Journal of Applied Logics - IfCoLog Journal of Logics and their Applications

Volume 12, Number 5

June 2025

ISBN 978-1-84890-488-0
ISSN (E) 2631-9829
ISSN (P) 2631-9810

College Publications
Scientific Director: Dov Gabbay
Managing Director: Jane Spurr

http://www.collegepublications.co.uk

SCOPE AND SUBMISSIONS

This journal considers submission in all areas of pure and applied logic, including:

pure logical systems
proof theory
constructive logic
categorical logic
modal and temporal logic
model theory
recursion theory
type theory
nominal theory
nonclassical logics
nonmonotonic logic
numerical and uncertainty reasoning
logic and AI
foundations of logic programming
belief change/revision
systems of knowledge and belief
logics and semantics of programming
specification and verification
agent theory
databases

dynamic logic
quantum logic
algebraic logic
logic and cognition
probabilistic logic
logic and networks
neuro-logical systems
complexity
argumentation theory
logic and computation
logic and language
logic engineering
knowledge-based systems
automated reasoning
knowledge representation
logic in hardware and VLSI
natural language
concurrent computation
planning

This journal will also consider papers on the application of logic in other subject areas: philosophy, cognitive science, physics etc. provided they have some formal content.

Submissions should be sent to Jane Spurr (jane@janespurr.net) as a pdf file, preferably compiled in LaTeX using the IFCoLog class file.

Contents

Proceedings of the 2nd Third Workshop: Introduction

Katalin Bimbó

Department of Philosophy, University of Alberta, Edmonton, Canada
`<bimbo@ualberta.ca>`

The *2nd Third Workshop* was held at the University of Alberta, in Edmonton, Canada, on May 13–14, 2024. The workshop was organized by me and several aspects of the workshop were financially supported by the *Insight Grant* entitled "From the Routley–Meyer semantics to gaggle theory and beyond: The evolution and use of relational semantics for substructural and other intensional logics."

The ordinal number "2nd" in the title of the workshop may suggest that a "1st Third Workshop" took place earlier; this is almost true. The *Third Workshop* was held eight years earlier, and its proceedings is Bimbó and Dunn [9]; we explained in the introduction how the title of the workshop came to be "Third." The Insight Grant that was connected to that workshop mainly dealt with the early development of the ternary relational semantics for relevance logics, which was the topic of one of the papers. The historical aspect of the more recent Insight Grant was intended to continue tracing the development of relational semantics into *gaggle theory*. The so-called *generalized Galois logic* (GGL) approach (or gaggle theory, for short) was invented by Dunn in the late 1980s, and he started to publish on gaggles in [15], in the early 1990s.[1] A paper of mine, [7] attempts to reconstruct — in the sense of the history of ideas — how gaggle theory emerged; however, conducting a more traditional historical investigation became impossible in the early 2020s. Thus, the research in the project shifted to obtaining new results that are in line with gaggle theory itself (rather than with its history).

It seems to me that there are at least three directions into which the research projects behind the *Third Workshop*s could be continued. Generalized Galois logic favors a particular type of semantics for substructural logics, which was emphasized

The support of the *Insight Grant #435–2019–0331*, awarded by the *Social Sciences and Humanities Research Council of Canada*, is gratefully acknowledged.

[1]Further papers and a book that added to gaggle theory include Dunn [16, 17, 18], Dunn and Meyer [21], Dunn [19], Dunn and Hardegree [20], and the book Bimbó and Dunn [8] aimed not only to expand gaggle theory but also to present systematically the state of the approach in the first decade of the 21st century.

in [7]. However, a historically-minded scholar could chart a map of what happened with other sorts of semantics during the same years, that is, during the 1970s–1990s. To mention some papers that opt for *operations* instead of relations in semantics, let me point to some publications (listed in the order of their appearance): Maksimova [28], Girard [24], Došen [12, 13], Rosenthal [31], Barwise [2], Došen and Schroeder-Heister [14], Ono [29], Girard [25] and van Benthem [3, 4]. I should emphasize that this is a more or less accidental list, that is, it is not based on systematic historical investigations (that I have not carried out).

In another direction, investigations could advance by extending the group of algebras that have been included into gaggle theory so far. Many logics do not algebraize simply into a gaggle, rather, algebras of logics are often "multi-gaggles" or "super-gaggles;" in other words, they are intricate combinations of algebras, each of which in itself is some sort of a gaggle. Although there are semantics that result by a piecemeal integration of semantics for component gaggles, it seems that no overarching treatment has been developed for *multi-gaggles*. Dunn originally defined gaggles to be certain algebras with a distributive lattice reduct; he soon expanded the class of algebras to be considered within gaggle theory. In some cases, there appears to be a consensus between logicians — expressed in the definition of semantics — about the "best way" to interpret a group of logics. However, an advantage afforded by the gaggle approach is that results about one kind of gaggles can be applied to another kind. An example is an operational semantics for logics (such as $\mathbf{T}_+$) with conjunction and disjunction distributing over each other where ideas for non-distributive gaggles are transposed to distributive gaggles resulting in a new (sound and complete) semantics. This completes what was initiated in [28], and then further elaborated on in [32] and [22]. It seems to us that other ways to tweak existing semantics might result in interpretations for additional logics, or they may yield semantics that fit specific technical goals or informal explications.

Lastly, on a speculative note, let us mention that connections between gaggle theory and *universal algebra* and *category theory* could be explored farther. There are algebraic studies of algebras of substructural logics (e.g., Galatos et al. [23]), and there are category-theoretic treatments of some substructural logics (such as intuitionistic and linear logics). There is a sizeable literature on duality between *topological structures* and algebras (including algebras of logics).[2] However, it appears that much more fruitful research could be conducted at the intersection of these areas.

---

[2]See e.g., Priestley [30], Urquhart [33], Goldblatt [26, 27], Allwein and Dunn [1], Clark and Davey [11], Bimbó and Dunn [8, 10], Bimbó [5, 6].

The papers in this issue are connected to the *2nd Third Workshop*, however, there is no 1–1 correspondence between the talks at the workshop and the papers. Some speakers chose not to submit a paper for the proceedings, and others wrote papers with co-authors who were not at the workshop or who contributed to the paper after the workshop. The table of contents shows the papers, thus, for the sake of comparison, we list the talks that were delivered at the workshop (in alphabetic order of the speakers).

(1) Guillermo Badia: *Definability of classes of frames and models in bi-intuitionistic logic*

(2) Katalin Bimbó: *Operational semantics for negation-free ticket entailment* $\mathbf{T}_+$

(3) Nicholas Ferenz: γ *and first-order relevant logics*

(4) Rohan French: *The constructive logic of paradox: On the Woodruff plan*

(5) Allen Hazen: *David Lewis and relational semantics*

(6) Shay Logan: *The philosophical significance of operational and stratified semantics*

(7) Edwin Mares: *Substructural philosophy*

(8) Joseph McDonald: *Monadic ortholattices: Completions and duality*

(9) David Ripley: *Recapture, ambiguity, and conflation*

(10) Igor Sedlár: *Algebras for relevant reasoners*

(11) Shawn Standefer: *Universal necessity, Scroggs properties, and deep classicality*

(12) Andrew Tedder: *Topical consequence relations*

(13) Zach Weber: *On strong and weak logics for non-classical computability*

(14) Yale Weiss: *A relevant framework for barriers to entailment*

In conclusion, I would like to acknowledge the people whose efforts helped to make this proceedings possible.[3] First of all, thanks to the authors of the papers, and thanks to the referees who provided feedback on the papers. I would like to thank Danielle Brown (currently, a PhD student) and Sobhan Jalilian (currently, an MA student) who provided copy-editing assistance as research assistants (i.e., RAs under my IG). I am grateful to Jane Spurr, Managing Director of College Publications for accepting the workshop proceedings for an issue of this journal.

---

[3] The typesetting of this proceedings utilizes the program TeX (originally designed by D. Knuth), the LaTeX format and various packages, some which were developed under the auspices of the American Mathematical Society.

References

[1] Gerard Allwein and J. Michael Dunn. Kripke models for linear logic. *Journal of Symbolic Logic*, 58(2):514–545, 1993.

[2] K. Jon Barwise. Constraints, channels and the flow of information. In P. Aczel, D. Israel, Y. Katagiri, and S. Peters, editors, *Situation Theory and its Applications*, volume 37 of *CSLI Lecture Notes*, pages 3–27. CSLI Publications, Stanford, CA, 1993.

[3] Johan van Benthem. *Language in Action: Categories, Lambdas and Dynamic Logic*. MIT Press, Cambridge, MA, 1995.

[4] Johan van Benthem. *Exploring Logical Dynamics*. Studies in Logic, Language and Information. CSLI Publications, Stanford, CA, 1996.

[5] Katalin Bimbó. Functorial duality for ortholattices and De Morgan lattices. *Logica Universalis*, 1(2):311–333, 2007.

[6] Katalin Bimbó. Dual gaggle semantics for entailment. *Notre Dame Journal of Formal Logic*, 50(1):23–41, 2009.

[7] Katalin Bimbó. On the origins of gaggle theory. In I. Sedlár, editor, *Logica Yearbook 2021*, pages 19–36. College Publications, London, UK, 2022.

[8] Katalin Bimbó and J. Michael Dunn. *Generalized Galois Logics: Relational Semantics of Nonclassical Logical Calculi*, volume 188 of *CSLI Lecture Notes*. CSLI Publications, Stanford, CA, 2008.

[9] Katalin Bimbó and J. Michael Dunn, editors. *Proceedings of the Third Workshop, May 16–17, 2016, Edmonton, Canada*, volume 4(3) of *IFCoLog Journal of Logics and Their Applications*, College Publications, London, UK, 2017.

[10] Katalin Bimbó and J. Michael Dunn. Larisa Maksimova's early contributions to relevance logic. In S. Odintsov, editor, *L. Maksimova on Implication, Interpolation and Definability*, volume 15 of *Outstanding Contributions to Logic*, pages 33–60. Springer Nature, Switzerland, 2018.

[11] David M. Clark and Brian A. Davey. *Natural Dualities for the Working Algebraist*, volume 57 of *Cambridge Studies in Advanced Mathematics*. Cambridge University Press, Cambridge, UK, 1998.

[12] Kosta Došen. Sequent-systems and groupoid models. I. *Studia Logica*, 47(4): 353–385, 1988.

[13] Kosta Došen. Sequent-systems and groupoid models. II. *Studia Logica*, 48(1): 41–65, 1989.

[14] Kosta Došen and Peter Schroeder-Heister, editors. *Substructural Logics*. Clarendon, Oxford, UK, 1993.

[15] J. Michael Dunn. Gaggle theory: An abstraction of Galois connections and residuation, with applications to negation, implication, and various logical operators. In J. van Eijck, editor, *Logics in AI: European Workshop JELIA '90*, number 478 in Lecture Notes in Computer Science, pages 31–51. Springer, Berlin, 1991.

[16] J. Michael Dunn. Star and perp: Two treatments of negation. *Philosophical Perspectives*, 7:331–357, 1993. (Language and Logic, 1993, J. E. Tomberlin (ed.)).

[17] J. Michael Dunn. Partial gaggles applied to logics with restricted structural rules. In K. Došen and P. Schroeder-Heister, editors, *Substructural Logics*, number 2 in Studies in Logic and Computation, pages 63–108. Clarendon, Oxford, UK, 1993.

[18] J. Michael Dunn. Gaggle theory applied to intuitionistic, modal and relevance logics. In I. Max and W. Stelzner, editors, *Logik und Mathematik. Frege-Kolloquium Jena 1993*, pages 335–368. W. de Gruyter, Berlin, 1995.

[19] J. Michael Dunn. A representation of relation algebras using Routley–Meyer frames. In C. A. Anderson and M. Zelëny, editors, *Logic, Meaning and Computation. Essays in Memory of Alonzo Church*, pages 77–108. Kluwer, Dordrecht, 2001.

[20] J. Michael Dunn and Gary M. Hardegree. *Algebraic Methods in Philosophical Logic*, volume 41 of *Oxford Logic Guides*. Oxford University Press, Oxford, UK, 2001.

[21] J. Michael Dunn and Robert K. Meyer. Combinators and structurally free logic. *Logic Journal of the IGPL*, 5(4):505–537, 1997.

[22] Kit Fine. Models for entailment. *Journal of Philosophical Logic*, 3(4):347–372, 1974.

[23] Nikolaos Galatos, Peter Jipsen, Tomasz Kowalski, and Hiroakira Ono. *Residuated Lattices: An Algebraic Glimpse at Substructural Logics*. Elsevier, Amsterdam, 2007.

[24] Jean-Yves Girard. Linear logic. *Theoretical Computer Science*, 50:1–102, 1987.

[25] Jean-Yves Girard. Linear logic: Its syntax and semantics. In J.-Y. Girard, Y. Lafont, and L. Regnier, editors, *Advances in Linear Logic*, number 222 in London Mathematical Society Lecture Note Series, pages 1–42. Cambridge University Press, Cambridge, UK, 1995.

[26] Robert Goldblatt. The Stone space of an ortholattice. *Bulletin of the London Mathematical Society*, 7:45–48, 1975.

[27] Robert Goldblatt. Varieties of complex algebras. *Annals of Pure and Applied Logic*, 44:173–242, 1989.

[28] Larisa L. Maksimova. Interpretatsiya i teoremy otdeleniya dlya ischisleniǐ E i R. *Algebra i logika*, 10(4):376–392, 1971. [An interpretation and separation theorems for the calculi E and R, (in Russian)].

[29] Hiroakira Ono. Semantics for substructural logics. In K. Došen and P. Schroeder-Heister, editors, *Substructural Logics*, number 2 in Studies in Logic and Computation, pages 259–291. Clarendon, Oxford (UK), 1993.

[30] Hilary A. Priestley. Representation of distributive lattices by means of ordered Stone spaces. *Bulletin of the London Mathematical Society*, 2:186–190, 1970.

[31] Kimmo I. Rosenthal. *Quantales and their Applications*. Number 234 in Pitman Research Notes in Mathematics. Longman and J. Wiley & Sons, Essex, UK and New York, NY, 1990.

[32] Alasdair Urquhart. Semantics for relevant logic. *Journal of Symbolic Logic*, 37 (1):159–169, 1972.

[33] Alasdair Urquhart. A topological representation theorem for lattices. *Algebra Universalis*, 8:45–58, 1978.

Received 22 April 2025

Definable Classes of Models and Frames in Bi-intuitionistic Logic

Guillermo Badia
University of Queensland, Australia
<g.badia@uq.edu.au>

Tomasz Kowalski[*]
Jagiellonian University, Poland,
La Trobe University and University of Queensland, Australia
<tomasz.s.kowalski@uj.edu.pl>,
<t.kowalski@latrobe.edu.au>, <uqtkowal@uq.edu.au>

Grigory Olkhovikov
Ruhr University Bochum, Germany
<grigory.olkhovikov@gmail.com>

Abstract

The question of the expressive power of a given logical language with Kripke relational semantics has at least two dimensions: (1) what the language can say about frames, and (2) what it can say about models. The Goldblatt–Thomason theorem provides a model-theoretic characterisation of modal axiomatisability for elementary classes of *frames* in terms of closure under taking generated subframes, disjoint unions, bounded morphic images, and reflection of ultrafilter extensions. Goldblatt also provides a similar characterisation for axiomatisability in intuitionistic logic of classes of *models* rather than frames. In this article we provide analogous results for bi-intuitionistic logic, a natural expressive extension of intuitionistic logic obtained by adding a binary connective dual to the intuitionistic implication, introduced in the 1970s independently by Dieter Klemke and Cecylia Rauszer. Together with previous results, such as a

We are grateful to Jim de Groot with whom we discussed several of the ideas that appear here. We are also grateful to Katalin Bimbó for many editing suggestions that improved the presentation greatly.

[*]Support from ERC HORIZON2020-MSCA-RISE project no. 101007627 (MOSAIC) is gratefully acknowledged.

van Benthem bisimulation characterisation theorem and a Lindström theorem, this provides a complete picture of the expressive power of propositional bi-intuitionistic logic.

2020 *Mathematics Subject Classification.* Primary: 03B20, Secondary: 03C52.

1 Introduction

The result of adding the algebraic dual operation of intuitionistic implication $\rightarrow$ (known as co-implication and denoted here by $\prec$) to intuitionistic logic is known as Heyting–Brouwer or bi-intuitionistic logic. It is a natural model-theoretic extension of the language of intuitionistic logic in the same sense as adding a backward looking $\lozenge^{-1}$ to modal logic with the primitive modality $\square$ is. Indeed, there is a strong analogy in a precise sense between temporal logic where we have both backward and forward looking operators and bi-intuitionistic logic (cf. [27]). In the 1970s, C. Rauszer started an intense study of various technical aspects of both propositional and predicate bi-intuitionistic logic in a series of interesting articles spanning over a decade [20, 21, 22, 23, 25, 24]. This work has been picked up in recent years by a number of scholars (e.g., [1, 2, 18, 12, 13, 9, 16, 19, 7]) and results on all sorts of proof-theoretic and model-theoretic properties of these systems have been produced. Historically, before Rauszer, at least one place where bi-intuitionistic logic was inadvertently introduced and studied is [15]; but since the article was in German and the main topic was not bi-intuitionistic logic, it did not attract much attention from the community of researchers in the area until recently.

The present contribution is focused on completing the picture of what exactly can be expressed model-theoretically in the language of propositional bi-intuitionistic logic. In [1], directed bisimulations were isolated as the correct model-theoretic relations to distinguish the standard translations of bi-intuitionistic formulas into first-order logic from other first-order formulas by means of a van Benthem characterization theorem. Following this work, in [18] these relations, in conjunction with other natural model-theoretic properties, were used to characterize the expressive power of bi-intuitionistic logic among all its possible expressive extensions through a Lindström-style theorem. Two interesting results that serve as inspiration for the work in the current paper are the main theorems from [11] and [10]. On the one hand, [11] contains the so called Goldblatt–Thomason theorem for modal logic which provides a model-theoretic characterization of the classes of frames that can be axiomatised by a modal theory among the classes that are already axiomatisable in first-order logic. On the other hand, in [10], Goldblatt provides a model-theoretic characterization of the classes of intuitionistic Kripke models axiomatisable in propositional intuitionistic logic. Our goal in the present work is to produce analogues of the theorems in [11] and [10] for bi-intuitionistic logic.

The article is arranged as follows. In Section 2, we give the basic definitions of the syntax and semantics of bi-intuitionistic logic, some algebraic background related to double-Heyting algebras, and the standard translation of bi-intuitionistic logic into first-order logic. In Section 3, we introduce the model-theoretic constructions and relations between structures that we will use in our main theorems. These include notions of countably saturated structures, bounded morphisms, subframes (and submodels), disjoint unions, directed bisimulations and prime filter extensions. Finally, we also provide some observations on algebraic duality between some of these relations and corresponding ones between algebras. In Section 4 we provide a model-theoretic characterization using some of the notions previously introduced of which classes of models exactly correspond to models of a set of bi-intuitionistic formulas (Theorem 27). In Section 5 we provide a direct analogue of the famous Goldblatt–Thomason theorem and in Section 6 we give a brief summary of what we have accomplished.

2 Preliminaries

We begin by defining the language of bi-intuitionistic logic and its standard Kripke semantics. Let Prop be an arbitrary but fixed set of proposition letters. Then the language $\mathbf{BI}(\mathrm{Prop})$ is the set of formulas generated by the grammar

$$\varphi ::= p \mid \top \mid \bot \mid \varphi \wedge \varphi \mid \varphi \vee \varphi \mid \varphi \to \varphi \mid \varphi \prec \varphi,$$

where p ranges over Prop. We will write $\mathbf{BI}$ instead of $\mathbf{BI}(\mathrm{Prop})$, unless the set of propositional letters changes to something other than Prop, as for example, in Lemma 28.

Definition 1. *A* Kripke frame *is any poset* $(W, \leq)$. *Elements of W are often called* worlds.

Let $(W, \leq)$ be a Kripke frame and let $a \subseteq W$. We write $\uparrow a$ for the upward closure of a and $\downarrow a$ for the downward closure of a. A subset equal to its upward closure is an *upset,* and a subset equal to its downward closure is a *downset.*

A *valuation* for a Kripke frame $(W, \leq)$ is a map V that assigns to each proposition letter $p \in \mathrm{Prop}$ an upset $V(p)$ of $(W, \leq)$. A *Kripke model* is a tuple $\mathfrak{M} = (W, \leq, V)$ consisting of a Kripke frame $(W, \leq)$ and a valuation V. We say that the Kripke model $\mathfrak{M}$ is *based on* the Kripke frame $(W, \leq)$, and that $(W, \leq)$ is the *underlying Kripke frame* of $\mathfrak{M}$. We will often use $\mathfrak{F}$ to stand for an arbitrary Kripke frame, and write $\mathfrak{M} = (\mathfrak{F}, V)$ for a model based on $\mathfrak{F}$.

1127

Definition 2. *Truth of a formula φ at a world w of a model $\mathfrak{M} = (\mathfrak{F}, V)$ based on a frame $\mathfrak{F} = (W, \leq)$ is denoted symbolically by $\mathfrak{M}, w \Vdash \varphi$ and defined recursively as follows:*

$$
\begin{aligned}
\mathfrak{M}, w &\Vdash p &\ &\textit{iff} &\ &x \in V(p) \\
\mathfrak{M}, w &\Vdash \top & &\textit{always} \\
\mathfrak{M}, w &\Vdash \bot & &\textit{never} \\
\mathfrak{M}, w &\Vdash \varphi \wedge \psi & &\textit{iff} & &\mathfrak{M}, w \Vdash \varphi \textit{ and } \mathfrak{M}, w \Vdash \psi \\
\mathfrak{M}, w &\Vdash \varphi \vee \psi & &\textit{iff} & &\mathfrak{M}, w \Vdash \varphi \textit{ or } \mathfrak{M}, w \Vdash \psi \\
\mathfrak{M}, w &\Vdash \varphi \to \psi & &\textit{iff} & &\forall v \in W (\textit{if } w \leq v \textit{ and } \mathfrak{M}, v \Vdash \varphi \textit{ then } \mathfrak{M}, v \Vdash \psi) \\
\mathfrak{M}, w &\Vdash \varphi \prec \psi & &\textit{iff} & &\exists v \in W (v \leq w \textit{ and } \mathfrak{M}, v \Vdash \varphi \textit{ and not } \mathfrak{M}, v \Vdash \psi)
\end{aligned}
$$

*For a set Γ of **BI**-formulas we write $\mathfrak{M}, w \Vdash \Gamma$, if $\mathfrak{M}, w \Vdash \varphi$ for all $\varphi \in \Gamma$.*

As usual, we extend the function V from Prop to the collection of all formulas by defining

$$V(\varphi) := \{w \in W \mid \mathfrak{M}, w \Vdash \varphi\}.$$

This set is often called the *truth set* of φ. If $V(\varphi) = W$, then we write $\mathfrak{M} \Vdash \varphi$.

The notion of truth at a world naturally extends to frames. For a frame $\mathfrak{F} = (W, \leq)$ and $w \in W$, we write $\mathfrak{F}, w \Vdash \varphi$ if $\mathfrak{M}, w \Vdash \varphi$ for every $\mathfrak{M}$ based on $(W, \leq)$. Further still, if $\mathfrak{F}, w \Vdash \varphi$ holds for every $w \in W$, or, equivalently, if $\mathfrak{M} \Vdash \varphi$ for every $\mathfrak{M}$ based on $(W, \leq)$, we say that φ is *valid* on $\mathfrak{F}$, and write $\mathfrak{F} \Vdash \varphi$ relying on context to distinguish between models and frames.

If φ is valid on every frame $\mathfrak{F}$ from some collection K of frames, we say that φ is *valid in K* and write $K \Vdash \varphi$.

2.1 Upset Algebras and Double-Heyting Algebras

For any frame $\mathfrak{F}$ and any valuation V, the truth sets naturally form an algebra of upsets. To begin with, we have $V(\varphi \wedge \psi) = V(\varphi) \cap V(\psi)$, $V(\varphi \vee \psi) = V(\varphi) \cup V(\psi)$ and $V(\top) = W$, $V(\bot) = \emptyset$, so the truth sets ordered by inclusion form a bounded distributive lattice of upsets of $\mathfrak{F}$. In general this lattice is a sublattice of the (bounded, distributive) lattice of all upsets of $\mathfrak{F}$. Furthermore, if we unfold the definitions of $\to$ and $\prec$ we see that the truth sets of $\varphi \to \psi$ and $\varphi \prec \psi$ in a Kripke model $\mathfrak{M} = (W, \leq, V)$ can be defined as follows:

$$
\begin{aligned}
V(\varphi \to \psi) &= W \setminus {\downarrow} (V(\varphi) \setminus V(\psi)) \\
V(\varphi \prec \psi) &= {\uparrow} (V(\varphi) \setminus V(\psi))
\end{aligned}
$$

Importantly, these sets are both upsets of $\mathfrak{F}$. The algebra defined this way will be called the *algebra of definable upsets* in Section 3.5. Applying exactly the same definitions to arbitrary upsets of $\mathfrak{F}$ (not necessarily values of any valuation), we obtain the *upset algebra* of $\mathfrak{F}$, denoted $\mathrm{Up}(\mathfrak{F})$. It is easy to verify that $\mathrm{Up}(\mathfrak{F})$ satisfies the following equivalences:

$$a \cap b \subseteq c \quad \text{iff} \quad b \subseteq a \to c,$$
$$a \cup b \supseteq c \quad \text{iff} \quad b \supseteq c \prec a.$$

It turns out that these are precisely the definitional properties of *double-Heyting algebras*.

Definition 3. *An algebra* $\mathbf{A} = (A; \wedge, \vee, \to, \prec, 0, 1)$, *such that* $(A; \wedge, \vee, 0, 1)$ *is a bounded distributive lattice and the equivalences*

$$x \wedge y \leq z \quad \textit{iff} \quad y \leq x \to z,$$
$$x \vee y \geq z \quad \textit{iff} \quad y \geq z \prec x,$$

hold for all $x, y, z \in A$, *is called a double-Heyting algebra.*

These properties are in fact equivalent to equations, so the class DH of all double-Heyting algebras is an *equational class*, hence, by the easy half of Birkhoff Theorem (see, e.g., [4] Lem. 4.36 and Thm. 4.41), a *variety*, that is, a class closed under homomorphic images, subalgebras and direct products. The connection between double-Heyting algebras and bi-intuitionistic logic is very tight, in fact **BiInt** is *algebraizable with equivalent algebraic semantics* DH.

Details of algebraizability are beyond the scope of this article, but to keep it as self-contained as sensible we give a sketch. Let Σ be a purely functional signature, say, $\Sigma = \{\wedge, \vee, \to, \prec\}$. Let $\mathbf{L}$ be a logic in the language of Σ, given as a deductive system $(\Gamma_\Sigma, \vdash)$ where Γ_Σ is a set of formulas over Σ and $\vdash$ a Tarskian consequence relation. Let $\mathcal{K}$ be a class of algebras of Σ, and let E_Σ be the set of all equations over Σ (formally $E_\Sigma = \Gamma_\Sigma \times \Gamma_\Sigma$). We say that $\mathbf{L}$ is algebraizable with equivalent algebraic semantics $\mathcal{K}$, if there are two translation maps $\varepsilon \colon \Gamma_\Sigma \to E_\Sigma$ and $\delta \colon E_\Sigma \to \Gamma_\Sigma$, which behave as the diagrams below indicate.

$$\delta(\varepsilon(\Gamma)) \dashv\vdash \Gamma \qquad
\begin{array}{ccc}
\Gamma \vdash \Pi & \xleftrightarrow{\text{iff}} & \varepsilon(\Gamma) \models \varepsilon(\Pi) \\
{\scriptstyle\text{case of}}\big\uparrow & & \big\downarrow{\scriptstyle\text{case of}} \\
\delta(\Psi) \vdash \delta(\Phi) & \xleftrightarrow{\text{iff}} & \Psi \models \Phi
\end{array}
\qquad \varepsilon(\delta(\Psi)) \equiv\!\models \Psi$$

Here, $\models$ is the usual equational (first-order) consequence relation, $\Gamma \vdash \Pi$ means $\Gamma \vdash \pi$ for all $\pi \in \Pi$, similarly for $\Psi \models \Phi$, and $\dashv\vdash$, $=\!\!\models\!\!=$ mean that the relevant deducibility holds both ways.

Proposition 1. *The logic* **BiInt** *is algebraizable with equivalent algebraic semantics* DH. *The translations are given by:*

$$\varphi \overset{\varepsilon}{\mapsto} (\varphi = 1) \qquad and \qquad (\tau_1 = \tau_2) \overset{\delta}{\mapsto} (\tau_1 \to \tau_2) \wedge (\tau_2 \to \tau_1).$$

Subvarieties of DH *correspond in a similar manner to axiomatic extensions of* **BiInt**.

For more on double-Heyting algebras, including a very nice crash course on the duality theory for them, see [26].

2.2 Standard Translation

Consider a first order language with a binary relation symbol R, and a unary predicate P for each $p \in \text{Prop}$. Following the tradition in modal logic [5], we call it the *correspondence language* $\mathbf{BI}^{corr}$ for **BI**. Note that each model $\mathfrak{M}$ for **BI** is also a model for $\mathbf{BI}^{corr}$ in the usual first-order sense, namely, W is the domain, $\leq$ is the interpretation of R, and $V(p)$ is the interpretation of P. The usual first-order satisfaction relation will be denoted by $\models$, and by $\mathfrak{M} \models \varphi[a]$ we mean that $a \in W$ satisfies the first-order formula φ.

The *standard translation* of bi-intuitionistic formulas to $\mathbf{BI}^{corr}$ is defined inductively:

$$ST_x(\bot) = \bot$$
$$ST_x(\top) = \top$$
$$ST_x(p) = Px$$
$$ST_x(\varphi \wedge \psi) = ST_x(\varphi) \wedge ST_x(\psi)$$
$$ST_x(\varphi \vee \psi) = ST_x(\varphi) \vee ST_x(\psi)$$
$$ST_x(\varphi \to \psi) = \forall y(\neg(Rxy \wedge ST_y(\varphi)) \vee ST_y(\psi))$$
$$ST_x(\varphi \prec \psi) = \exists y(Ryx \wedge ST_y(\varphi) \wedge \neg ST_y(\psi)),$$

with the usual proviso of choosing y carefully to avoid free variable capture.

Proposition 2. *For any Kripke model* $\mathfrak{M} = (W, \leq, V)$, *any* $w \in W$, *and any bi-intuitionistic formula* φ, *we have:*

$$\mathfrak{M}, w \Vdash \varphi \qquad if\ and\ only\ if \qquad \mathfrak{M} \models ST_x(\varphi)[w].$$

Proof. We omit the proof, which would proceed by straightforward induction on the complexity of formulas. $\square$

3 Some Model and Frame Constructions

The results below will involve some machinery from the classical model theory of first-order logic and thus we will start the section by recalling a few useful notions which can be found in any of the standard references, such as [14].

3.1 Countably Saturated Structures

We will use $\overline{a}$, $\overline{b}$, and $\overline{x}$, $\overline{y}$ to denote sequences of elements of a model and variables, respectively. For a model $\mathfrak{M}$, by $\mathrm{dom}(\mathfrak{M})$ we denote the domain of $\mathfrak{M}$. In particular, if $\mathfrak{M}$ is a model based on a Kripke frame $(W, \leq)$, then $\mathrm{dom}(\mathfrak{M}) = W$. If $X \subseteq \mathrm{dom}(\mathfrak{M})$, then $(\mathfrak{M}, a)_{a \in X}$ is the *expansion* of $\mathfrak{M}$ obtained by adding a constant c_a for each $a \in X$.

We say that a set of first order formulas $\Phi(\overline{x})$ is *realisable* in a model $\mathfrak{M}$ if there is some sequence $\overline{a}$ of elements of W such that $\mathfrak{M} \vDash \Phi[\overline{a}]$. $\Phi(\overline{x})$ is said to be *refutable* in $\mathfrak{M}$ if $\Phi'(\overline{x}) = \{\neg\varphi : \varphi \in \Phi(\overline{x})\}$ is realisable in $\mathfrak{M}$.

Definition 4. *Let λ be a cardinal. A model $\mathfrak{M}$ of first order logic is said to be λ-saturated if whenever $X \subseteq \mathrm{dom}(\mathfrak{M})$ and $|X| < \lambda$, then some expansion $(\mathfrak{M}, a)_{a \in X}$ of $\mathfrak{M}$ realises every set of formulas $\Phi(x)$ of the language of $(\mathfrak{M}, a)_{a \in X}$ which is consistent with the set of all first order sentences true in $(\mathfrak{M}, a)_{a \in X}$.*

Note that if $\kappa < \lambda$, then λ-saturation implies κ-saturation.

Definition 5. *Let $\mathfrak{M}$ and $\mathfrak{N}$ be two first order models. $\mathfrak{N}$ is an elementary extension of $\mathfrak{M}$ if $\mathfrak{M}$ is (isomorphic to) a submodel $\mathfrak{N}'$ of $\mathfrak{N}$ such that where $\overline{a}$ is a sequence of elements of $\mathfrak{N}'$, for each first order formula φ,*

$$\mathfrak{N}' \vDash \varphi[\overline{a}] \quad \text{iff} \quad \mathfrak{N} \vDash \varphi[\overline{a}].$$

The following proposition is a corollary of Theorems 6.1.4 and 6.1.8 in [6]. The method of obtaining such saturated models is often referred to as the Keisler method.

Proposition 3. *Each model $\mathfrak{M}$ for $\mathbf{BI}^{corr}$ has an ω-saturated elementary extension, obtainable as an ultrapower of $\mathfrak{M}$.*

3.2 Bounded Morphisms, Generated Subframes, Disjoint Unions

Bounded morphisms between Kripke frames and models are the maps that have the correct functorial properties in category theory sense. For models, in particular, they preserve and reflect truth of **BI**-formulas. They map valuations to valuations, and if w is a world and f is a bounded morphism, then every formula true at w is also true at $f(w)$ (preservation), and every formula true at $f(w)$ is also true at w as well (reflection).

Definition 6. *Let $\mathfrak{F} = (W, \leq)$ and $\mathfrak{F}' = (W', \leq')$ be two Kripke frames. A function $f \colon W \to W'$ is a* bounded morphism *from $\mathfrak{F}$ to $\mathfrak{F}'$ if it is order-preserving and moreover, for all $w, v \in W$ and $u' \in W'$:*

(ub) if $f(w) \leq' u'$ then there exists a $u \in W$ such that $w \leq u$ and $f(u) = u'$;

(lb) if $u' \leq' f(w)$ then there exists a $u \in W$ such that $u \leq w$ and $f(u) = u'$.

If f is surjective, we say that $\mathfrak{F}'$ is a bounded-morphic image *of $\mathfrak{F}$.*

Definition 7. *Let $\mathfrak{M} = (\mathfrak{F}, V)$ and $\mathfrak{M}' = (\mathfrak{F}', V')$ be two Kripke models. A function $f \colon W \to W'$ is a* bounded morphism *from $\mathfrak{M}$ to $\mathfrak{M}'$ if it is a bounded morphism from $\mathfrak{F} \to \mathfrak{F}'$ and for all $w \in W$ and $p \in \mathrm{Prop}$:*

(prop) $w \in V(p)$ if and only if $f(w) \in V'(p)$.

Proposition 4. *Let $\mathfrak{M} = (\mathfrak{F}, V)$ and $\mathfrak{M}' = (\mathfrak{F}', V')$ be two Kripke models and $f \colon \mathfrak{M} \to \mathfrak{M}'$ a bounded morphism. Then for all $w \in W$ and $\varphi \in \mathbf{BI}$ we have*

$$\mathfrak{M}, w \Vdash \varphi \quad \textit{iff} \quad \mathfrak{M}', f(w) \Vdash \varphi.$$

Proof. Induction on the complexity of φ. We leave it to the reader as an exercise. $\square$

Definition 8. *Let $\mathfrak{M}$ and $\mathfrak{N}$ be Kripke models such that $\mathfrak{M}$ is a substructure of $\mathfrak{N}$ is the first-order sense. If the natural inclusion map is a bounded morphism, then we say that $\mathfrak{M}$ is a* generated submodel *of $\mathfrak{N}$. The same,* mutatis mutandis, *applies to Kripke frames, the resulting notion being that of a* generated subframe.

Lemma 5. *Let $\mathfrak{F} = (W, \leq)$ and $\mathfrak{F}' = (W', \leq')$ be Kripke frames. Then, $\mathfrak{F}'$ is a generated subframe of $\mathfrak{F}$ if and only if (i) $W' \subseteq W$, (ii) $\leq'$ is the restriction of $\leq$ to W', and (iii) W' is both upward and downward closed in W.*

If $\mathfrak{M} = (\mathfrak{F}, V)$ and $\mathfrak{M}' = (\mathfrak{F}', V')$ are Kripke models, then $\mathfrak{M}'$ is a generated submodel of $\mathfrak{M}$ if and only if (i)–(iii) hold and moreover (iv) V' is the restriction of V to W'.

Proof. Straightforward from the definitions. $\square$

Yet another frame construction we will need is the disjoint union of frames and of models.

Definition 9. *Let* $\mathcal{F} = \{\mathfrak{F}_i \colon i \in I\}$ *be a set of Kripke frames with* $\mathfrak{F}_i = (W_i, \leq_i)$, *and let* $\mathcal{M} = \{(\mathfrak{F}_i, V_i) \colon i \in I\}$ *be a set of Kripke models. The* disjoint union *of* $\mathcal{F}$ *is defined by*

$$\coprod_{i \in I} \mathfrak{F}_i = (\biguplus W_i, \biguplus \leq_i),$$

where $\biguplus W_i$ *and* $\biguplus \leq_i$ *are the disjoint union of the sets* W_i *and the disjoint union of the relations* $\leq_i$, *respectively. Similarly, the disjoint union of* $\mathcal{M}$ *is defined by*

$$\coprod_{i \in I} (\mathfrak{F}_i, V_i) = (\coprod_{i \in I} \mathfrak{F}_i, V),$$

where for $w \in W_i$, $w \in V(p)$ *if* $w \in V_i(p)$.

Note that for sets $\mathcal{F}$ and $\mathcal{M}$ above, the identity function $\iota_j(w) = w$, for each $w \in W_j$, is a bounded morphism between Kripke frames and models, respectively. The proposition below follows easily from the definitions involved.

Proposition 6. *Let* $\{(\mathfrak{F}_i, V_i) \colon i \in I\}$ *be a set of Kripke models based on frames* $\mathfrak{F}_i = (W_i, \leq_i)$. *Let* $(\mathfrak{A}, V^A)$ *be a Kripke model based on a frame* $\mathfrak{A} = (A, \leq^A)$ *and let* $(\mathfrak{B}, V^B)$ *be a Kripke model based on a frame* $\mathfrak{B} = (B, \leq^B)$. *Then, for any* $\varphi \in \mathbf{BI}$, *the following hold:*

1. $(\mathfrak{F}_i, V_i) \Vdash \varphi$ *for all* $i \in I$, *if and only if* $\coprod_{i \in I}(\mathfrak{F}_i, V_i) \Vdash \varphi$.

2. $\mathfrak{F}_i \Vdash \varphi$ *for all* $i \in I$, *if and only if* $\coprod_{i \in I} \mathfrak{F}_i \Vdash \varphi$.

3. *If* $(\mathfrak{A}, V^A)$ *is a generated submodel of* $(\mathfrak{B}, V^B)$ *and* $\mathfrak{B} \Vdash \varphi$, *then* $\mathfrak{A} \Vdash \varphi$.

4. *If* $\mathfrak{A}$ *is a generated subframe of* $\mathfrak{B}$ *and* $\mathfrak{B} \Vdash \varphi$, *then* $\mathfrak{A} \Vdash \varphi$.

5. *If* $(\mathfrak{A}, V^A)$ *is a bounded-morphic image of* $(\mathfrak{B}, V^B)$ *and* $(\mathfrak{B}, V^B) \Vdash \varphi$ *then* $(\mathfrak{A}, V^A) \Vdash \varphi$.

6. *If* $\mathfrak{A}$ *is a bounded-morphic image of* $\mathfrak{B}$ *and* $\mathfrak{B} \Vdash \varphi$ *then* $\mathfrak{A} \Vdash \varphi$.

3.3 Shadows of Duality

Duality theory for double-Heyting algebras piggybacks on Priestley duality, of which we assume the reader to have some knowledge. However, since the whole topological machinery will not be used, we will recall the concepts and results absolutely necessary to follow our arguments. For a very readable account of Priestley duality we refer the reader to [8]. For the specific case of double-Heyting algebras Ch. 2 of [26] serves as a good reference.

We mentioned upset algebras in Section 2.1 briefly, now we will give proper definitions.

Definition 10. *Let* $\mathfrak{F} = (W, \leq)$ *be a Kripke frame. The* upset algebra $\mathfrak{F}^+$ *of* $\mathfrak{F}$ *is the structure* $\langle \mathrm{Up}(W); \cap, \cup, \to, \prec, 0, 1 \rangle$, *where* $\mathrm{Up}(W)$ *is the set of all upsets of* $(W, \leq)$, *the operations* $\cap$ *and* $\cup$ *are set-theoretical,* $0 = \emptyset$, $1 = W$, *and the remaining operations are defined by*

$$X \prec Y = \{w \in W : \exists v \in W(v \leq w \,\&\, v \in X \,\&\, v \notin Y)\},$$

$$X \to Y = \{w \in W : \forall v \in W((w \leq v \,\&\, v \in X) \Rightarrow v \in Y)\}.$$

The reader is invited to check that the definitions of $\to$ and $\prec$ given above coincide with these from Section 2.1. The algebraic structure of $\mathfrak{F}^+$ does not depend on valuations for $\mathfrak{F}$, and therefore, for any Kripke *model* $\mathfrak{M}$ we will write $\mathfrak{M}^+$ for the upset algebra of the underlying frame.

The next proposition is not difficult to verify using observations from Section 2.1. We leave it as an exercise to the reader.

Proposition 7. *Let* $\mathfrak{M}$ *be a Kripke model for bi-intuitionistic logic. Then* $\mathfrak{M}^+$ *is a double-Heyting algebra.*

Note that for a specific model $\mathfrak{M}$ the algebra $\mathfrak{M}^+$ can have too many elements, as not every upset is guaranteed to be a value of a formula. However, for a frame $\mathfrak{F}$, the upset algebra is exactly what we need, each upset is a value of a formula in *some* valuation.

Having thus constructed algebras from frames, we want to go back and construct frames from algebras. As usual, this is trickier. First, let us recall the notion of a *prime filter* in a lattice L. Namely, a subset F of L is a *filter* if it is a nonempty upset, and moreover $x, y \in F$ implies $x \wedge y \in F$. A filter F is *prime* if $x \vee y \in F$ implies $x \in F$ or $y \in F$.

Definition 11. *Let* $\mathbf{A}$ *be a double-Heyting algebra, and let* $\mathcal{F}_p(\mathbf{A})$ *be the set of all prime filters of the underlying lattice of* $\mathbf{A}$. *The* prime filter frame $\mathbf{A}_+$ *of* $\mathbf{A}$ *is the poset* $(\mathcal{F}_p(\mathbf{A}), \subseteq)$.

Naturally, for any frame $\mathfrak{F}$ we can form the frame $(\mathfrak{F}^+)_+$, and for any algebra $\mathbf{A}$ we can form the algebra $(\mathbf{A}_+)^+$. For finite frames and finite algebras, we have a perfect situation:

Proposition 8. *Let* $\mathfrak{F}$ *be a finite frame, and let* $\mathbf{A}$ *be a finite double-Heyting algebra. Then,* $(\mathfrak{F}^+)_+ \cong \mathfrak{F}$ *and* $(\mathbf{A}_+)^+ \cong \mathbf{A}$.

For infinite frames and algebras isomorphism no longer holds: algebras have too many prime filters, and frames have too many upsets. Topology comes to the rescue, but as we mentioned already, we will not need it. We will, however, need the concept of a double-dual model, in Section 3.5 and later.

Definition 12. *For a model $\mathfrak{M} = (\mathfrak{F}, V)$, with $\mathfrak{F} = (W, \leq)$, the double-dual model $(\mathfrak{M}^+)_+ = ((\mathfrak{F}^+)_+, V^*)$ of $\mathfrak{M}$, is defined by setting $V^*(p) = \{w \in \mathcal{F}_p(\mathfrak{F}^+) : V(p) \in w\}$.*

Lemma 9. *Let $\mathfrak{F} = (W, \leq)$. Then, in $(\mathfrak{F}^+)_+$ we have*

$$w \subseteq v \quad \text{iff} \quad \forall x, y \in \mathrm{Up}(W)(x \to y \in w \,\&\, x \in v \;\Rightarrow\; y \in v).$$

Proof. First, observe that the following holds for any $x, y \in \mathrm{Up}(W)$:

$$(\dagger) \qquad\qquad (W \setminus {\downarrow} (x \setminus y)) \cap x \subseteq y.$$

To see it, take any $a \notin {\downarrow} (x \setminus y)$ such that $a \in x$. Then ${\uparrow} a \cap (x \setminus y) = \emptyset$. Since $a \in x$, this implies ${\uparrow} a \cap (W \setminus y) = \emptyset$, that is, ${\uparrow} a \subseteq y$, hence $a \in y$.

Now, assume $w \subseteq v$ and take $x, y \in \mathrm{Up}(W)$ such that $x \to y \in w$ and $x \in v$. By definition of $\to$, we have $W \setminus {\downarrow} (x \setminus y) \in w$. So, by assumption $W \setminus {\downarrow} (x \setminus y) \in v$. Since v is a filter, $(W \setminus {\downarrow} (x \setminus y)) \cap x \in v$ so by $(\dagger)$ $y \in v$.

Conversely, assume that for all $x, y \in \mathrm{Up}(W)$ such that $x \to y \in w$ and $x \in v$ we have $y \in v$. Take any $z \in w$. Since w is a filter, for an arbitrary x we have $x \to z \in w$. Pick any $x \in v$. Then, by assumption $z \in v$, proving $w \subseteq v$. $\qquad\square$

We end this section by two propositions stating the results we will need in Section 5.

Proposition 10. *Let $\{\mathfrak{F}_i : i \in I\} \cup \{\mathfrak{G}, \mathfrak{H}\}$ be a set of Kripke frames. The following hold:*

1. $\left(\coprod_{i \in I} \mathfrak{F}_i\right)^+ \cong \prod_{i \in I} \mathfrak{F}_i^+$.

2. If $\mathfrak{G}$ is a generated subframe of $\mathfrak{H}$, then $\mathfrak{H}^+$ is a homomorphic image of $\mathfrak{G}^+$.

3. If $\mathfrak{G}$ is a bounded-morphic image of $\mathfrak{H}$, then $\mathfrak{H}^+$ is a subalgebra of $\mathfrak{G}^+$.

Proposition 11. *Let $\mathbf{A}$ and $\mathbf{B}$ be double-Heyting algebras. The following hold:*

1. $\mathbf{A}$ is a subalgebra of $(\mathbf{A}_+)^+$.

2. If $\mathbf{A}$ is a homomorphic image of $\mathbf{B}$, then $\mathbf{B}_+$ is a generated subframe of $\mathbf{A}_+$.

3. If $\mathbf{A}$ is a subalgebra of $\mathbf{B}$, then $\mathbf{B}_+$ is a bounded-morphic image of $\mathbf{A}_+$.

The double dual algebra $(\mathbf{A}_+)^+$ is known as the *canonical extension* of $\mathbf{A}$.

3.4 Directed Bisimulations

The relations (directed bisimulations) that we introduce in this section originally appeared in [17] in a different form. The current version has essentially been used in [1, 18].

Definition 13. *Let $\mathfrak{M} = (W, \leq, V)$ and $\mathfrak{M}' = (W', \leq', V')$ be two Kripke models. A* directed bisimulation *between $\mathfrak{M}$ and $\mathfrak{M}'$ is a pair (Z_1, Z_2) of relations $Z_1 \subseteq W \times W'$ and $Z_2 \subseteq W' \times W$ such that:*

(D_1) *if wZ_1w' and $w \in V(p)$, then $w' \in V'(p)$, for all $p \in \mathrm{Prop}$;*

(D_2) *if $w'Z_2w$ and $w' \in V'(p)$, then $w \in V(p)$, for all $p \in \mathrm{Prop}$;*

(D_3) *if wZ_1w' and $w' \leq' v'$, then $\exists v \in W$ such that $w \leq v$ and vZ_1v' and $v'Z_2v$;*

(D_4) *if $w'Z_2w$ and $w \leq v$, then $\exists v' \in W'$ such that $w' \leq' v'$ and $v'Z_2v$ and vZ_1v';*

(D_5) *if wZ_1w' and $v \leq w$, then $\exists v' \in W'$ such that $v' \leq' w'$ and vZ_1v' and $v'Z_2v$;*

(D_6) *if $w'Z_2w$ and $v' \leq' w'$, then $\exists v \in W$ such that $v \leq w$ and $v'Z_2v$ and vZ_1v'.*

Two worlds $w \in W$ and $w' \in W'$ are called directed bisimilar *if there exists a directed bisimulation (Z_1, Z_2) from $\mathfrak{M}$ to $\mathfrak{M}'$ such that wZ_1w'. We denote this by $\mathfrak{M}, w \rightharpoonup \mathfrak{M}', w'$.*

Directed bisimilar states preserve truth, as shown in the following proposition. (For a proof see [1].)

Proposition 12. *Let $\mathfrak{M} = (W, \leq, V)$ and $\mathfrak{M}' = (W', \leq', V')$ be Kripke models, $w \in W$ and $w' \in W'$. If $\mathfrak{M}, w \rightharpoonup \mathfrak{M}', w'$, then for any formula $\varphi \in \mathbf{BI}$ we have that $\mathfrak{M}, w \Vdash \varphi$ only if $\mathfrak{M}', w' \Vdash \varphi$.*

Bounded morphisms give rise to directed bisimulations.

Proposition 13. *Let $\mathfrak{M}_1 = (W_1, \leq_1, V_1)$ and $\mathfrak{M}_2 = (W_2, \leq_2, V_2)$ be two Kripke models for bi-intuitionistic logic. Furthermore, let $f \colon W_1 \longrightarrow W_2$ be a bounded morphism. Then, the pair $\langle Z_1, Z_2 \rangle$ is a directed bisimulation, where:*

$$xZ_1y \quad \text{iff} \quad f(x) \leq_2 y,$$
$$xZ_2y \quad \text{iff} \quad x \leq_2 f(y).$$

Moreover, if f is surjective, then the directed bisimulation $\langle Z_1, Z_2 \rangle$ is surjective with respect to both $\mathfrak{M}_1$ and $\mathfrak{M}_2$.

Proof. Let $i, j \in \{1, 2\}$, $i \neq j$. First, if xZ_iy, we immediately have $\mathfrak{M}_i, x \Vdash p$ only if $\mathfrak{M}_j, y \Vdash p$ by persistence of atomic formulas across the partial ordering $\leq_i$.

If xZ_1y (i.e., $f(x) \leq_2 y$) and $y \leq_2 b$ for some $b \in W_2$, then $f(x) \leq_2 b$ and we have that there exists $b' \in W_1$ such that $x \leq_1 b'$ and $f(b') = b$ by properties of bounded morphisms. Thus bZ_2b' since $b \leq_2 f(b')$ and $b'Z_1b$ as $f(b') \leq_2 b$.

If xZ_1y (i.e., $f(x) \leq_2 y$) and $b \leq_1 x$ for some $b \in W_1$, so $f(b) \leq_2 f(x)$ by properties of bounded morphisms, and thus $f(b) \leq_2 y$. But then $bZ_1f(b)$ since $f(b) \leq_2 f(b)$ and $f(b)Z_2b$ given that $f(b) \leq_2 f(b)$.

The remaining conditions are shown similarly. Finally, suppose that f is surjective. Then if $y \in W_1$, we have that $f(y)Z_2y$ since $f(y) \leq_2 f(y)$. Moreover, if $y \in W_2$ there must be x such that $f(x) = y$ by surjectivity, so $f(x) \leq_2 y$ and hence xZ_1y as desired. $\square$

Proposition 14. *Let $\mathfrak{M} = (W, \leq, V)$ and $\mathfrak{M}' = (W', \leq', V')$ be two Kripke models and $\langle Z_1, Z_2 \rangle$ (where $Z_1 \subseteq W \times W'$ and $Z_2 \subseteq W' \times W$) be a directed bisimulation surjective with respect to $\mathfrak{M}'$. Then $\mathfrak{M} \Vdash \varphi$ only if $\mathfrak{M}' \Vdash \varphi$, for every formula φ of* **BI**.

Proof. This is a corollary of the preservation under directed bisimulations of the formulas of **BI**. Suppose that $\mathfrak{M}' \nVdash \varphi$, so there is $w' \in W'$ such that $\mathfrak{M}', w' \nVdash \varphi$. Given that we have assumed $\langle Z_1, Z_2 \rangle$ to be surjective with respect to $\mathfrak{M}'$, there must be $w \in W$ such that wZ_1w'. By the preservation of the formulas of **BI** under directed bisimulations, we must have that $\mathfrak{M}, w \nVdash \varphi$ and thus $\mathfrak{M} \nVdash \varphi$. $\square$

Let us write $\mathsf{bitp}_{\mathfrak{M}}(e)$ for the 'bi-intuitionistic type' of a point e in a model $\mathfrak{M}$, i.e., the set of all first order translations of bi-intuitionistic formulas that e satisfies. If $\mathfrak{M}_1, \mathfrak{M}_2$ are models, we will write $\mathfrak{M}_1, w \Rightarrow_{\mathbf{BI}} \mathfrak{M}_2, v$ if $\mathfrak{M}_1, w \Vdash \varphi$ implies $\mathfrak{M}_2, v \Vdash \varphi$, for every formula φ of **BI**. If $\mathfrak{M}_1, w \Rightarrow_{\mathbf{BI}} \mathfrak{M}_2, v$ and $\mathfrak{M}_2, v \Rightarrow_{\mathbf{BI}} \mathfrak{M}_1, w$ both hold, we write $\mathfrak{M}_1, w \equiv_{\mathbf{BI}} \mathfrak{M}_2, v$ and say that $\mathfrak{M}_1, w$ and $\mathfrak{M}_2, v$ are directed bisimulation equivalent. Clearly, if $\mathfrak{M}_1, w \equiv_{\mathbf{BI}} \mathfrak{M}_2, v$ then they satisfy precisely the same bi-intuitionistic formulas.

Proposition 15. *Let $\mathfrak{M}_1$ and $\mathfrak{M}_2$ be two Kripke models. Suppose that $\mathfrak{M}_1$ and $\mathfrak{M}_2$ are ω-saturated as first order models. Then the relation $\Rightarrow_{\mathbf{BI}}$ induces a directed bisimulation $\langle Z_1, Z_2 \rangle$ between $\mathfrak{M}_1$ and $\mathfrak{M}_2$ defined as follows:*

$$xZ_1y \quad \textit{iff} \quad \mathsf{bitp}_{\mathfrak{M}_1}(x) \subseteq \mathsf{bitp}_{\mathfrak{M}_2}(y),$$

$$xZ_2y \quad \textit{iff} \quad \mathsf{bitp}_{\mathfrak{M}_2}(x) \subseteq \mathsf{bitp}_{\mathfrak{M}_1}(y).$$

Proof. In what follows let $\{i, j\} = \{1, 2\}$. By definition, if xZ_iy, i.e., $\mathsf{bitp}_{\mathfrak{M}_i}(x) \subseteq \mathsf{bitp}_{\mathfrak{M}_j}(y)$, we have that $\mathfrak{M}_i, w \Vdash \varphi$ only if $\mathfrak{M}_j, u \Vdash \varphi$ for every formula φ of **BI**, and, in particular, $\mathfrak{M}_i, w \Vdash p$ only if $\mathfrak{M}_j, u \Vdash p$ for every propositional variable.

Now suppose that xZ_iy, i.e., $\mathsf{bitp}_{\mathfrak{M}_i}(x) \subseteq \mathsf{bitp}_{\mathfrak{M}_j}(y)$, and $y \leq_j b$ for some b of $\mathfrak{M}_j$. Consider

$$\mathsf{nbitp}_{\mathfrak{M}_j}(y) = \{\neg ST_x(\psi) \colon \mathfrak{M}_j, y \not\Vdash \psi, \psi \in \mathbf{BI}\}.$$

We claim that the set of formulas $\mathsf{bitp}_{\mathfrak{M}_j}(b) \cup \mathsf{nbitp}_{\mathfrak{M}_j}(b)$ is satisfiable in $\mathfrak{M}_i$ by an element b' such that $x \leq_i b'$. Take any finite subset S of $\mathsf{bitp}_{\mathfrak{M}_j}(b) \cup \mathsf{nbitp}_{\mathfrak{M}_j}(b)$. Say, $\{ST_z(\delta_1), \ldots, ST_z(\delta_n)\} \subseteq \mathsf{bitp}_{\mathfrak{M}_j}(b)$, $\{\neg ST_z(\sigma_1), \ldots, \neg ST_z(\sigma_m)\} \subseteq \mathsf{nbitp}_{\mathfrak{M}_j}(b)$, and

$$S = \{ST_z(\delta_1), \ldots, ST_z(\delta_n)\} \cup \{\neg ST_z(\sigma_1), \ldots, \neg ST_z(\sigma_m)\}.$$

It is then clear that $\mathfrak{M}_j, y \not\Vdash \bigwedge\{\delta_1, \ldots, \delta_n\} \to \bigvee\{\sigma_1, \ldots, \sigma_m\}$, and thus, $\mathfrak{M}_j \not\models ST_z(\bigwedge\{\delta_1, \ldots, \delta_n\} \to \bigvee\{\sigma_1, \ldots, \sigma_m\})[y]$. Given that $\mathsf{bitp}_{\mathfrak{M}_i}(x) \subseteq \mathsf{bitp}_{\mathfrak{M}_j}(y)$, we get $\mathfrak{M}_i \not\models ST_z(\bigwedge\{\delta_1, \ldots, \delta_n\} \to \bigvee\{\sigma_1, \ldots, \sigma_m\})[x]$. It follows that S is satisfiable in $\mathfrak{M}_i$ by an element b_0 such that $x \leq_i b_0$. By the ω-saturation of $\mathfrak{M}_i$, there must be an element b' such that $x \leq_i b'$ realising the whole of $\mathsf{bitp}_{\mathfrak{M}_j}(b) \cup \mathsf{nbitp}_{\mathfrak{M}_j}(b)$. Since $\mathsf{bitp}_{M_j}(b)$ is realised by b', we have that bZ_jb'. Since b' realises $\mathsf{nbitp}_{\mathfrak{M}_j}(b)$, that is, $\mathfrak{M}_j, b \not\Vdash \psi$ only if $\mathfrak{M}_i, b' \not\Vdash \psi$, by contraposing we obtain that $\mathsf{bitp}_{\mathfrak{M}_i}(b') \subseteq \mathsf{bitp}_{\mathfrak{M}_j}(b)$, that is, $b'Z_ib$.

On the other hand, suppose that xZ_iy, that is, $\mathsf{bitp}_{\mathfrak{M}_i}(x) \subseteq \mathsf{bitp}_{\mathfrak{M}_j}(y)$, and $b \leq_i x$, for some b from $\mathfrak{M}_i$. We claim that the set of formulas $\mathsf{bitp}_{\mathfrak{M}_i}(b) \cup \mathsf{nbitp}_{\mathfrak{M}_i}(b)$ is satisfiable in $\mathfrak{M}_j$ by an element b' such that $b' \leq_j y$. As before, take any finite subset S of $\mathsf{bitp}_{\mathfrak{M}_i}(b) \cup \mathsf{nbitp}_{\mathfrak{M}_i}(b)$. Say, $\{ST_z(\delta_1), \ldots, ST_z(\delta_n)\} \subseteq \mathsf{bitp}_{\mathfrak{M}_i}(b)$, $\{\neg ST_z(\sigma_1), \ldots, \neg ST_z(\sigma_m)\} \subseteq \mathsf{nbitp}_{\mathfrak{M}_i}(b)$, and

$$S = \{ST_z(\delta_1), \ldots, ST_z(\delta_n)\} \cup \{\neg ST_z(\sigma_1), \ldots, \neg ST_z(\sigma_m)\}.$$

It is clear that $\mathfrak{M}_i, x \Vdash \bigwedge\{\delta_1, \ldots, \delta_n\} \prec \bigvee\{\sigma_1, \ldots, \sigma_m\}$, so given that $\mathsf{bitp}_{\mathfrak{M}_i}(x) \subseteq \mathsf{bitp}_{\mathfrak{M}_j}(y)$, we obtain $\mathfrak{M}_j, y \Vdash \bigwedge\{\delta_1, \ldots, \delta_n\} \prec \bigvee\{\sigma_1, \ldots, \sigma_m\}$. It follows that S is satisfiable in $\mathfrak{M}_j$ by an element b_0 such that $b_0 \leq_j y$. By the ω-saturation of $\mathfrak{M}_j$, there must be an element b' such that $b' \leq_j y$ realising the whole of $\mathsf{bitp}_{\mathfrak{M}_i}(b) \cup \mathsf{nbitp}_{\mathfrak{M}_i}(b)$. Since $\mathsf{bitp}_{\mathfrak{M}_i}(b)$ is realised by b', we have that bZ_ib' and since b' realises $\mathsf{nbitp}_{\mathfrak{M}_i}(b)$, we have $b'Z_jb$. $\qquad\square$

3.5 Definable Prime Filter Extensions

Recall from Section 3.3 the double-dual model $(\mathfrak{M}^+)_+ = ((\mathfrak{F}^+)_+, V^*)$, with $\mathfrak{F} = (W, \leq)$. We will call $(\mathfrak{M}^+)_+$ a *prime filter extension* of $\mathfrak{M}$ and denote it by $\mathfrak{pe}(\mathfrak{M})$. This is because the mapping $\pi \colon W \longrightarrow \mathcal{F}_p(\mathrm{Up}(W))$ defined by $\pi(w) = \{x \in \mathrm{Up}(W) \colon w \in x\}$ is an embedding, in the standard model-theoretic sense, of the structure $\mathfrak{M}$ into $\mathfrak{pe}(\mathfrak{M})$.

For $x, y \subseteq \mathrm{Up}(W)$ we will say that x *is separated from* y if for any finite subsets $x' \subseteq x$ and $y' \subseteq y$ we have that $\bigcap x' \not\subseteq \bigcup y'$. The following version of the prime filter theorem is well known.

Lemma 16. *If x is separated from y, then x is included in a prime filter of $\mathrm{Up}(W)$ that is disjoint from y.*

Proposition 17. *Let $\mathfrak{M} = (W, \leq, V)$ be a Kripke model for bi-intuitionistic logic. Then for any $u \in \mathcal{F}_p(\mathrm{Up}(W))$ and any formula φ of* **BI** *the following equivalences hold:*

1. $V(\varphi) \in u \quad$ iff $\quad \mathbf{pe}(\mathfrak{M}), u \Vdash \varphi$.

*2. $\mathfrak{M}, w \Vdash \varphi \quad$ iff $\quad \mathbf{pe}(\mathfrak{M}), \pi(w) \Vdash \varphi$ for any $\varphi \in$ **BI**.*

*3. $\mathfrak{M} \Vdash \varphi \quad$ iff $\quad \mathbf{pe}(\mathfrak{M}) \Vdash \varphi$ for any $\varphi \in$ **BI**.*

Proof. We establish (1) by induction on the complexity of φ. The case when φ is a propositional variable is immediate by definition of $V^{\mathbf{pe}(\mathfrak{M})}$. The case when $\varphi = \bot$ is trivial since $V(\bot) = \emptyset$, which can never belong to a filter u. The cases for $\vee$ and $\wedge$ are obvious.

Let $\varphi = \psi \to \chi$. Suppose that $\mathbf{pe}(\mathfrak{M}), u \nVdash \psi \to \chi$. Since the order in $\mathbf{pe}(\mathfrak{M})$ is prime-filter inclusion, we get that there is u_1 such that $u \subseteq u_1$ while $\mathbf{pe}(\mathfrak{M}), u_1 \Vdash \psi$ and $\mathbf{pe}(\mathfrak{M}), u_1 \nVdash \chi$. By inductive hypothesis, $V(\psi) \in u_1$ and $V(\chi) \notin u_1$. Hence $V(\psi) \to V(\chi) = V(\psi \to \chi) \notin u$ by Lemma 9. On the other hand, suppose that $\mathbf{pe}(\mathfrak{M}), u \Vdash \psi \to \chi$. Next, we claim that $u \cup \{V(\psi)\}$ cannot be separated from $\{V(\chi)\}$. For otherwise, there would be $v \in \mathcal{F}_p(\mathrm{Up}(W))$ such that $u \subseteq v, V(\psi) \in v$, and $V(\chi) \notin v$. By inductive hypothesis, this would imply that $\mathbf{pe}(\mathfrak{M}), v \Vdash \psi$ and $\mathbf{pe}(\mathfrak{M}), v \nVdash \chi$, contradicting $\mathbf{pe}(\mathfrak{M}), u \Vdash \psi \to \chi$. Thus $u \cup \{V(\psi)\}$ cannot be separated from $\{V(\chi)\}$ and since u is closed under finite intersections there must be some $y \in u$ with $y \cap V(\psi) \subseteq V(\chi)$, which entails that $y \subseteq V(\psi \to \chi)$ and hence $V(\psi \to \chi) \in u$ as u is a filter.

Suppose $V(\psi \prec \chi) \notin u$. Let $u' \in \mathcal{F}_p(\mathrm{Up}(W))$ be any prime filter such that $u' \subseteq u$ and suppose $\mathbf{pe}(\mathfrak{M}), u' \Vdash \psi$. Then by the induction hypothesis we have $V(\psi) \in u'$. Since $V(\psi) \subseteq V(\chi) \cup V(\psi \prec \chi)$ we find that $V(\chi) \cup V(\psi \prec \chi) \in u'$. We have $V(\psi \prec \chi) \notin u'$, because $u' \subseteq u$, so since u' is prime it must be the case that $V(\chi) \in u'$. The induction hypothesis then implies $\mathbf{pe}(\mathfrak{M}), u' \Vdash \chi$. By Lemma 9, it implies that no prime filter u' below u can satisfy ψ but not χ, and therefore $\mathbf{pe}(\mathfrak{M}), u \nVdash \psi \prec \chi$. For the converse, suppose $\mathbf{pe}(\mathfrak{M}), u \nVdash \psi \prec \chi$, and let $I = \mathrm{Up}(W) \setminus u$. Then, by well-known properties of distributive lattices, I is a prime ideal. Suppose towards a contradiction that there is no $a \in I$ such that

$V(\psi) \subseteq V(\chi) \cup a$. Then $\{V(\psi)\}$ is separated from $I \cup \{V(\chi)\}$, so there exists a prime filter u' containing $V(\psi)$ which is disjoint from $I \cup \{V(\chi)\}$. This implies $V(\psi) \in u'$ and $V(\chi) \notin u'$, so by the induction hypothesis we have $\mathbf{pe}(\mathfrak{M}), u' \Vdash \psi$ while $\mathbf{pe}(\mathfrak{M}), u' \nVdash \chi$. But since $u' \subseteq \mathrm{Up}(W) \setminus I = u$, this contradicts the assumption that $u \nVdash \psi \prec \chi$. So there must be some $a \in I$ such that $V(\psi) \subseteq V(\chi) \cup a$. This implies $V(\psi \prec \chi) \subseteq a$. Since $a \in I$ we have $a \notin u$, and hence $V(\psi \prec \chi) \notin u$.

For part (2) note that we have $\mathfrak{M}, w \Vdash \varphi$ iff $w \in V(\varphi)$ iff $V(\varphi) \in \{x \in \mathrm{Up}(W) \colon w \in x\} = \pi(w)$ iff, by part (1), $\mathbf{pe}(\mathfrak{M}), \pi(w) \Vdash \varphi$.

For part (3) suppose that $\mathfrak{M} \Vdash \varphi$. Then $V(\varphi) = W$ and hence $V(\varphi) \in u$ for every $u \in \mathcal{F}_p(\mathrm{Up}(W))$, so $\mathbf{pe}(\mathfrak{M}), u \Vdash \varphi$. Thus $\mathbf{pe}(\mathfrak{M}) \Vdash \varphi$ since u was arbitrary. Conversely if $\mathfrak{M} \nVdash \varphi$, there is some $w \in W$ such that $\mathfrak{M}, w \nVdash \varphi$. Thus, $\mathbf{pe}(\mathfrak{M}), \pi(w) \nVdash \varphi$, and $\mathbf{pe}(\mathfrak{M}) \nVdash \varphi$ as desired. $\square$

Definition 14. *Let $\mathfrak{M} = (W, \leq, V)$ be a model. The* definable dual algebra $\mathfrak{M}^{+\delta}$ *of $\mathfrak{M}$ is the structure $\langle U(\mathfrak{M}), \to, \prec, \cap, \cup, 1, 0 \rangle$, where $U(\mathfrak{M}) = \{V(\varphi) \colon \varphi \in \mathbf{BI}\}$ and the operations are defined as in Definition 10. In particular, $V(\top) = W$ and $V(\bot) = \emptyset$.*

Proposition 18. *Let $\mathfrak{M}$ be a Kripke model for bi-intuitionistic logic. Then $\mathfrak{M}^{+\delta}$ is a double Heyting algebra. Indeed, $\mathfrak{M}^{+\delta}$ is a subalgebra of $\mathfrak{M}^+$ generated by the set $\{V(p) \colon p \in \mathrm{Prop}\}$.*

Definition 15. *Let $\mathfrak{M} = (W, \leq, V)$ be a model. The* dual model $(\mathfrak{M}^{+\delta})_+$ *of $\mathfrak{M}^{+\delta}$ is the structure $\langle \mathcal{F}_p(U(\mathfrak{M})), \leq^*, V^* \rangle$, where*

1. *$\mathcal{F}_p(U(\mathfrak{M}))$ is the set of all prime filters of $U(\mathfrak{M})$ in the algebra $\mathfrak{M}^{+\delta}$,*

2. *$w \leq^* v$ iff for all $x, y \in U(\mathfrak{M})$, if $x \to y \in w$ and $x \in v$ then $y \in v$,*

3. *$V^*(p) = \{w \in \mathcal{F}_p(U(\mathfrak{M})) \colon V(p) \in w\}$.*

We will write $\mathfrak{M}^\delta$ for $(\mathfrak{M}^{+\delta})_+$, and call it the *definable prime filter extension* of $\mathfrak{M}$. It is easy to see that $\mathfrak{M}$ embeds into $\mathfrak{M}^\delta$ via the map $\pi \colon W \longrightarrow \mathcal{F}_p(U(\mathfrak{M}))$ given by

$$\pi(w) = \{x \in U(\mathfrak{M}) \colon w \in x\}.$$

By Proposition 11(3) it follows that $\mathfrak{M}^\delta$ is a bounded-morphic image of $\mathbf{pe}(\mathfrak{M})$, but we will prove it directly below.

Proposition 19. *Let $\mathfrak{M}$ be a Kripke model for bi-intuitionistic logic. The map $x \mapsto x \cap U(\mathfrak{M})$ is a bounded morphism from $\mathbf{pe}(\mathfrak{M})$ onto $\mathfrak{M}^\delta$. Moreover, for each $x \in \mathcal{F}_p(U(\mathfrak{M}))$ there is $y \in \mathcal{F}_p(\mathrm{Up}(W))$ such that $x = y \cap U(\mathfrak{M})$.*

Proof. Consider the inclusion homomorphism i from $\mathfrak{M}^{+\delta}$ into $\mathfrak{M}^+$. A dual mapping $i_+ \colon (\mathfrak{M}^+)_+ \longrightarrow (\mathfrak{M}^{+\delta})_+$ is defined by the equation:

$$i_+(x) = i^{-1}(x).$$

But $i^{-1}(x) = \{y \in U(\mathfrak{M}) \colon y \in x\}$, so i_+ is indeed the map mentioned in the statement of the proposition. Furthermore, i_+ is a bounded morphism. The only thing that we need to notice is that $\mathfrak{pe}(\mathfrak{M}), x \Vdash p$ iff $V(p) \in x$ iff $V(p) \in x \cap U(\mathfrak{M})$ iff $\mathfrak{M}^\delta, x \cap U(\mathfrak{M}) \Vdash p$, for any propositional variable p of **BI**.

Finally, let $x \in \mathcal{F}_p(U(\mathfrak{M}))$. Since x is a prime filter of $U(\mathfrak{M})$ it is not difficult to see that x is separated from $U(\mathfrak{M}) \setminus x$. Hence, by Lemma 16, there is a prime filter $y \subseteq U(\mathfrak{M})$, such that $y \supseteq x$ and $y \cap (U(\mathfrak{M}) \setminus x) = \emptyset$. It follows that $x = y \cap U(\mathfrak{M})$, as desired. A similar argument establishes that the mapping under consideration is surjective. $\square$

We will say that a class K is *closed under surjective directed bisimulations* if whenever $\mathfrak{M} = (W, \leq, V)$ and $\mathfrak{M}' = (W', \leq', V')$ are two Kripke models and $\langle Z_1, Z_2 \rangle$ (where $Z_1 \subseteq W \times W'$ and $Z_2 \subseteq W' \times W$) is a directed bisimulation surjective with respect to $\mathfrak{M}'$, then $\mathfrak{M} \in K$ only if $\mathfrak{M}' \in K$.

Corollary 20. *Suppose K is a class of Kripke models for bi-intuitionistic logic closed under surjective directed bisimulations. Then $\mathfrak{pe}(\mathfrak{M}) \in K$ if and only if $\mathfrak{M}^\delta \in K$.*

Proof. Let g be the bounded morphism $x \mapsto x \cap U(\mathfrak{M})$, given in Proposition 19. By Proposition 13, the pair $\langle Z_1, Z_2 \rangle$ is a directed bisimulation which is surjective with respect to both $\mathfrak{M}^\delta$ and $\mathfrak{pe}(\mathfrak{M})$, where:

$$xZ_1y \quad \text{iff} \quad g(x) \leq^{\mathfrak{M}^\delta} y,$$
$$xZ_2y \quad \text{iff} \quad x \leq^{\mathfrak{M}^\delta} g(y).$$

Consequently, $\mathfrak{pe}(\mathfrak{M}) \in K$ iff $\mathfrak{M}^\delta \in K$ by the closure assumption on K. $\square$

Corollary 21. *Let $\mathfrak{M}$ be a Kripke model for bi-intuitionistic logic. Then for any prime filter u of $U(\mathfrak{M})$ and formula φ of **BI** we have that $V(\varphi) \in u$ if and only if $\mathfrak{M}^\delta, u \Vdash \varphi$. Moreover, $\mathfrak{M} \Vdash \varphi$ if and only if $\mathfrak{M}^\delta \Vdash \varphi$ for any such φ.*

Proof. Consider the mapping given in Proposition 19. We have that $\mathfrak{pe}(\mathfrak{M}), x \Vdash \varphi$ iff $\mathfrak{M}^\delta, x \cap U(\mathfrak{M}) \Vdash \varphi$ and that indeed all elements of $\mathfrak{M}^\delta$ are of the form $x \cap U(\mathfrak{M})$. Therefore, using Proposition 17, we see that if $u = x \cap U(\mathfrak{M})$ is a prime filter of $U(\mathfrak{M})$, $V(\varphi) \in u$ iff $V(\varphi) \in x$ iff $\mathfrak{pe}(\mathfrak{M}), x \Vdash \varphi$ iff $\mathfrak{M}^\delta, u \Vdash \varphi$.

The last part of the result follows from Proposition 17 again using Corollary 20. $\square$

According to Corollaries 14, 21 and Proposition 17, we can see that a class of models K definable by a theory of **BI** is always going to be closed under surjective directed bisimulations, (definable) prime extensions and disjoint unions. Indeed, both K and its complement $\overline{K}$ will be closed under (definable) prime extensions.

Definition 16. *Let $\mathfrak{M}$ be a model. A submodel (in the classical first-order sense) $\mathfrak{M}'$ of $\mathfrak{M}$ will be called an* inner submodel *of $\mathfrak{M}$ if the pair $\langle I, I \rangle$, where I is the identity relation on $\mathfrak{M}'$, is a directed bisimulation between $\mathfrak{M}$ and $\mathfrak{M}'$.*

Clearly, if $\mathfrak{M}'$ is an inner submodel of $\mathfrak{M}$, then $\langle I, I \rangle$ is a directed bisimulation surjective with respect to $\mathfrak{M}'$. Note that, using Corollary 14, it follows that a definable class of models is always going to be closed under inner submodels, that is, given a class K, if $\mathfrak{M} \in K$ and $\mathfrak{M}'$ is an inner submodel of $\mathfrak{M}$, then $\mathfrak{M}' \in K$.

Lemma 22. *Let $\{\mathfrak{M}_i : i \in I\}$ be a family of Kripke models and $\prod_{i \in I} \mathfrak{M}_i / U$ an ultraproduct. Then $\prod \mathfrak{M}_i / U$ is isomorphic to an inner submodel of the ultrapower $(\coprod_{i \in I} \mathfrak{M}_i)^I / U$. Hence, a class closed under disjoint unions, inner submodels and ultrapowers is closed under ultraproducts.*

Proof. Simply consider the mapping $f/U \mapsto f'/U$ where $f' : I \longrightarrow \bigcup_{i \in I} W_i$ is defined by $f'(i) = f(i)$. $\square$

Lemma 23. *Let $\mathbf{BI}^{corr+}$ be obtained by adding a list of constants to $\mathbf{BI}^{corr}$. Suppose K is a class of Kripke models which is closed under ultraproducts. Then for any set Θ of formulas of $\mathbf{BI}^{corr+}$, if Θ is finitely satisfiable in K then Θ is satisfiable in K.*

Proof. This is simply the proof of the compactness theorem using ultraproducts which can be found for example in [3] (Theorem 4.1). $\square$

If $\mathfrak{M} = (W, \leq, V)$ is a model and $x \in W$, we consider the inner submodel $\mathfrak{M}_x$ with the domain

$$W_x = \{y \mid \exists n \in \mathbb{N}\, \exists z_0, \ldots, z_n (\bigwedge_{i < n} (z_{i+1} \leq z_i \ \lor \ z_i \leq z_{i+1}) \land (z_n = y) \land (z_0 = x))\},$$

where both the ordering and the valuation are induced from $\mathfrak{M} = (W, \leq, V)$.

Lemma 24. *Let $\mathfrak{M} = (W, \leq, V)$ be a Kripke model for bi-intuitionistic logic and K a class of models closed under directed bisimulations and disjoint unions, then*

$$\mathfrak{M} \in K \quad iff \quad \mathfrak{M}_x \in K, \text{ for all } x \in W.$$

Proof. First, we can take isomorphic copies of the models $\mathfrak{M}_x$ ($x \in W$) to make sure that they are pairwise disjoint. For definiteness, we can simply let the domain of $\mathfrak{M}_x$ be the set $\{\langle w, x \rangle \mid w \in W_x\}$. Notice that for the disjoint union $\coprod_{x \in W} \mathfrak{M}_x$, the mapping $\langle w, x \rangle \mapsto w$ is a surjective bounded morphism (cf. the remark after Definition 9). This suffices to establish our result. $\square$

Proposition 25. *Let $\mathfrak{M} = (W, \leq, V)$ be a Kripke model and $\mathfrak{N} = (W', \leq', V')$ an ω-saturated Kripke model such that*

$$\mathfrak{N} \Vdash \varphi \quad \textit{iff} \quad \mathfrak{M} \Vdash \varphi$$

for any formula $\varphi \in \mathbf{BI}$. Then there is a directed bisimulation surjective with respect to both $\mathfrak{N}$ and $\mathfrak{M}^\delta$ between these two models.

Proof. Given $x \in W'$, let

$$f(x) = \{V(\varphi) \colon \varphi \in \mathbf{BI} \,\&\, \mathfrak{N}, x \Vdash \varphi\}.$$

It is not difficult to verify that $f(x)$ is indeed in the domain of $\mathfrak{M}^\delta$. Now define the following pair of relations:

$$x Z_1 y \quad \text{iff} \quad x \subseteq f(y),$$
$$x Z_2 y \quad \text{iff} \quad f(x) \subseteq y.$$

Next we show that $\langle Z_1, Z_2 \rangle$ is a directed bisimulation surjective with respect to both $\mathfrak{N}$ and $\mathfrak{M}^\delta$.

Assume that $x Z_1 y$, that is, $x \subseteq f(y)$. Now, $\mathfrak{M}^\delta, x \Vdash p$ implies that $V(p) \in x$, so by assumption, $V(p) \in f(y)$ as well. The latter means that $\mathfrak{N}, y \Vdash p$ by definition of f. On the other hand if $x Z_2 y$, i.e., $f(x) \subseteq y$ and $\mathfrak{N}, x \Vdash p$, then $V(p) \in f(x)$. Consequently, $V(p) \in y$, by our assumption, and that means that $\mathfrak{M}^\delta, y \Vdash p$ as desired.

For the next condition, suppose that $x Z_1 y$, that is, $x \subseteq f(y)$ and $y \leq' b$. Take $f(b)$. It is easy to see that $x \leq^\delta f(b)$. For assume that $V(\varphi) \to V(\psi) = V(\varphi \to \psi) \in x$ while $V(\varphi) \in f(b)$. Then $\mathfrak{N}, b \Vdash \varphi$ whereas $V(\varphi \to \psi) \in f(y)$, which means that $\mathfrak{N}, y \Vdash \varphi \to \psi$, so $\mathfrak{N}, b \Vdash \psi$ given that $y \leq' b$. Thus, $V(\psi) \in f(b)$ as desired, and $x \leq^\delta f(b)$ follows by Lemma 9. Moreover, $b Z_2 f(b)$ while $f(b) Z_1 b$. On the other hand assume that $x Z_2 y$, that is, $f(x) \subseteq y$ and $y \leq^\delta b$. Consider the set Δ defined as

$$\{ST_x(\varphi) \colon \varphi \in \mathbf{BI} \,\&\, \mathfrak{M}^\delta, b \Vdash \varphi\} \cup \{\neg ST_x(\psi) \colon \psi \in \mathbf{BI} \,\&\, \mathfrak{M}^\delta, b \not\Vdash \psi\}.$$

Take any finite $\Delta_0 \subseteq \Delta$. We may assume that

$$\Delta_0 = \{ST_x(\varphi_0), \ldots, ST_x(\varphi_j)\} \cup \{\neg ST_x(\psi_0), \ldots, \neg ST_x(\psi_k)\}.$$

Now, $\mathfrak{M}^\delta, y \nVdash \bigwedge_{i<j+1} \varphi_i \to \bigvee_{i<k+1} \psi_i$, so $V(\bigwedge_{i<j+1} \varphi_i \to \bigvee_{i<k+1} \psi_i) \notin y$, and therefore $V(\bigwedge_{i<j+1} \varphi_i \to \bigvee_{i<k+1} \psi_i) \notin f(x)$. This means that $\mathfrak{N}, x \nVdash \bigwedge_{i<j+1} \varphi_i \to \bigvee_{i<k+1} \psi_i$, hence, there must be b_0 such that $x \leq' b_0$ and $\mathfrak{N}, b_0 \Vdash \bigwedge_{i<j+1} \varphi_i$ while $\mathfrak{N}, b_0 \nVdash \bigvee_{i<k+1} \psi_i$. By the ω-saturation of $\mathfrak{N}$ we must have b' such that $x \leq' b'$ while also b' satisfies $\{ST_x(\varphi) \colon \varphi \in \mathbf{BI}\ \&\ \mathfrak{M}^\delta, b \Vdash \varphi\}$ (that is, $\mathfrak{M}^\delta, b \Rightarrow_{\mathbf{BI}} \mathfrak{N}, b'$) and b' satisfies $\{\neg ST_x(\psi) \colon \psi \in \mathbf{BI}\ \&\ \mathfrak{M}^\delta, b \nVdash \psi\}$ (that is, $\mathfrak{N}, b' \Rightarrow_{\mathbf{BI}} \mathfrak{M}^\delta, b$). Finally, $b \subseteq f(b')$, that is $bZ_1 b'$, since $V(\varphi) \in b$ implies that $\mathfrak{M}^\delta, b \Vdash \varphi$ which in turn means that $\mathfrak{N}, b' \Vdash \varphi$, so indeed $V(\varphi) \in f(b')$. Similarly, $f(b') \subseteq b$, that is $b'Z_2 b$, since $\mathfrak{N}, b' \Rightarrow_{\mathbf{BI}} \mathfrak{M}^\delta, b$.

For the next condition, suppose that $xZ_2 y$, that is, $f(x) \subseteq y$ and $b \leq' y$. Take $f(b)$. It is easy to see that $f(b) \leq^\delta f(x)$. For assume that $V(\varphi) \to V(\psi) = V(\varphi \to \psi) \in f(b)$ while $V(\varphi) \in f(x)$. Then $\mathfrak{N}, x \Vdash \varphi$ whereas $\mathfrak{N}, b \Vdash \varphi \to \psi$, so $\mathfrak{N}, x \Vdash \psi$ given that $b \leq' x$. Thus, $V(\psi) \in f(x)$ as desired, and $f(b) \leq^\delta f(x)$ follows by Lemma 9, whence also $f(b) \leq^\delta f(y)$ by transitivity of $\subseteq$. Moreover, $bZ_2 f(b)$ while $f(b)Z_1 b$.

Now suppose that $xZ_1 y$, that is, $x \subseteq f(y)$ and $b \leq x$ for some b of $\mathfrak{M}^\delta$. We must find a $b' \in W'$ such that $b' \leq' y$, $bZ_1 b'$ and $b'Z_2 b$. Consider the set Δ defined as

$$\{ST_x(\varphi) \colon \varphi \in \mathbf{BI}\ \&\ \mathfrak{M}^\delta, b \Vdash \varphi\} \cup \{\neg ST_x(\psi) \colon \varphi \in \mathbf{BI}\ \&\ \mathfrak{M}^\delta, b \nVdash \psi\}.$$

Take any finite $\Delta_0 \subseteq \Delta$. We may assume that

$$\Delta_0 = \{ST_x(\varphi_0), \ldots, ST_x(\varphi_j)\} \cup \{\neg ST_x(\psi_0), \ldots, \neg ST_x(\psi_k)\}.$$

Now, $\mathfrak{M}^\delta, x \Vdash \bigwedge_{i<j+1} \varphi_i \prec \bigvee_{i<k+1} \psi_i$, so $V(\bigwedge_{i<j+1} \varphi_i \prec \bigvee_{i<k+1} \psi_i) \in x$, so $V(\bigwedge_{i<j+1} \varphi_i \prec \bigvee_{i<k+1} \psi_i) \in f(y)$. The latter means that $\mathfrak{N}, y \Vdash \bigwedge_{i<j+1} \varphi_i \prec \bigvee_{i<k+1} \psi_i$, hence, there must be b_0 such that $b_0 \leq' y$ and $\mathfrak{N}, b_0 \Vdash \bigwedge_{i<j+1} \varphi_i$ while $\mathfrak{N}, b_0 \nVdash \bigvee_{i<k+1} \psi_i$. By the ω-saturation of $\mathfrak{N}$ we must have b' such that $b' \leq' y$ while also b' satisfies $\{ST_x(\varphi) \colon \varphi \in \mathbf{BI}\ \&\ \mathfrak{M}^\delta, b \Vdash \varphi\}$ (that is, $\mathfrak{M}^\delta, b \Rightarrow_{\mathbf{BI}} \mathfrak{N}, b'$) and b' satisfies $\{\neg ST_x(\psi) \colon \psi \in \mathbf{BI}\ \&\ \mathfrak{M}^\delta, b \nVdash \psi\}$ (that is, $\mathfrak{N}, b' \Rightarrow_{\mathbf{BI}} \mathfrak{M}^\delta, b$). Finally, $b \subseteq f(b')$, that is, $bZ_1 b'$ since $V(\varphi) \in b$ implies that $\mathfrak{M}^\delta, b \Vdash \varphi$ which means that $\mathfrak{N}, b' \Vdash \varphi$, so indeed $V(\varphi) \in f(b')$, and $f(b') \subseteq b$, that is, $b'Z_2 b$ similarly since $\mathfrak{N}, b' \Rightarrow_{\mathbf{BI}} \mathfrak{M}^\delta, b$.

Finally, we show that $\langle Z_1, Z_2 \rangle$ is surjective with respect to both $\mathfrak{N}$ and $\mathfrak{M}^\delta$. Assuming first that $x \in W'$, we have that $f(x)Z_1 x$ trivially. On the other hand, if x is a world in $\mathfrak{M}^\delta$ then it suffices to find $y \in \mathfrak{N}$ such that $\mathfrak{M}^\delta, x \nVdash \varphi$ implies that $\mathfrak{N}, y \nVdash \varphi$ for all formulas φ of $\mathbf{BI}$. This will show that $yZ_2 x$. Consider the set $\{\neg ST_x(\varphi) \colon \varphi \in \mathbf{BI}\ \&\ \mathfrak{M}^\delta, x \nVdash \varphi\}$ and take some finite subset $\{\neg ST_x(\varphi_0), \ldots, \neg ST_x(\varphi_n)\}$. We know that $\mathfrak{M}^\delta, x \nVdash \bigvee_{j<n+1} \varphi_j$, so $\mathfrak{M}^\delta \nVdash \bigvee_{j<n+1} \varphi_j$, so $\mathfrak{M} \nVdash \bigvee_{j<n+1} \varphi_j$ and by the hypothesis of the proposition, $\mathfrak{N} \nVdash \bigvee_{j<n+1} \varphi_j$. Consequently, there is $z \in \mathfrak{N}$ such that $\mathfrak{N}, z \nVdash \bigvee_{j<n+1} \varphi_j$, so z satisfies $\{\neg ST_x(\varphi_0), \ldots, \neg ST_x(\varphi_n)\}$. By the ω-saturation of $\mathfrak{N}$, there must be y satisfying $\{\neg ST_x(\varphi) \colon \varphi \in \mathbf{BI}\ \&\ \mathfrak{M}^\delta, x \nVdash \varphi\}$ as desired. $\qquad\square$

Lemma 26. *Let K be a class of Kripke models for bi-intuitionistic logic. If K is closed under surjective directed bisimulations and both K and $\overline{K}$ are closed under definable prime filter extensions, then both K and its complement are closed under ultrapowers.*

Proof. First suppose $\mathfrak{M} \in K$ and $\prod \mathfrak{M}/U$ is an ultrapower of $\mathfrak{M}$. Consider next an ultrapower $\mathfrak{N} = \prod(\prod \mathfrak{M}/U)/D$ obtained by the Keisler method which is ω-saturated. Applying Łoś theorem twice, we get that $\mathfrak{M} \Vdash \varphi$ if and only if $\mathfrak{N} \Vdash \varphi$. Thus, using Proposition 25, there is a surjective (with respect to $\mathfrak{N}$) directed bisimulation between $\mathfrak{M}^\delta$ and $\mathfrak{N}$.

Now $\mathfrak{M}^\delta \in K$ by the closure of K under definable prime filter extensions. By the closure under surjective directed bisimulations, we also see that $\mathfrak{N} \in K$. Again applying Proposition 25 with $\mathfrak{N}$ and $\prod \mathfrak{M}/U$ this time, we get that there is a directed bisimulation surjective with respect to $\prod \mathfrak{M}/U$ between $\mathfrak{N}$ and $(\prod \mathfrak{M}/U)^\delta$, so indeed the latter is in K and since $\overline{K}$ is closed under definable prime filter extensions, we must have that $\prod \mathfrak{M}/U \in K$, as desired.

On the other hand if $\mathfrak{M} \in \overline{K}$, since also $\mathfrak{M}^\delta \in \overline{K}$ then, by Proposition 25, we have that $\prod(\prod \mathfrak{M}/U)/D \in \overline{K}$ by the closure of K under surjective directed bisimulations. Moreover, $\prod \mathfrak{M}/U \in \overline{K}$ by the closure of K under ultrapowers established above. $\square$

4 Classes of Models Axiomatisable in BiInt

In this section we are going to characterise classes of models *axiomatisable* by theories in **BI**. The exact meaning of axiomatisability that we employ here will be clarified in the following definition, where, as usual, for a set $\Theta \subseteq \mathbf{BI}$, we write $\mathrm{Mod}(\Theta)$ for $\{\mathfrak{M} \mid \mathfrak{M} \Vdash \Theta\}$.

Definition 17. *A class K of Kripke models is said to be* axiomatisable *or* definable *in* **BI** *if there is a set Θ of formulas of* **BI** *such that $\mathrm{Mod}(\Theta) = K$.*

Theorem 27. *Let K be a class of Kripke models for bi-intuitionistic logic. Then the following are equivalent:*

1. *K is bi-intuitionistically definable, that is, $K = \mathrm{Mod}(\Theta)$ for some collection Θ of formulas of* **BI**.

2. *K is closed under surjective directed bisimulations and disjoint unions while both K and its complement $\overline{K}$ are closed under prime filter extensions.*

3. *K is closed under surjective directed bisimulations and disjoint unions while both K and its complement $\overline{K}$ are closed under definable extensions.*

    4. K is closed under surjective directed bisimulations and disjoint unions while both K and its complement $\overline{K}$ are closed under ultrapowers.

Proof. $(1) \Rightarrow (2)$: Closure under surjective directed bisimulations comes from Corollary 14. Closure under disjoint unions is a consequence of previous results. Finally, the remaining closure properties follow from Proposition 17.

$(2) \Rightarrow (3)$: By Corollary 20, we see that if K is closed under prime filter extensions, then it is also closed under definable extensions.

$(3) \Rightarrow (4)$: By Lemma 26.

$(4) \Rightarrow (1)$: Suppose that K and $\overline{K}$ are closed as indicated. Observe that K is closed under ultraproducts according to Lemma 22. Let

$$\mathrm{Th}(K) = \{\varphi \in \mathbf{BI} \colon \mathfrak{M} \Vdash \varphi, \text{ for all } \mathfrak{M} \in K\}.$$

All we need to do is show that $K = \mathrm{Mod}(\mathrm{Th}(K))$. The direction $K \subseteq \mathrm{Mod}(\mathrm{Th}(K))$ is obvious, so let $\mathfrak{M} = (W, \leq, V) \in \mathrm{Mod}(\mathrm{Th}(K))$ to establish that $\mathfrak{M} \in K$. To this end, we fix an arbitrary $x \in W$.

Next, consider the following set Δ of formulas of $\mathbf{BI}^{corr+}$:

$$\{ST_x(\varphi) \colon \varphi \in \mathbf{BI} \ \& \ \mathfrak{M}, x \Vdash \varphi\} \cup \{\neg ST_x(\varphi) \colon \varphi \in \mathbf{BI} \ \& \ \mathfrak{M}, x \nVdash \varphi\}.$$

We will show that Δ is finitely satisfiable in K, which, together with the assumption that K is closed under ultraproducts, will yield, by appealing to Lemma 23, that indeed Δ is satisfiable in K. To this end, take any finite $\Delta_0 \subseteq \Delta$. Without loss of generality, we may assume that

$$\Delta_0 = \{ST_x(\varphi_1), \ldots, ST_x(\varphi_n), \neg ST_x(\psi_1), \ldots, \neg ST_x(\psi_m)\}.$$

But then $\mathfrak{M}, x \nVdash \bigwedge_{i \leq n} \varphi_i \to \bigvee_{i \leq m} \psi_i$, so $\bigwedge_{i \leq n} \varphi_i \to \bigvee_{i \leq m} \psi_i \notin \mathrm{Th}(K)$ since $\mathfrak{M} \in \mathrm{Mod}(\mathrm{Th}(K))$. Therefore, there must be $\mathfrak{M}' \in K$ such that $\mathfrak{M}' \nVdash \bigwedge_{i \leq n} \varphi_i \to \bigvee_{i \leq m} \psi_i$. Thus Δ is finitely satisfiable in K, and so Δ is satisfiable in K by some point y in a model $\mathfrak{N} = (W', \leq', V')$. Hence, $\mathfrak{N}, y \equiv_{\mathbf{BI}} \mathfrak{M}, x$.

    Now we may assume that both $\mathfrak{N}$ and $\mathfrak{M}$ are ω-saturated since if we take ω-saturated ultrapowers $\mathfrak{N}'$ and $\mathfrak{M}'$ of $\mathfrak{N}$ and $\mathfrak{M}$ respectively, we have that $\mathfrak{N}' \in K$ iff $\mathfrak{N} \in K$ and that $\mathfrak{M}' \in K$ iff $\mathfrak{M} \in K$ (by closure of both K and $\overline{K}$ under ultrapowers). By Proposition 15, we have that $\langle Z_1, Z_2 \rangle$ is a directed bisimulation between $\mathfrak{M}$ and $\mathfrak{N}$, defined as follows:

$$zZ_1u \quad \text{iff} \quad \mathsf{bitp}_{\mathfrak{M}}(z) \subseteq \mathsf{bitp}_{\mathfrak{N}}(u),$$
$$zZ_2u \quad \text{iff} \quad \mathsf{bitp}_{\mathfrak{N}}(z) \subseteq \mathsf{bitp}_{\mathfrak{M}}(u).$$

We now establish the following two claims:

Claim 1. The restriction of $\langle Z_1, Z_2 \rangle$ to W_x is a directed bisimulation between $\mathfrak{M}_x$ and $\mathfrak{N}$. More precisely, this directed bisimulation can be given as $\langle Y_1, Y_2 \rangle$ such that:

$$
\begin{aligned}
zY_1 u \quad &\text{iff} \quad z \in W_x \ \& \ zZ_1 u, \\
zY_2 u \quad &\text{iff} \quad zZ_2 u \ \& \ u \in W_x.
\end{aligned}
$$

Proof (of Claim 1). Conditions (D_1) and (D_2) hold for $\langle Y_1, Y_2 \rangle$ trivially since it is a restriction of a directed bisimulation.

As for (D_3), if $w \in W_x$ and $w', v' \in W'$ are such that both $wY_1 w'$ and $w' \leq' v'$, then we must have $wZ_1 w'$. Since $\langle Z_1, Z_2 \rangle$ is a directed bisimulation, we can choose a $v \in W$ such that $w \leq v$, $vZ_1 v'$, and $v'Z_2 v$. Since W_x is closed under both $\leq$ and $\geq$, it follows that $v \in W_x$, so that both $vY_1 v'$, and $v'Y_2 v$, as desired.

As for (D_4), if $w, v \in W_x$ and $w' \in W'$ are such that both $wY_1 w'$ and $w \leq v$, then we must have $wZ_1 w'$. Since $\langle Z_1, Z_2 \rangle$ is a directed bisimulation, we can choose a $v' \in W'$ such that $w' \leq v'$, $vZ_1 v'$, and $v'Z_2 v$. But then clearly also $vY_1 v'$, and $v'Y_2 v$, as desired.

We argue similarly for the conditions (D_5) and (D_6).

Claim 1 is proven.

Claim 2. $\langle Y_1, Y_2 \rangle$ surjective with respect to $\mathfrak{M}_x$.

Indeed, take an arbitrary $w \in W_x$; for some $n \in \mathbb{N}$ and some $z_0, \ldots, z_n$ we must have

$$
\bigwedge_{i < n} (z_{i+1} \leq z_i \ \vee \ z_i \leq z_{i+1}) \wedge (z_n = w) \wedge (z_0 = x).
$$

Proof (of Claim 2). We show by induction on n that, for some $v \in W'$, we must have both $wY_1 v$ and $vY_2 w$.

Basis. $n = 0$. Then $w = x$ and we have both $xZ_1 y$ and $yZ_2 x$ by the choice of y.

Induction step. Suppose we have shown our claim for every $m \leq k$, and assume that $n = k + 1$. Then $z_k \in W_x$ and, for some $v' \in W'$ we must have both $z_k Y_1 v'$ and $v'Y_2 z_k$. By the choice of z_k, either $z_k \leq w$ or $w \leq z_k$ must hold. If $z_k \leq w$, then $v'Y_2 z_k$ and Claim 1 imply, in virtue of (D_4), the existence of a $v \in W$ such that both $wY_1 v$ and $vY_2 w$, as desired. On the other hand, if $w \leq z_k$, then $z_k Y_1 v'$ and Claim 1 together entail the same conclusion by (D_5).

Claim 2 is proven.

It follows from Claims 1 and 2 that, in virtue of the closure of K under surjective directed bisimulations, we must have $\mathfrak{M}_x \in K$. Since $x \in W$ was chosen arbitrarily, Lemma 24 implies that also $\mathfrak{M} \in K$. $\square$

5 Goldblatt–Thomason Theorem

We end by proving a bi-intuitionistic analogue of the Goldblatt–Thomason Theorem. Apart from the result itself, this section also serves an illustration of algebraic methods. Most of the preliminary results have already been stated, we only use one additional lemma.

Lemma 28. *Let $\mathfrak{F}$ be a Kripke frame. There exists an ultrapower $\mathfrak{F}^I/U$ such that $\mathfrak{pe}(\mathfrak{F})$ is a bounded-morphic image of $\mathfrak{F}^I/U$.*

Proof. We only sketch a proof. Expand the first-order language of $\mathfrak{F}$ by adding as many unary predicates P_j as there are upsets of $\mathfrak{F}$. Let $\widehat{\mathfrak{F}}$ be $\mathfrak{F}$ expanded by interpreting each P_j as a distinct upset of $\mathfrak{F}$. By Proposition 3 there exists an ultrapower $\widehat{\mathfrak{F}}^I/U$ which is ω-saturated. Now expanding (if necessary) Prop to Prop^+ so that there is a distinct propositional variable p_j for each P_j, and reinterpreting P_j as the value of p_j, we have that $\widehat{\mathfrak{F}}^I/U$ realises all sets of $\mathbf{BI}(\mathrm{Prop}^+)$ formulas maximally consistent over $\mathfrak{F}$. Define a map $f\colon \widehat{\mathfrak{F}}^I/U \to \mathfrak{pe}(\mathfrak{F})$ by setting

$$f(a) := \{P_j \in \mathrm{Up}(\mathfrak{F})\colon \widehat{\mathfrak{F}}^I/U \models P_j(a)\}$$

This map is order preserving, for if $a \leq b$, then since P_j are upsets, we have $f(a) \subseteq f(b)$. It also satisfies the bounded morphism conditions. For suppose $f(a) \subseteq Z$, then since Z is a prime-filter of upsets of $\mathfrak{F}$, we have $Z = \{P_j^F\colon j \in J, \text{ for some } J\}$, that is, Z is a type. Moreover, since Z is a prime filter, Z is maximally consistent and hence, by saturation, $Z = f(b)$ for some b. This proves the upward bounded-morphism condition, the proof of the downward condition is analogous. The map f does not depend on the language expansion, so in fact it is a bounded morphism from $\mathfrak{F}^I/U$ to $\mathfrak{pe}(\mathfrak{F})$. $\qquad\square$

Theorem 29. *Let K be a class of Kripke frames closed under ultrapowers. Then K is bi-intuitionistically definable if and only if K is closed under generated subframes, bounded-morphic images and disjoint unions, and reflects prime-filter extensions.*

Proof. Assume K is bi-intuitionistically definable. Closure under generated subframes, bounded-morphic images and disjoint unions follows immediately from Proposition 6. By Proposition 17(3) it also follows that K reflects prime-filter extensions.

For the converse, assume K is an elementary class closed under generated subframes, bounded-morphic images and disjoint unions, and reflecting prime-filter extensions. Let $\Sigma(K)$ be the set of all formulas valid in K. Take any frame $\mathfrak{F}$ such that $\mathfrak{F} \Vdash \Sigma(K)$. To prove the theorem it suffices to show that $\mathfrak{F}$ belongs to K.

By Proposition 1 we have that $\mathfrak{F}^+ \models \{\varphi = 1 \colon \varphi \in \Sigma(K)\}$. By subdirect representation theorem, $\mathfrak{F}^+ \in HSP\{\mathfrak{P}^+ \colon \mathfrak{P} \in K\}$. Put $\mathbf{A} = \mathfrak{F}^+$, and $\mathfrak{P} = \coprod_{i \in I} \mathfrak{P}_i$. Hence, there is a set $\{\mathfrak{P}_i \colon i \in I\}$ of frames from K such that $\mathbf{A}$ is a homomorphic image of an algebra $\mathbf{B}$ which is a subalgebra of $\mathbf{C} = \prod_{i \in I} \mathfrak{P}_i^+$. By Proposition 10(1),

$$\mathbf{C} = \prod_{i \in I} \mathfrak{P}_i^+ \cong \left(\coprod_{i \in I} \mathfrak{P}_i \right)^+ = \mathfrak{P}^+$$

and since K is closed under disjoint unions, we have $\mathfrak{P} \in K$. Using Proposition 11 (2, 3), we obtain the following: (i) $\mathbf{A}_+$ is a generated subframe of $\mathbf{B}_+$ and (ii) $\mathbf{B}_+$ is a bounded-morphic image of $\mathbf{C}_+ \cong (\mathfrak{P}^+)_+ = \mathfrak{pe}(\mathfrak{P})$. Since K is closed under ultrapowers, by Lemma 28 we get that $\mathfrak{pe}(\mathfrak{P}) \in K$. But then, $\mathfrak{P} \in K$, so by closure properties of K we get that $\mathbf{B}_+$ and $\mathbf{A}_+$ are in K. Finally, since $\mathbf{A}_+ = (\mathfrak{F}^+)_+ \cong \mathfrak{pe}(\mathfrak{F})$, and K reflects prime-filter extensions, $\mathfrak{F} \in K$, as required. $\square$

6 Conclusion

In this paper we have settled two questions. On the one hand, we have provided a characterization of the classes of models of bi-intuitionistic logic that can be axiomatised by a theory in the language of the logic (Theorem 27). On the other hand, we have obtained an analogous result for the classes of frames closed under ultrapowers for which a similar thing can be done (Theorem 29). These two new results contribute to building a fuller picture of the expressivity of propositional bi-intuitionistic logic, adding to the results in [1, 18].

References

[1] G. Badia. Bi-simulating in bi-intuitionistic logic. *Studia Logica*, 104(5):1037–1050, 2016.

[2] G. Badia. A remark on Maksimova's variable separation property in super-bi-intuitionistic logics. *Australasian Journal of Logic*, 14(1):46–53, 2017.

[3] J. L. Bell and A B. Slomson. *Models and Ultraproducts: An Introduction.* North-Holland, Amsterdam, 1969.

[4] C. Bergman. *Universal algebra. Fundamentals and Selected Topics*, volume 301 of *Pure and Applied Mathematics*. CRC Press, Boca Raton, FL, 2012.

[5] P. Blackburn, M. de Rijke, and Y. Venema. *Modal Logic*. Cambridge University Press, Cambridge, UK, 2001.

[6] C. C. Chang and H. J. Keisler. *Model Theory*. North-Holland, Amsterdam, 1973.

[7] T. Crolard. Subtractive logic. *Theoretical Computer Science*, 254(1–2):151–185, 2001.

[8] B. A. Davey and H. A. Priestley. *Introduction to Lattices and Order*. Cambridge University Press, Cambridge, UK, 2nd edition, 2002.

[9] J. de Groot and D. Pattinson. Hennessy–Milner properties for (modal) bi-intuitionistic logic. In R. Iemhoff, M. Moortgat, and R. de Queiroz, editors, *Logic, Language, Information, and Computation. WoLLIC 2019*, number 11541 in Lecture Notes in Computer Science, Berlin, 2019. Springer.

[10] R. Goldblatt. Axiomatic classes of intuitionistic models. *Journal of Universal Computer Science*, 11(12):1945–1962, 2005.

[11] R. Goldblatt and S. K. Thomason. Axiomatic classes in propositional modal logic. In J. N. Crossley, editor, *Algebra and Logic*, number 450 in Lecture Notes in Mathematics, pages 163–173. Springer-Verlag, Berlin, 1975.

[12] R. Goré and L. Postniece. Combining derivations and refutations for cut-free completeness in bi-intuitionistic logic. *Journal of Logic and Computation*, 20 (1):233–260, 2010.

[13] R. Goré and I. Shillito. Bi-intuitionistic logics: A new instance of an old problem. *Advances in Modal Logic*, 13:269–288, 2020.

[14] W. Hodges. *Model Theory*. Cambridge University Press, Cambridge, UK, 1993.

[15] Dieter Klemke. Ein Henkin-Beweis für die Vollständigkeit eines Kalküls relativ zur Grzegorczyk-Semantik. *Archiv für Mathematische Logik und Grundlagenforschung*, 14:148–161, 1971.

[16] T. Kowalski and H. Ono. Analytic cut and interpolation for bi-intuitionistic logic. *Review of Symbolic Logic*, 10(2):259–283, 2017.

[17] N. Kurtonina and M. de Rijke. Simulating without negation. *Journal of Logic and Computation*, 7(4):501–522, 1997.

[18] G. Olkhovikov and G. Badia. Maximality of bi-intuitionistic propositional logic. *Journal of Logic and Computation*, 32(1):1–31, 2022.

[19] L. Pinto and T. Uustalu. A proof-theoretic study of bi-intuitionistic propositional sequent calculus. *Journal of Logic and Computation*, 28(1):165–202, 2018.

[20] C. Rauszer. Representation theorem for semi-Boolean algebras. I, II. *Bulletin L'Académie Polonaise des Science. Série des Sciences Mathématiques, Astronomiques et Physiques*, 19:881–887, 889–892, 1971, 1972.

[21] C. Rauszer. Semi-Boolean algebras and their applications to intuitionistic logic with dual operations. *Fundamenta Mathematicae*, 83(3):219–249, 1974.

[22] C. Rauszer. A formalization of the propositional calculus of H–B logic. *Studia Logica*, 33:23–34, 1974.

[23] C. Rauszer. Applications of Kripke models to Heyting–Brouwer logic. *Studia Logica*, 36(1–2):61–71, 1977.

[24] C. Rauszer. Model theory for an extension of intuitionistic logic. *Studia Logica*, 36(1–2):73–87, 1977.

[25] C. Rauszer. An algebraic and Kripke-style approach to a certain extension of intuitionistic logic. Warszawa: Instytut Matematyczny Polskiej Akademi Nauk, 1980. (http://eudml.org/doc/268511).

[26] C. J. Taylor. *Double-Heyting Algebras*. Phd thesis, La Trobe University, 2017. (URL: opal.latrobe.edu.au/articles/thesis/Double_Heyting_algebras/21841596?file=38762820).

[27] F. Wolter. On logics with coimplication. *Journal of Philosophical Logic*, 27(4):353–387, 1998.

Received 25 February 2025

γ-ADMISSIBILITY IN FIRST-ORDER MODAL RELEVANT LOGICS: METHOD OF NORMAL MODELS

NICHOLAS FERENZ

Centre of Philosophy, School of Arts and Humanities, University of Lisbon
`<nicholas.ferenz@gmail.com>`

THOMAS M. FERGUSON

Department of Cognitive Science, Rensselaer Polytechnic Institute
`<fergut5@rpi.edu>`

Abstract

The admissibility of the rule γ in relevant logics has been a significant problem. Omitted from the first statements of relevant logics, γ (admissibility) has played a significant role in relevant logics. Several methods have been developed to assess γ in propositional relevant logics. Seki [27, 28, 29], building on existing methods, shows γ admissible in modal relevant logics using the methods of normal models, algebraic semantics, and metavaluations. In this paper, we show the γ-admissibility of many first-order extensions of Seki's **L3** logics (with fusion, left implication, and the Ackermann truth constant). We follow Ferenz and Ferguson [6] in using the normal models method on ternary relational models with the Mares–Goldblatt interpretation of the quantifiers, extending the arguments to include modalities.

2020 *Mathematics Subject Classification.* Primary: 03B47, Secondary: 03B45, 03B62.

1 Introduction

Ackermann [1] gave a list of rules for a system *strenge Implikation*, which avoids the paradoxes of strict and material conditionals, of which γ is this third rule. Essentially, γ is a form of disjunctive syllogism, itself equivalent in classical logic to modus ponens on the material conditional. In defining the first axiomatic presentations of relevant logics, Anderson and Belnap rejected the rule γ, because Ackermann's understanding of γ is a primitive rule which can be used to derive sentences from non-theorem assumptions. In particular, in relevant logic theories should not be taken to be closed under γ in this sense. To do so would entail that anything and everything follows from an inconsistent theory. The denial of this explosion from inconsistent theories is a core part of the relevantist project.

On the other hand, we could take γ as a rule under which the set of theorems are closed. Formally, we will write

$$\neg \mathcal{A} \vee \mathcal{B}, \ \mathcal{A} \Rightarrow \mathcal{B}$$

where the '$\Rightarrow$' symbol denotes that the rule is a *rule of proof* in the sense of [13, 31]. In particular, a rule of proof is a rule that preserves theorem-hood and not merely truth in a theory: $\mathcal{A}_1, \ldots, \mathcal{A}_n \Rightarrow \mathcal{B}$ means that if each $\mathcal{A}_i$ is a theorem, then so is $\mathcal{B}$. The rule γ is omitted in the definition of relevant logics, but there are cases where, like cut, we can show the rule is admissible (and typically not derivable). That is, there are relevant logics where the addition of γ does not result in new theorems.

The rule γ has been shown admissible for large classes of propositional relevant logics via a number of methods. The first proof of γ-admissibility was given in [18], wherein it is shown that **R**, **E**, and **T**—the favorite children of Anderson and Belnap—admit γ using *algebraic techniques*. Later, further techniques were developed, such as *normal models* [23], *metavaluations* [17], and *reduced frames* [30]. A detailed account of the history of γ in relevant logics is found in [33].

For modal (propositional) relevant logics, a similar plethora of results (and many open questions) is found. Meyer and Mares [16] and Sylvan (né Routley) and Meyer [22] had shown some modal relevant logics admit γ. However, Seki has done much work providing the most general and comprehensive proofs of the admissibility of γ in modal relevant logics using the normal models method [27], metavaluations [28], and algebraic methods [29]. Notably, the method of normal models Seki employs works for a large class of base (modal and regular) modal relevant logics, and with some Sahlqvist axiom and rule schemes. The results of this paper aim to handle first-order extensions of the former: the base modal relevant logics considered by Seki.

The history of admissibility of γ in first-order relevant logics has historically been more impoverished, likely due in part to the difficulties in producing a nice semantics for these logics (or because of the degree of complication of the extant semantics). The first of such proofs is found in [19, Theorem 6.]. The authors use an algebraic method to prove γ admissible in the logic **RQ** (defined below).

A handful of other results concern the logics generated by using the Tarskian interpretation of the quantifiers on ternary relational semantics with a constant universal domain. The set of validities on this semantics we will call $\overline{\mathbf{RQ}}$. This set of validities, however, does not correspond to the logic $\mathbf{RQ}$.[1] This is the incompleteness result given in [8]. However, we do know that this logic admits γ (see, Weiss [34] and Kripke [14]).[2]

The last result for first-order logics is given in [6]. These results build on the semantics presented in [5], which generalizes the semantics of [15]. These models employ the Mares–Goldblatt interpretation of the quantifiers, which itself requires a general frame semantics. In the terminology of [27], Ferenz and Ferguson prove γ admissible for first-order extensions of a class of propositional relevant logics. We aim here to combine the methods of [6] and [27] to prove γ admissible in classes of first-order modal logics based on L3 logics.

As an aside, there are relevant logics and theories that lack γ-admissibility. A notable example is found in [20]: several relevant logics extended by Boolean negation (and lacking contraction) fail to admit γ. If we consider particular theories for which γ-admissibility fails, the naïve set theory of Brady [3] does not admit γ, though the motivation for the presence of theorems whose negations are also theorems necessitates against γ. Moreover, the failure of γ in the relevant arithmetic $\mathbf{R}^{\sharp}$ dealt a serious blow to the whole project [9].

We will consider logics as sets of theorems given by a Hilbert-style axiom system presented with axiom and rule schemes. Thus, we can and will throughout this paper use $\Rightarrow$ for all rules. The usual notion of a proof is employed. Given this focus, we can consider all rules as rules of proof. The subtlety required for a consequence relation is beyond the scope of this paper. A logic will be defined with a *primitive* set of rules. A second class of rules is the *derived* rules, whose form can be captured exactly by a series of axioms and primitive rule applications. For example, a rule form of conjunction elimination of the form $\mathcal{A} \wedge \mathcal{B} \Rightarrow \mathcal{A}$ can be shown to be derivable using modus ponens and the axiom $\mathcal{A} \wedge \mathcal{B} \to \mathcal{A}$. An *admissible* rule is one where, if there exists a proof of each of the premises, then there is a proof of the consequence. However, the proofs need not be related in any way.

The admissibility of γ is of interest not merely to relevant logicians. The rule bears a remarkable similarity to *cut*, and insights into γ and *cut* may lead to insights in the other.[3] This may be relevant for the complexity of proofs, for, as noted in [33], the speedup theorem for *cut* suggests a speedup theorem for γ.[4]

---

[1]It is still an open question if and what the axiomatization of such a logic would be.

[2]Kripke does not make it clear whether his results concern $\mathbf{RQ}$ or $\overline{\mathbf{RQ}}$. Conversations with Yale Weiss seem to confirm that Kripke's results are for the latter.

[3]Meyer, Dunn and Leblanc [19, p. 120] put as follows: "the cut theorem … is for classical theories simply γ in peculiar notation."

[4]The speedup theorem for cut in classical logic can be found, e.g., in [21]. The theorem essen-

Here, we combine the methods of [6] and [27] to prove a class of first-order modal relevant logics admitting the rule γ. The specific class of logics in question are first-order logics extended by modalities with non-Sahlqvist formulas. In [27], several classes of propositional modal relevant logics are shown to admit γ. One sub-class are called L3 logics, which extend a base logic with axiom and rules schemes from a specified list which includes only a few modal axioms and rules. We consider first-order extensions of the logics in this class. Moreover, we discuss the Barcan formulas and conjecture that the normal models method (at least as presented here) is incapable handling certain Barcan formulas. In particular, the frame conditions are not invariant under frame normalization.

2 First-order Modal Relevant Logics

2.1 Language

A first-order modal language is built up from a set of symbols divided as follows. We are given a denumerable set of variables $Var = \{x_0, x_1, \dots\}$. Here we assume a fixed but arbitrary ordering of the elements of Var, which is tracked by variable subscripts. A *signature* $\mathbb{S}$ consists of (i) a set of constant symbols $Con^{\mathbb{S}} = \{c_0, c_1, \dots\}$, and (ii) a non-empty set of predicate symbols $Pred^{\mathbb{S}}$, where $P^n \in Pred^{\mathbb{S}}$ is an n-ary predicate. The set of n-ary predicates shall be written as $Pred^n \subseteq Pred^{\mathbb{S}}$. We further have the constant symbol t, the binary operators $\land, \lor, \rightarrow, \circ, \leftarrow$, the unary operators $\neg, \Box$, and $\Diamond$, and the quantifier symbols $\forall, \exists$.

We now take an arbitrary signature to be fixed for the remainder of the paper.[5] We take the usual definition of a term, with respect to a signature, and use τ (with or without decoration) as varying over the set of terms. Given a set of individuals U, a *variable assignment* we take to be a countably infinite sequence of individuals, $f \in U^\omega$: the nth element, written fn is the individual assigned to the variable x_n. Given variable assignments f and g, we say that g is an x_n-variant of f when g differs from f in at most the nth position. We denote this (symmetric) relation by $f \sim_n g$ (or $f \sim_{x_n} g$).

Definition 2.1. The *first-order modal language* $\mathfrak{L}$ is defined in BNF as follows:

$$A ::= P^n(\tau_1, \dots, \tau_n) \mid t \mid \neg A \mid A \land A \mid A \lor A \mid A \rightarrow A \mid A \circ A \mid A \leftarrow A \mid \forall x_n A \mid \exists x_n A \mid \Box A \mid \Diamond A.$$

---

tially means that the size of a proof of a theorem without *cut* has a significantly larger lower bounds than when allowing for proofs with *cut*.

[5]In the canonical model constructions, the signature is assumed to contain sufficiently many constant symbols.

We suppress parentheses (implicit in the construction for each binary connective) by assuming that each and every unary connective (including quantifiers) bind stronger than the binary connectives, that the implications (right and left) bind equally and weaker than fusion, and that fusion binds weaker than the extensional conjunction and disjunction. Note that, when a merely modal, propositional, or first-order fragment of the first-order modal language is intended, this will be made clear by context and defined in the usual way.

We deploy several notions of substitution. We will write $f[j/n]$ (or $f[j/x_n]$), with $j \in U$ to denote the result of replacing the n-th element of f with the individual j. We will use $\mathcal{A}[\tau/x]$ to denote the result of replacing every free occurrence of x in $\mathcal{A}$ with τ, and similarly write $\mathcal{A}[\tau_0/v_0, \ldots, \tau_n/v_n]$ for a simultaneous substitution. We employ the usual definitions of *bound* and *free* variables. A term τ is *free for x* in $\mathcal{A}$ if τ does not become bound in the resulting formula $\mathcal{A}[\tau/x]$.

When we write a formula with a variable superscript, such as $\mathcal{A}^x$, this means that x does not occur free in $\mathcal{A}$.

In the context of some of the weaker logics, the modalities of $\Box$ and $\Diamond$ are not dual (in the usual way via negation). Seki gives a pair of negation dual modalities for our primitives which, due to the duality being definitional, can be taken as defined modalities in our systems. These are defined as follows:

$$\boxdot\mathcal{A} =_{df} \neg\Diamond\neg\mathcal{A} \qquad\qquad \Diamonddot\mathcal{A} =_{df} \neg\Box\neg\mathcal{A}$$

2.2 Axiomatic Presentation

We start with the axiomatic presentation of propositional modal relevant logics, as given in Seki [27]. We end the section with the first-order extensions of all logics constructed using the axioms and rule schemes of Seki reproduced here. The least logic for which the results of this paper apply are first-order extensions of what Seki calls the L3 logics. The most basic modal L3 logic is the logic $\mathbf{G.C}^{g,d}$, which extends the regular modal relevant logic $\mathbf{B.C}$ with the principle of excluded middle (A10), the disjunctive rules (R6), (Q1), and (Q2).[6] Note that, in order to facilitate easy comparisons with Seki, we adopt the same labels for axioms and rules.

Definition 2.2 (Propositional Logics). The base propositional logic $\mathbf{G.C}^{g,d}$ is defined by the following axioms and rules:

(A1) $\mathcal{A} \to \mathcal{A}$

(A2) $\mathcal{A} \wedge \mathcal{B} \to \mathcal{A}$

---

[6]It is not only suggested, but actually the case that one obtains $\mathbf{B.C}$ by eliminating these extra rules, the principle of excluded middle (A10), the disjunctive rules (R6), (Q1) and (Q2).

(A3) $\mathcal{A} \wedge \mathcal{B} \to \mathcal{B}$

(A4) $(\mathcal{A} \to \mathcal{B}) \wedge (\mathcal{A} \to \mathcal{C}) \to (\mathcal{A} \to (\mathcal{B} \wedge \mathcal{C}))$

(A5) $\mathcal{A} \to \mathcal{A} \vee \mathcal{B}$

(A6) $\mathcal{B} \to \mathcal{A} \vee \mathcal{B}$

(A7) $(\mathcal{A} \to \mathcal{C}) \wedge (\mathcal{B} \to \mathcal{C}) \to ((\mathcal{A} \vee \mathcal{B}) \to \mathcal{C})$

(A8) $\mathcal{A} \wedge (\mathcal{B} \vee \mathcal{C}) \to (\mathcal{A} \wedge \mathcal{B}) \vee (\mathcal{A} \wedge \mathcal{C})$

(A9) $\neg\neg\mathcal{A} \to \mathcal{A}$

(A10) $A \vee \neg A$

(A11) $\Box\mathcal{A} \wedge \Box\mathcal{B} \to \Box(\mathcal{A} \wedge \mathcal{B})$

(A12) $\Diamond(\mathcal{A} \vee \mathcal{B}) \to \Diamond\mathcal{A} \vee \Diamond\mathcal{B}$

(R1) $\mathcal{A}, \mathcal{A} \to \mathcal{B} \Rightarrow \mathcal{B}$

(R2) $\mathcal{A}, \mathcal{B} \Rightarrow \mathcal{A} \wedge \mathcal{B}$

(R3) $\mathcal{A} \to \mathcal{B} \Rightarrow (\mathcal{B} \to \mathcal{C}) \to (\mathcal{A} \to \mathcal{C})$

(R4) $\mathcal{A} \to \mathcal{B} \Rightarrow (\mathcal{C} \to \mathcal{A}) \to (\mathcal{C} \to \mathcal{B})$

(R5) $\mathcal{A} \to \mathcal{B} \Rightarrow \neg\mathcal{B} \to \neg\mathcal{A}$

(R6) $\mathcal{C} \vee \mathcal{A} \Rightarrow \mathcal{C} \vee \neg(\mathcal{A} \to \neg\mathcal{A})$

(R7) $\mathcal{A} \to \mathcal{B} \Rightarrow \Box\mathcal{A} \to \Box\mathcal{B}$

(R8) $\mathcal{A} \to \mathcal{B} \Rightarrow \Diamond\mathcal{A} \to \Diamond\mathcal{B}$

(R9) $\mathcal{A} \vee \mathcal{B} \Rightarrow \mathcal{A} \vee \varodot\mathcal{B}$

(R10) $\mathcal{A} \vee \mathcal{B} \Rightarrow \mathcal{A} \vee \Diamond\mathcal{B}$

(Q1) $\mathcal{C} \vee (\mathcal{A} \to \mathcal{B}), \mathcal{C} \vee \mathcal{A} \Rightarrow \mathcal{C} \vee \mathcal{B}$

(Q2) $\mathcal{C} \vee (\mathcal{A} \to \neg\mathcal{B}) \Rightarrow \mathcal{C} \vee (\mathcal{B} \to \neg\mathcal{A})$

(R$\circ$) $\mathcal{A} \to (\mathcal{B} \to \mathcal{C}) \Longleftrightarrow (\mathcal{A} \circ \mathcal{B}) \to \mathcal{C}$

(R$\leftarrow$) $\mathcal{A} \to (\mathcal{B} \to \mathcal{C}) \Longleftrightarrow \mathcal{B} \to (\mathcal{C} \leftarrow \mathcal{A})$

(Rt) $t \to \mathcal{A} \Longleftrightarrow \mathcal{A}$

The logic $\mathbf{G.C}^{g,d}$ may be extended by any of the following axioms and rules.

(B1) $\mathcal{A} \wedge (\mathcal{A} \to \mathcal{B}) \to \mathcal{B}$

(B2) $(\mathcal{A} \to \mathcal{B}) \wedge (\mathcal{B} \to \mathcal{C}) \to (\mathcal{A} \to \mathcal{C})$

(B3) $(\mathcal{A} \to (\mathcal{A} \to \mathcal{B})) \to (\mathcal{A} \to \mathcal{B})$

(B4) $\mathcal{A} \to ((\mathcal{A} \to \mathcal{B}) \to \mathcal{B})$

(B5) $\mathcal{A} \to (\mathcal{B} \to \mathcal{B})$

(B6) $\mathcal{A} \to (\mathcal{B} \to \mathcal{A})$

(B7) $(\mathcal{A} \to \mathcal{B}) \to ((\mathcal{A} \to \mathcal{C}) \to (\mathcal{A} \to \mathcal{B} \wedge \mathcal{C}))$

(B8) $\mathcal{A} \to (\mathcal{A} \to \mathcal{A})$

(B9) $\mathcal{A} \vee \mathcal{B} \to ((\mathcal{A} \to \mathcal{B}) \to \mathcal{B})$

(B10) $(\mathcal{A} \wedge \mathcal{B} \to \mathcal{C}) \to (\mathcal{A} \wedge \neg \mathcal{C} \to \neg \mathcal{B})$

(B11) $\mathcal{A} \to \neg(\mathcal{A} \to \mathcal{A})$

(B12) $(\mathcal{A} \to \neg \mathcal{A}) \to \neg \mathcal{A}$

(B13) $(\mathcal{A} \to \neg \mathcal{B}) \to (\mathcal{B} \to \neg \mathcal{A})$

(B14) $\mathcal{A} \to \mathcal{B} \vee \neg \mathcal{B}$

(B15) $\mathcal{A} \to (\neg \mathcal{A} \to \mathcal{B})$

(B16) $(\mathcal{A} \to \mathcal{B}) \to ((\mathcal{B} \to \mathcal{C}) \to (\mathcal{A} \to \mathcal{C}))$

(B17) $(\mathcal{A} \to \mathcal{B}) \to ((\mathcal{C} \to \mathcal{A}) \to (\mathcal{C} \to \mathcal{B}))$

(B18) $(\mathcal{A} \to (\mathcal{B} \to \mathcal{C})) \to (\mathcal{B} \to (\mathcal{A} \to \mathcal{C}))$

(B19) $(\mathcal{A} \to (\mathcal{B} \to \mathcal{C})) \to ((\mathcal{A} \to \mathcal{B}) \to (\mathcal{A} \to \mathcal{C}))$

(B20) $(\mathcal{A} \to \mathcal{B}) \to ((\mathcal{A} \to (\mathcal{B} \to \mathcal{C})) \to (\mathcal{A} \to \mathcal{C}))$

(B21) $(\mathcal{A} \wedge \mathcal{B} \to \mathcal{C}) \to (\mathcal{A} \to (\mathcal{B} \to \mathcal{C}))$

(B22) $\Box(\mathcal{A} \to \mathcal{B}) \to (\Box \mathcal{A} \to \Box \mathcal{B})$

(B23) $\boxdot(\mathcal{A} \to \mathcal{B}) \to (\Diamond\mathcal{A} \to \Diamond\mathcal{B})$

(B24) $\Box(\mathcal{A} \to \mathcal{B}) \to (\Diamonddot\mathcal{A} \to \Diamonddot\mathcal{B})$

(Q3) $\mathcal{C} \lor (\neg\mathcal{A} \to \mathcal{A}) \Rrightarrow \mathcal{C} \lor \mathcal{A}$

(Q4) $\mathcal{A} \Rrightarrow (\mathcal{A} \to \mathcal{B}) \to \mathcal{B}$

(Q5) $\mathcal{E} \lor (\mathcal{A} \to \mathcal{B}), \mathcal{E} \lor (\mathcal{C} \to \mathcal{D}) \Rrightarrow \mathcal{E} \lor ((\mathcal{B} \to \mathcal{C}) \to (\mathcal{A} \to \mathcal{D}))$

(Q6) $\mathcal{C} \lor (\mathcal{A} \to \mathcal{B}) \Rrightarrow \mathcal{C} \lor (\Box\mathcal{A} \to \Box\mathcal{B})$

(Q7) $\mathcal{C} \lor (\mathcal{A} \to \mathcal{B}) \Rrightarrow \mathcal{C} \lor (\Diamond\mathcal{A} \to \Diamond\mathcal{B})$

(Q8) $\mathcal{B} \lor \Box\mathcal{A} \Rrightarrow \mathcal{B} \lor \mathcal{A}$

(Q9) $\mathcal{B} \lor \boxdot\mathcal{A} \Rrightarrow \mathcal{B} \lor \mathcal{A}$

(Q10) $\mathcal{B} \lor \mathcal{A} \Rrightarrow \mathcal{B} \lor \Box\mathcal{A}$

(Q11) $\mathcal{A} \Rrightarrow \Box\mathcal{A}$

(Q12) $\mathcal{A} \Rrightarrow \boxdot\mathcal{A}$

Technically speaking, our base logic builds into it the rules (Q1) and (Q2), and is thus an L3 logic. We nonetheless introduce Seki's distinction. In the propositional cases, L1 logics are obtained from (initially) dropping (Q1) and (Q2) and adding a subset of axioms (B1)–(B15) and rules (Q1)–(Q4). The L2 logics are those obtained from an L1 logic such that (Q1) is derivable and we add a subset of axioms (B16)–(B21) and rule (Q5). Then, L3 requires (Q1) and (Q2) derivable and any of the axioms (B1)–(B21) and rules (Q3)–(Q5).

In the modal cases, the logics $\mathbf{L}_R$ are extensions of L1 logics with (A11), (A12), (R7)–(R10), and a subset of the axiom (B22) and rules (Q6)–(Q12), which is restricted so that a logic has (B22) or (Q7) only if it has derivable (Q1) or (Q2), respectively. The minimal logic in the $\mathbf{L}_R$ category or regular modal relevant logics is $\mathbf{G.C}^g$.[7] The basic normal modal relevant logics $\mathbf{G.K}^g$, defined as $\mathbf{G.C}^g$ with (Q1) and (Q2)—a modal logic on the base L3 logic—with additionally the axioms (A13) and (B22), and the rule (Q11), where the former is:

(A13) $\Box(\mathcal{A} \to \mathcal{B}) \to (\Diamond\mathcal{A} \to \Diamond\mathcal{B})$

---

[7]Note that the uses of the terms *normal* and *regular* are drawn from Seki [26]. Ultimately, the definitions are due to Segerberg in [25], but are often attributed to Chellas in [4].

The class of normal relevant logics L_N may also be extended by the axioms (B23) and (B24).

Note that the class L_N contains axiom (A13), which is prohibited from the logics in the class L_R. At least, that is, in terms of axiom schemes. It will be noted that the two classes of logics, L_R and L_N must be dealt with separately, as in [27]. That is, at least following the proofs within.

We note that our results only apply to logics with (Q1) and (Q2)—L3 logics— and therefore that the extension by any subset of the (Bi) and (Qi) axioms and rules is covered by our results. We also reiterate that our logics contain fusion and left implication and the Ackermann truth constant, and thus the derivability of some of the axioms and rules presented in this hierarchy may be subject to change when the underlying logic is not conservatively extended by these operations.

In the above axiom and rule schemes there are not many with modalities. In Seki [27], there is first the proof of γ admissibility for the L1, L2, and L3 modal logics, and additional proof for some modal axioms and rules in certain Sahlqvist forms. Due to space considerations, these modal extensions are not dealt with in this paper, but we note that many of the familiar conditions on the accessibility relation $S_\Box$ should be preserved under the method of normalization and we expect that γ-admissibility ought to hold for corresponding extensions of L3 logics. It is not obvious in *all* cases; e.g., insofar as the linear ordering of a relation $S_\Box$ may not survive normalization, γ may prove challenging for relevant versions of **S4.3**. We leave such investigations for future work, however.

2.2.1 First-order Logics

Definition 2.3 (First-Order Logics). Let **L.M** be an extension of **B.C** as defined above with non-modal part **L** and modal part **M**.[8] The logic **LQ.M** is defined by adding the following axioms and rule schemes, where τ is free for x in $\mathcal{X}$, in the first-order language.[9]

($\forall$E)	$\forall x \mathcal{X} \to \mathcal{X}[\tau/x]$	(dEC)	$\mathcal{A}^x \wedge \exists x \mathcal{B} \to \exists x (\mathcal{A}^x \wedge \mathcal{B})$
($\exists$I)	$\mathcal{X}[\tau/x] \to \exists x \mathcal{X}$	(R$\forall$I)	$\mathcal{A}^x \to \mathcal{B} \Rightarrow \mathcal{A}^x \to \forall x \mathcal{B}$
(EC)	$\forall x (\mathcal{A} \vee \mathcal{B}^x) \to \forall x \mathcal{A} \vee \mathcal{B}^x$	(R$\exists$E)	$\mathcal{A} \to \mathcal{B}^x \Rightarrow \exists x \mathcal{A} \to \mathcal{B}^x$

---

[8] This dot notation is fairly standard in the literature on modal relevant logic. Ferenz [5] explains that the dot tracks, in many cases, whether the $\{\neg, \wedge, \vee, \Box, \Diamond\}$-fragment is "sufficiently classical." We therefore acknowledge that the dot notation usefully tracks an important notion, and that the notion can be overbearing. In this paper, we aim to simplify notation by quantifying over the intended class of logics with a single notation.

[9] The reader is reminded that a super-scripted x means that x does not occur free in the decorated (sub)formula.

We define the logic **QL.M** similarly, but without the so-called *extensional confinement* axiom (EC) and its dual (dEC).

Because we have chosen to include both $\circ$ and $\leftarrow$ and their governing rules, the following formulas are theorem(-schemes):[10]

(A∃E) $\forall x(\mathcal{A} \to \mathcal{B}^x) \to (\exists x \mathcal{A} \to \mathcal{B}^x)$ (A∀I) $\forall x(\mathcal{A}^x \to \mathcal{B}) \to (\mathcal{A}^x \to \forall x \mathcal{B})$

In the relational setting, at least with respect to the Mares–Goldblatt interpretation of the quantifier on a ternary relational semantics with a universal domain, (EC) and (dEC) and result in the more interesting first-order extensions. Thus we consider logics with and without these axiom schemes. More interesting toggles can be found in the neighbourhood setting, but such is beyond the scope of this paper and approaches to γ-admissibility using normal models.

In this paper our results will be for first-order extensions of modal extensions of L3 logics. Seki defines the class of L_R logics as a certain class of regular modal logics extending L1 logics, and he defines L_N as normal extensions of L3 logics. We define our class of interest here. We denote by L_{QR} a first-order extension (with or without extensional confinement) of $\mathbf{G.C}^{g,d}$ by any subset of (B1)–(B24) and (Q3)–(Q12), and by L_{QN} we denote a similar extension that includes both (A13) and (Q11). Moreover, we also use the notations L_{QR} and L_{QN} to refer to the class of logics so-described. Context should disambiguate between this uses-as-class and uses-as-individual.

2.2.2   Barcan and Buridan Formulas

There are, however, many interesting axiom and rule schemes when one considers mixing modalities and quantifiers. The most famous of these axiom schemes are the Barcan formulas. We consider the addition of Barcan formulas and the related Buridan formulas to the systems defined above. The Barcan formulas are named after Ruth Barcan Marcus, who discussed the formal and philosophical significance of some of these formulas in [2]. The Buridan formulas are so-called by Garson [10] though this naming is not a standard in the literature.

(BF) $\forall x \Box \mathcal{A} \to \Box \forall x \mathcal{A}$ (BF$_\Diamond$) $\Diamond \exists x \mathcal{A} \to \exists x \Diamond \mathcal{A}$

(CBF) $\Box \forall x \mathcal{A} \to \forall x \Box \mathcal{A}$ (CBF$_\Diamond$) $\exists x \Diamond \mathcal{A} \to \Diamond \exists x \mathcal{A}$

(BuF) $\exists x \Box \mathcal{A} \to \Box \exists x \mathcal{A}$ (BuF$_\Diamond$) $\Diamond \forall x \mathcal{A} \to \forall x \Diamond \mathcal{A}$

---

[10]Note that, as pointed out in [32], there are first-order relevant logics that are not conservatively extended by $\circ$ and $\leftarrow$ because these formulas become derivable when they were not theorems in non-extended logic.

(CBuF) $\Box\exists x\mathcal{A} \to \exists x\Box\mathcal{A}$ $\qquad\qquad$ (CBuF$_\Diamond$) $\forall x\Diamond\mathcal{A} \to \Diamond\forall x\mathcal{A}$

The following is easy to show.

Lemma 2.4. The logic $\mathbf{QG.C}^{g,d}$ has (CBF), (BuF), (CBF$_\Diamond$), and (BuF$_\Diamond$) as theorems.

2.3 Semantics

Definition 2.5. A *(general frame) ternary relational frame for* $\mathbf{G.C}^{g,d}$ is a tuple $\mathfrak{F} = \langle W, N, R, {}^*, S_\Box, S_\Diamond, e, Prop\rangle$ where $\emptyset \neq N \subseteq W$, $R \subseteq W^3$, ${}^*\colon W \longrightarrow W$, $S_\Box, S_\Diamond \subseteq W^2$, $e \in W$ (and define $u = e^*$), and $Prop$ is a collection of $\leq$-upsets where the $\leq$ relation is defined by, for each $a, b \in W$, $a \leq b =_{df} \exists x \in N(Rxab)$.[11] In addition, we require the conditions below to be satisfied:[12]

($c1$) $a \leq a$;[13]

($c2$) If $a \leq a', b \leq b', c' \leq c$, and $Ra'b'c'$, then $Rabc$, $\qquad$ (i.e., $R\!\downarrow\!\downarrow\!\uparrow$);

($c3$) $b \leq c$ implies $c^* \leq b^*$;

($c4$) $a^{**} = a$;

($c5$) $S_\Box ab$ and $c \leq a$ implies $S_\Box cb$;

($c6$) $S_\Diamond ab$ and $a \leq c$ implies $S_\Diamond cb$;

($c7$) N is an $\leq$-upset;

($c8$) $a \in N$ implies $a^* \leq a$;

($c9$) $a \in N$ implies Ra^*aa^*;

($c10$) $a \in N$ implies $S_\Diamond aa$;

($c11$) $a \in N$ implies $S_\Diamond aa$;

($c12$) $Ruab$ implies $a = e$ or $b = u$;

---

[11]Formally speaking, we define and denote the set of $\leq$-upsets as $\wp(W)^\uparrow = \{X \in \wp(W)\colon \forall a, b \in W(a \in X \ \& \ a \leq b \Rightarrow b \in X)\}$.

[12]Note that condition (CX) corresponds (roughly) to Seki's condition (pX).

[13]As remarked in Seki [27], the relation $\leq$ is not always transitive. However, with the tonicity of R we obtain transitivity. Suppose that $a \leq b$ and $b \leq c$. Then $Rdbc$ for some $d \in N$. By the tonicity condition, then $Rdac$, which entails $a \leq c$.

(c13) $Reue$;

(c14) $e \neq u$;

(c15) $S_\square ee$;

(c16) $S_\square ua$ implies $a = u$;

(c17) $S_\diamond ea$ implies $a = e$;

(c18) $S_\diamond uu$;

(q1) $a \in N$ implies $Raaa$;

(q2) $a \in N$ implies $Rabc \Rightarrow Rac^*b^*$;

(p1) $Prop$ contains N, and is closed under $\cap, \cup, \neg, \rightarrow, \circ, \leftarrow, \square$ and $\diamond$ defined by:

    (i) $\neg X =_{df} \{a \in W : \alpha^* \notin X\}$

    (ii) $X \rightarrow Y =_{df} \{a \in W : \forall b, c \in W (Rabc \ \& \ b \in X \Rightarrow c \in Y)\}$

    (iii) $X \circ Y =_{df} \{a \in W : \exists b, c \in W (Rbca \ \& \ b \in X \ \& \ c \in Y)\}$

    (iv) $X \leftarrow Y =_{df} \{a \in W : \forall b, c \in W (Rbac \ \& \ b \in X \Rightarrow c \in Y)\}$

    (v) $\square X =_{df} \{a \in W : \forall b \in W (S_\square ab \Rightarrow b \in X)\}$

    (vi) $\diamond X =_{df} \{a \in W : \exists b \in W (S_\diamond ab \ \& \ b \in X)\}$

A *model for* $\mathbf{G.C}^{g,d}$ on ternary relational frame $\mathfrak{F}$ is a pair $\mathfrak{M} = \langle \mathfrak{F}, ||-|| \rangle$ where $||-||$ is a valuation function that assigns an $\leq$-upset $||p|| \subseteq W$ to each propositional variable p. This assignment is extended to all formulas by having $||t|| = N$ and the following:

$$||\mathcal{A} \wedge \mathcal{B}|| = ||\mathcal{A}|| \cap ||\mathcal{B}|| \qquad\qquad ||\neg\mathcal{A}|| = \neg||\mathcal{A}||$$
$$||\mathcal{A} \rightarrow \mathcal{B}|| = ||\mathcal{A}|| \rightarrow ||\mathcal{B}|| \qquad\qquad ||\mathcal{A} \vee \mathcal{B}|| = ||\mathcal{A}|| \cup ||\mathcal{B}||$$
$$||\mathcal{A} \leftarrow \mathcal{B}|| = ||\mathcal{A}|| \leftarrow ||\mathcal{B}|| \qquad\qquad ||\mathcal{A} \circ \mathcal{B}|| = ||\mathcal{A}|| \circ ||\mathcal{B}||$$
$$||\square\mathcal{A}|| = \square||\mathcal{A}|| \qquad\qquad ||\diamond\mathcal{A}|| = \diamond||\mathcal{A}||$$

For a logic L_{QR} and L_{QN}, addition axiom and rule schemes will require additional frame conditions. The frame conditions listed below are all modular over $\mathbf{G.C}^{g,d}$. As noted above, in logics weaker than L3, there are a couple of non-modular relations (where we understand "non-modularity" to apply to those semantic conditions that cannot be characterized by a single axiom). Additionally, the split between L_R and L_N requires the additional axiom (A13), with condition (C19) below. For simplicity, for labeling conditions by their correspondences, we take (bi) for the axiom (Bi) (for $1 \leq i \leq 24$), and (qj) for (Qj) (for $3 \leq j \leq 12$).[14]

$(c19)$ $Rabc$ & $S_\Diamond bd \Rightarrow \exists x, y \in W(S_\Box ax$ & $S_\Diamond cy$ & $Rxdy)$

$(b1)$ $Raaa$

$(b2)$ $Rabc \Rightarrow \exists x \in W(Rabx$ & $Raxc)$

$(b3)$ $Rabc \Rightarrow \exists x \in W(Rabx$ & $Rxbc)$

$(b4)$ $Rabc \Rightarrow Rbac$

$(b5)$ $a \neq e$ & $Rabc \Rightarrow b \leq c$

$(b6)$ $b \neq e$ & $Rabc \Rightarrow a \leq c$

$(b7)$ $Rabc$ & $Rcdf \Rightarrow Radf$ & $Rbdf$

$(b8)$ $Rabc \Rightarrow a \leq c$ or $b \leq c$

$(b9)$ $Rabc \Rightarrow Rbac$ & $a \leq c$

$(b10)$ $Rabc \Rightarrow \exists x \in W(b \leq x$ & $c^* \leq x$ & $Raxb^*)$

$(b11)$ $Ra^* aa^*$

$(b12)$ $Raa^* a$

$(b13)$ $Rabc \Rightarrow Rac^* b^*$

$(b14)$ $a \neq e \Rightarrow a^* \leq a$

$(b15)$ $c \neq e$ & $Rabc \Rightarrow a \leq b^*$

$(b16)$ $Rabc$ & $Rcdf \Rightarrow \exists x \in W(Radx$ & $Rbxf)$

$(b17)$ $Rabc$ & $Rcdf \Rightarrow \exists x \in W(Rbdx$ & $Raxf)$

$(b18)$ $Rabc$ & $Rcdf \Rightarrow \exists x \in W(Radx$ & $Rxbf)$

$(b19)$ $Rabc$ & $Rcdf \Rightarrow \exists x, y \in W(Radx$ & $Rbdy$ & $Rxyf)$

$(b20)$ $Rabc$ & $Rcdf \Rightarrow \exists x, y \in W(Radx$ & $Rbdy$ & $Ryxf)$

---

[14]Some of these conditions use the defined relations for the negation-dual modalities, which we define by (i) $S_\Box ab$ iff $S_\Diamond a^* b^*$ and (ii) $S_\Diamond ab$ iff $S_\Box a^* b^*$.

$(b21)$ $Rabc$ & $Rcdf$ $\Rightarrow$ $\exists x \in W(b \leq x$ & $d \leq x$ & $Raxf)$

$(b22)$ $Rabc$ & $S_\Box cd$ $\Rightarrow$ $\exists x, y \in W(S_\Box ax$ & $S_\Box by$ & $Rxyd)$

$(b23)$ $Rabc$ & $S_\Diamond bd$ $\Rightarrow$ $\exists x, y \in W(S_{\boxdot} ax$ & $S_\Diamond cy$ & $Rxdy)$

$(b24)$ $Rabc$ & $S_{\talloblong} bd$ $\Rightarrow$ $\exists x, y \in W(S_\Box ax$ & $S_{\talloblong} cy$ & $Rxdy)$

$(q3)$ $a \in N \Rightarrow Raa^*a$

$(q4)$ $\exists x \in N(Raxa)$

$(q5)$ $a \in N$ & $Rabc$ & $Rcdf$ $\Rightarrow$ $\exists x, y \in W(Radx$ & $Rbxy$ & $Rayf)$

$(q6)$ $a \in N$ & $Rabc$ & $S_\Box cd \Rightarrow \exists x \in W(Raxd$ & $S_\Box bx)$

$(q7)$ $a \in N$ & $Rabc$ & $S_\Diamond bd \Rightarrow \exists x \in W(Radx$ & $S_\Diamond cx)$

$(q8)$ $a \in N \Rightarrow S_\Box aa$

$(q9)$ $a \in N \Rightarrow S_{\boxdot} aa$

$(q10)$ $a \in N$ & $S_\Box ab \Rightarrow a \leq b$

$(q11)$ $a \in N$ & $S_\Box ab \Rightarrow b \in N$

$(q12)$ $a \in N$ & $S_{\boxdot} ab \Rightarrow b \in N$

Definition 2.6 (Models for **LQ.M**). Let **L.M** be a modal relevant logic extending $\mathbf{G.C}^{g,d}$. A *frame for **LQ.M*** (an **LQ.M**-frame) is a tuple

$$\mathfrak{F} = \langle W, N, R, ^*, S_\Box, S_\Diamond, e, U, Prop, PropFun \rangle,$$

where $\langle W, N, R, ^*, S_\Box, S_\Diamond, e, Prop \rangle$ is an **L.M**-frame, U is a non-empty set of individuals, and $PropFun \subseteq \{\varphi \colon U^\omega \longrightarrow Prop\}$. Moreover, the following conditions are satisfied:

$(cq1)$ $PropFun$ contains a constant function φ_N $(\varphi_N f = N)$, and is closed under $\cap, \cup, \neg \rightarrow, \circ, \leftarrow, \Box, \Diamond, \forall_n$ and $\exists_n$, for every $n \in \omega$, where

   (a) $(\oplus\varphi)f = \oplus(\varphi f)$, for each $\oplus \in \{\neg, \Box, \Diamond\}$;

   (b) $(\varphi \otimes \psi)f = \varphi f \otimes \psi f$, for each $\otimes \in \{\cap, \cup, \rightarrow, \circ\}$;

   (c) $(\forall_n\varphi)f = \displaystyle\bigcap_{g \sim_{x_n} f} \varphi g = \bigcup\{X \in Prop \colon X \subseteq \bigcap_{g \sim_{x_n} f} \varphi g\}$;

(d) $(\exists_n \varphi)f = \bigsqcup_{g \sim_{x_n} f} \varphi g = \bigcap \{X \in Prop: \bigcup_{g \sim_{x_n} f} \varphi g \subseteq X\}.$

$(cq2)$ For every $\varphi \in PropFun$, $X, Y \in Prop$, $n \in \omega$, and $f \in U^\omega$,[15]

$\quad$ (cEC) $X \smallsetminus Y \subseteq \bigcap_{j \in U} \varphi(f[j/n])$ only if $X \smallsetminus Y \subseteq (\forall_n \varphi)f$;

$\quad$ (cdEC) $\bigcup_{j \in U} \varphi(f[j/n]) \subseteq X \cup \overline{Y}$ only if $(\exists_n \varphi)f \subseteq X \cup \overline{Y}$.

Note that the interpretation of the quantifiers here is the Mares–Goldblatt interpretation introduced in a pair of papers [15, 12].

A *pre-model for* $\boldsymbol{LQ.M}$ is a tuple $\mathfrak{M} = \langle \mathfrak{F}, |-| \rangle$ such that $\mathfrak{F}$ is a frame for $\boldsymbol{LQ.M}$ and $|-|$ is a valuation function that assigns:

(i) an individual $|c| \in U$ to each constant symbol c;

(ii) a function $|P^n|: U^n \longrightarrow \wp(W)$ to each n-ary predicate symbol P^n; $\quad$ and

(iii) a propositional function $|\mathcal{A}|: U^\omega \longrightarrow \wp(W)$ to each formula $\mathcal{A}$ such that, when $\mathcal{A}$ is atomic, for every $f \in U^\omega$: $|P^n \tau_1, \ldots, \tau_n|f = |P^n|(|\tau_1|f, \ldots, |\tau_n|f)$ where "$|\tau|f$" is fn when τ is the variable x_n, and $|c|$ when τ is the constant symbol c. Moreover, when $\mathcal{A}$ is not atomic (or $\boldsymbol{t}$), the valuation is extended as follows:

$$|\boldsymbol{t}| = \varphi_N \qquad\qquad |\neg\mathcal{A}| = \neg|\mathcal{A}|$$
$$|\mathcal{A} \wedge \mathcal{B}| = |\mathcal{A}| \cap |\mathcal{B}| \qquad\qquad |\mathcal{A} \circ \mathcal{B}| = |\mathcal{A}| \circ |\mathcal{B}|$$
$$|\mathcal{A} \vee \mathcal{B}| = |\mathcal{A}| \cup |\mathcal{B}| \qquad\qquad |\mathcal{A} \rightarrow \mathcal{B}| = |\mathcal{A}| \rightarrow |\mathcal{B}|$$
$$|\forall x_n \mathcal{A}| = \forall_n |\mathcal{A}| \qquad\qquad |\exists x_n \mathcal{A}| = \exists_n |\mathcal{A}|$$
$$|\Box\mathcal{A}| = \Box|\mathcal{A}| \qquad\qquad |\Diamond\mathcal{A}| = \Diamond|\mathcal{A}|$$

A *model for* $\boldsymbol{LQ}$ is a pre-model for $\boldsymbol{LQ}$ that assigns an element of $Prop$ to each atomic formula.

A formula $\mathcal{A}$ is *satisfied* by a variable assignment f in a model $\mathfrak{M}$, written $\mathfrak{M}, f \vDash \mathcal{A}$, when $N \subseteq |\mathcal{A}|f$. A formula is *valid in a model* $\mathfrak{M}$ ($\mathfrak{M} \vDash \mathcal{A}$) when it is satisfied by every variable assignment in that model; *valid in a frame* $\mathfrak{F}$ ($\mathfrak{F} \vDash \mathcal{A}$) when it is valid in every model based on that frame; *valid in a class of frames* $\mathbb{C}$ ($\mathbb{C} \vDash \mathcal{A}$) when it is valid in every frame in that class. Admissibility of a rule in the semantic context is understood as the preservation of validity.

---

[15]Note that $X \smallsetminus Y$ and $\overline{Y}$ are defined in the usual set-theoretic sense, but that $Prop$ is not necessarily closed under either of these operations.

Theorem 2.7 (Soundness and Completeness for **LQ.M**). *The logics* **QB.C** *and the extensions by sets of the axiom and rule schemes above are sound and complete w.r.t. the corresponding class of frames.*

Remark 2.8. The proof is as in Ferenz [5]. However, we note three differences. First, in [5] there is no left implication. This will not affect any relevant extensions of **QB.C** that are conservatively extended by left implication. We will note this fact where relevant and in the summary of the main theorem of this paper. Secondly, there is no e (or u) element in the frames considered there. This is easily rectified, and the inclusion of e in Seki is only to deal with certain axiom schemes not dealt with in [5]. Third, as just mentioned, not all of the axioms and rule schemes listed above were explicitly considered. However, the verification of modular soundness and completeness over **QB.C** using the frame conditions in [27] (for logics conservatively extended by left implication) is straightforward, and left to the reader.

We are particularly interested in the class of frames corresponding to the class L_{QR} or L_{QN} of first-order modal logics. For any logic L_{QR} or L_{QN}, a corresponding frame will be called an $\mathfrak{F}_{QR}$ or $\mathfrak{F}_{QN}$, respectively. Similarly, we use $\mathfrak{F}_{QR}$ or $\mathfrak{F}_{QN}$ to both denote the class of frames corresponding to the class of logics L_{QR} or L_{QN}, respectively, and to range over concrete frames in the class. Moreover, we use $\mathfrak{F}_Q$ to refer to either $\mathfrak{F}_{QR}$ or $\mathfrak{F}_{QN}$.

3 Gamma Admissibility for 'Easy' First-order Modal Relevant Logics

In this section, we deal with γ-admissibility for L_Q logics. The modal propositional base of these first-order logics (without fusion, left implication, and the intensional truth constant t) was shown to admit γ by the methods of Seki [27]. We remind the reader that the first-order logics in question have both disjunctive rules (Q1) and (Q2), making the logic what Seki calls an L3 logic. Moreover, they contain the principle of excluded middle (A10), and the disjunctive rule (R6). Our results, that is, strictly only apply to first-order extensions of L3 logics, since, due to the tonicity conditions required for fusion and left implication, require (Q1) and (Q2). The normal models method employed in this paper shows that certain logics are sound and complete w.r.t. normal models in such a way that γ can be shown admissible.

Definition 3.1 (Normal Models). For any logic **LQ.M** defined above (and similarly for **QL.M**), a **QL.M**-model (-frame) is *normal* if it satisfies the following:

(Norm) $a = a^*$, for some $a \in N$.

Note that while, say, the logic **QB.C** might have some normal models, it cannot be proved sound and complete w.r.t. the class of its normal models by the constructions of this paper, due to requiring the axiom of the excluded middle.

Definition 3.2 (Normalization of a frame $\mathfrak{F}$ (at 0 for o)). For a logic L_Q (which thus has a modal part that admits γ by Seki's arguments), $0 \notin W$ and $o \in N$, the *normalization of an L_Q-frame at 0 for o* $\mathfrak{F}_Q = \langle W, N, R, {}^*, S_\Box, S_\Diamond, e, U, Prop, PropFun \rangle$ is a frame

$$\mathfrak{F}'_Q = \langle W', N', R', {}^{*'}, S'_\Box, S'_\Diamond, e, U, Prop', PropFun' \rangle$$

defined by:[16]

1. $W' = W \cup \{0\}$;

2. $N' = N \cup \{0\}$;

3. R' is given by $R'abc$ iff $Rabc$, whenever $a, b, c \in W$, and when $a, b \in W$, R' satisfies the following:

   (a) $R'000$; (d) $R'a00$ iff $Raoo^*$; (g) $R'ab0$ iff $Rabo^*$.

   (b) $R'00a$ iff $Rooa$; (e) $R'0ab$ iff $Roab$;

   (c) $R'0a0$ iff $Roao^*$; (f) $R'a0b$ iff $Raob$;

4. ${}^{*'}$ is defined by:

   (a) $a^{*'} = a^*$ when $a \in W$;   (b) $0^{*'} = 0$.

5. $S'_\Box$ is given by $S_\Box ab$ iff $S'_\Box ab$, whenever $a, c \in W$, and for all $a \in W$:

   (a) $S'_\Box 00$; (b) $S'_\Box 0a$ iff $S_\Box oa$; (c) $S'_\Box a0$ iff $S_\Box ao^*$.

6. $S'_\Diamond$ is given by $S_\Diamond ab$ iff $S'_\Diamond ab$, whenever $a, c \in W$, and for all $a \in W$:

   (a) $S'_\Diamond 00$; (b) $S'_\Diamond 0a$ iff $S_\Diamond o^* a$; (c) $S'_\Diamond a0$ iff $S_\Diamond ao$.

7. We define $Prop'$ so that for each $X \in Prop$,

   (u1) add $CL^{\leq'}(X)$ to $Prop'$, where $CL^{\leq'}(X) = \{b \in W' : \exists a \in W' \text{ s.t. } a \leq' b\}$;

   (u2) add $X \cup \{0\}$ to $Prop'$ when $o \in X$.

8. We define $PropFun'$ so that for each $\varphi \in PropFun$, add φ' and φ'' to $PropFun'$, where these new functions are defined by the following:[17]

---

[16] The o and 0 are used as in [27], and correspond to T and T' of [24, p. 387].

[17] We note that the resulting $PropFun'$ is not, in the terminology of [15], *full*. That is, it does not contain every function from U^ω into $Prop'$. Notably, it is missing what we might call *mixed* functions, which sometimes return an element of X of $Prop$ with $o \in X$, and sometimes return an element Y of $PropFun'$ which contains 0.

$$\forall f \in U^{\omega}: \qquad \varphi' f = CL^{\leq'}(\varphi f) \qquad \text{and} \qquad \varphi'' f = \begin{cases} \varphi f \cup \{0\}, & \text{if } o \in \varphi f; \\ CL^{\leq'}(\varphi f), & \text{otherwise.} \end{cases}$$

We called this normalized frame *the normalization of* $\mathfrak{F}$ *at 0 for* $o \in N$.

The next few lemmas have proofs that are standard in the literature, and, as noted in [6] for logics with the Mares–Goldblatt interpretation of the quantifiers, do not rely on the extra first-order machinery given by $U, Prop$, and $PropFun$.

Lemma 3.3. If $\mathfrak{F}_Q$ is an L_{QR}-frame or L_{QN}-frame and $\mathfrak{F}'$ is the normalization of $\mathfrak{F}_Q$ at 0 for o:

1. the relations $R', S'_\square$ and $S'_\diamond$ are well defined;

2. the ordering $\leq'$ and relations $S'_\square$ and $S'_\diamond$ are such that, for all $a, b \in W$:

   (a) $a \leq' b$ iff $a \leq b$ (d) $0 \leq' b$ iff $o \leq b$ (g) $a \leq' 0$ if $a \leq o^*$

   (b) $S'_\square ab$ iff $S_\square ab$ (e) $S'_\square 0b$ iff $S_\square ob$ (h) $S'_\square a0$ iff $S_\square ao^*$

   (c) $S'_\diamond ab$ iff $S_\diamond ab$ (f) $S'_\diamond 0b$ iff $S_\diamond o^*b$ (i) $S'_\diamond a0$ iff $S_\diamond ao$

Corollary 3.4. If $\mathfrak{F}_Q$ is an L_{QR}-frame or L_{QN}-frame and $\mathfrak{F}'$ is the normalization of $\mathfrak{F}_Q$ at 0 for o, then $o^* \leq' 0 \leq' o$.

Lemma 3.5. For a logic L_{QR} or L_{QN}, let

$$\mathfrak{F}' = \langle W', N', R', {}^{*'}, S'_\square, S'_\diamond, e, U, Prop', PropFun' \rangle$$

be the normalization at 0 for $o \in N$ of the frame $\mathfrak{L}$ for

$$\mathfrak{F}_Q = \langle W, N, R, {}^*, S_\square, S_\diamond, e, U, Prop, PropFun \rangle$$

for L_{QR} or L_{QN}. Then $\mathfrak{F}'$ is an $\mathfrak{F}_Q$-frame for L_{QR} or L_{QN}, as appropriate.

Proof. We must check every frame postulate, for both L_{QR} and L_{QN}. By our supposition that $\mathfrak{L}$ is based on an L3 logic that admits γ by the arguments of Seki [27] (and [24, pp. 389–390]), $(c1)$, $(c3)$–$(c18)$, $(q1)$–$(q2)$, are satisfied for an L_{QR} logic, and similarly that the conditions $(q11)$, $(b22)$, and $(c19)$ are satisfied for any L_{QN} logic. Here we show in detail that $(c2)$, $(p1)$, and $(cq1)$ hold. We refer the reader to [6] for the arguments that show that $(cq2)$ holds when the logic has (EC).

For $(c2)$, we show the case for the tonicity of the middle argument place. Suppose that $R'ab'c$ and that $b \leq' b'$. The interesting case is when $R'000$ and $b \leq' 0$. By $(q1)$

Rooo, and then and $(q2)$ *Roo*o**. From this and the previous frame satisfying $(c2)$, we have that *Roo*o**. By definition we then obtain that $R'0b0$, as required. The case for the third position of the ternary relation is similar.

For $(p1)$, we first show that every element of *Prop'* is an upset, i.e., *Prop'* is well defined. First, each $CL^{\leq}(X)$ is an upset by definition. Second, suppose that $a \leq b$ and $a \in X \cup \{0\}$. The only potentially problematic case is when $a = 0$, which gives $0 \leq' b$. The latter is iff $o \leq b$. By $(u2)$ $o \in X$, and so $o \in X \cup \{0\}$. Moreover, the new set N' is given by $N \cup \{0\}$, since $o \in N$.

We introduce some notation. For an $X \in Prop$, we denote by X' ambiguously either of $X \cup \{0\}$ or $CL^{\leq}(X)$ in *Prop'*. We are required to show that, e.g., for any $X', Y' \in Prop'$, there is a unique $Z' \in Prop'$ such that $X' \to Y' = Z'$. We show this for every operation required of *Prop'*. We show, e.g., that $X' \to Y'$, with X' from $X \in Prop$ and Y' from $Y \in Prop$ is either $CL^{\leq'}(X \to Y)$ or $(X \to Y) \cup \{0\}$. This has the further consequence that, for every $X' \in Prop'$ from $X \in Prop$, either $X' = X$ or $X' = X \cup \{0\}$.

Here we consider the cases for $\neg, \cap, \Box$, and $\to$.

Case $\neg$: We need to show that, for $X' \in Prop'$ and $X \in Prop$ its source, either $\neg(X') = CL^{\leq'}(\neg X)$ or $\neg(X') = \neg X \cup \{0\}$. Suppose that $CL^{\leq'}(\neg X) \neq \neg(X')$; we show that it must follow that $\neg(X') = \neg X \cup \{0\}$ by considering two cases:

- There is a $b \in CL^{\leq'}(\neg X)$ such that $b \notin \neg(X')$. Were $b \in W$, we would easily reason that $b \in \neg(X')$, so $b = 0$. But $0 \in CL^{\leq'}(\neg X)$ entails that $o^* \in \neg X$, whence $o \notin X$ and $o \notin X'$, whence $0 \notin X'$, whence $0 \in \neg(X')$, which contradicts the assumption that $0 \notin \neg(X')$.

- There is a $b \in \neg(X')$ such that $b \notin CL^{\leq'}(\neg X)$. As before, if $b \in W$, we would get that $b \in \neg X$ (and thus in its closure), so assume that $b = 0$ and consequently that $0 \notin X'$. Now, by the earlier lemma, this means that $X' = X$. Consequently, we can rewrite $\neg(X') = \{a \in W : a^* \notin X'\} \cup \{a \in W' \smallsetminus W : a^* \notin X'\}$ as $\{a \in W : a^* \notin X\} \cup \{a \in W' \smallsetminus W : a^* \notin X'\}$ and as $\neg X \cup \{0\}$.

Case $\cap$: We need to show that, for $X', Y' \in Prop'$ and $X, Y \in Prop$ their sources, either $X' \cap Y' = CL^{\leq'}(X \cap Y)$ or $X' \cap Y' = (X \cap Y) \cup \{0\}$. Suppose that $CL^{\leq'}(X \cap Y) \neq X' \cap Y'$. Then consider two cases:

- There is a $b \in CL^{\leq'}(X \cap Y)$ such that $b \notin X' \cap Y'$. If $b \in W$ then $b \in X, Y$, which would entail that $b \in X' \cap Y'$. So assume that $b = 0$ and that $0 \in CL^{\leq'}(X \cap Y)$. It follows that $o^* \in X \cap Y$. Now, assume w.l.o.g. that $0 \notin X'$. Then $o^* \notin X'$, whence $o^* \notin X$. Contradiction.

- There is a $b \in X' \cap Y'$ such that $b \notin CL^{\leq'}(X \cap Y)$. That $b \in W$ leads easily to a contradiction, so assume that $b = 0$. Then $X' = X \cup \{0\}$ and $Y' = Y \cup \{0\}$, whence it is a simple step to infer that $X' \cap Y' = (X \cap Y) \cup \{0\}$.

Case $\Box$: We need to show that, for $X' \in Prop'$ and $X \in Prop$ its source, either $\Box(X') = CL^{\leq'}(\Box X)$ or $\Box(X') = \Box X \cup \{0\}$. Suppose that $CL^{\leq'}(\Box X) \neq \Box(X')$. Then consider two cases:

• There is a $b \in CL^{\leq'}(\Box X)$ such that $b \notin \Box X'$, whence there exists a $c \in W'$ such that $S'_{\Box}bc$ for which $c \notin X'$. We consider four subcases, each of which entails that $c \in X'$:

○ $(b, c \in W)$ $S'_{\Box}bc$ entails that $S_{\Box}bc$; since $b \in \Box X$, $c \in X \subseteq X'$.

○ $(b \in W, c = 0)$ $S'_{\Box}b0$, whence $S_{\Box}bo^*$, $o^* \in X \subseteq X'$, and $c = 0 \in X'$ by heredity.

○ $(b = 0, c \in W)$ $S'_{\Box}0c$, whence $S_{\Box}oc$. Since $0 \in CL^{\leq'}(\Box X)$, $o^* \in \Box X$ and by heredity, $o \in \Box X$, so it follows that $c \in X \subseteq X'$.

○ $(b, c = 0)$ $S'_{\Box}00$, whence $S_{\Box}oo^*$. As $0 \in CL^{\leq'}(\Box X)$, $o^* \in \Box X$, whence $o \in \Box X$, whence $o^* \in X \subseteq X'$. As $o^* \in X'$, however, $c = 0 \in X'$ by heredity.

• There is a $b \in \Box(X')$ such that $b \notin CL^{\leq'}(\Box X)$. In this case, either $b = 0$ or $b \in W$:

○ If $b \in W$, then $b \notin CL^{\leq'}(\Box X)$ entails that $b \notin \Box X$ and that there exists a $c \in W$ such that $S_{\Box}bc$ and $c \notin X$. As $c \in W$, it follows that $c \notin X'$. This means that $S'_{\Box}bc$, so from $b \in \Box(X')$, it follows that $c \in X$.

○ If $b = 0$, we show that this case entails that $\Box(X') = \Box X \cup \{0\}$. Since $b = 0$ is a member of both sets, we need to focus on elements of W. Let $f \notin \Box X$. Then there is a $d \in W$ such that $S_{\Box}fd$ and $d \notin X$, whence $S_{\Box}fd$ and $d \notin X'$, i.e., $f \notin \Box(X')$. Let $f \notin \Box(X')$; then there exists $d \in W'$ such that $S'_{\Box}fd$ and $d \notin X'$. If $d = 0$, then $S'_{\Box}f0$ means that $S_{\Box}fo^*$ and that $d = 0 \notin X'$ means that $o^* \notin X$, whence $f \notin \Box X$. If $d \in W$, then we can similarly infer that $S_{\Box}fd$ and $d \notin X$, whence $f \notin \Box X$. Consequently, $\Box(X') = \Box X \cup \{0\}$.

Case $\rightarrow$: We need to show that, for $X', Y' \in Prop'$ and $X, Y \in Prop$ their sources, either $X' \rightarrow Y' = CL^{\leq'}(X \rightarrow Y)$ or $X' \rightarrow Y' = (X \cap Y) \rightarrow \{0\}$.

We first suppose that $X' \rightarrow Y' \neq CL^{\leq'}(X \rightarrow Y)$. There are two cases. First, suppose that $a \in X' \rightarrow Y'$, but $a \notin CL^{\leq'}(X \rightarrow Y)$. We first show that $a = 0$. For reductio, let $a \in W$. Then by $a \notin CL^{\leq'}(X \rightarrow Y)$, we have that $a \notin X \rightarrow Y$. Then there are $b, c \in W$ such that $Rabc$, $b \in X$, and $c \notin Y$. $b, c \in W$ then implies that $b \in X'$, and $c \notin Y'$. Moreover, we obtain $R'abc$. This all gives that $a \notin X' \rightarrow Y'$, which is a contradiction. So now that we have shown that $a = 0$, let us show that $X' \rightarrow Y' = (X \rightarrow Y) \cup \{0\}$.

It suffices to show that, for $a \in W$, $a \in X' \rightarrow Y'$ iff $a \in X \rightarrow Y$. First, if

$a \notin X' \to Y'$, then there are $b, c \in W'$ where $R'abc$. The interesting cases are when $b = 0$ or $c = 0$. We show one subcase. Let $b = c = 0$. Then we have $R'a00$, $0 \in X'$, $0 \notin Y'$. These give that $Raoo^*$, $o \in X$, and $o^* \notin Y$ straightforwardly, which entails that $a \notin X \to Y$. Next, if $a \notin X \to Y$, then it is easy to show that $a \notin X' \to Y'$.

The second case is when $a \notin X' \to Y'$, but $a \in CL^{\leq'}(X \to Y)$. From the latter, we have that $\exists b, c \in W'$ where $R'abc$, $b \in X'$, and $c \notin Y'$. We go through the possible cases for where a, b, c are in W or are 0, and we show each leads to contradiction.

Case $a = b = c = 0$: Then we have $R'000$. By $(c2)$, we then have that $Rooo^*$. Moreover, it is easily shown that $o \in X$ and $o^* \notin Y$. Therefore, $o \notin X \to Y$. This is in contradiction with $0 \in CL^{\leq'}(X \to Y)$.

Case $a = 0$, $b, c \in W$: Then we have $R'0bc$, $b \in X'$, and $c \notin X'$, which gives $Robc$, $b \in X$, $c \notin Y$. This entails that $o, 0 \notin CL^{\leq'}(X \to Y)$, which is a contradiction.

Case $a, c \in W$, $b = 0$: Then we have $R'a0c$, $0 \in X'$, and $c \notin Y'$, which gives $R'aoc$, $o \in X$, and $c \notin Y$, which entail that $a \notin X \to Y$, a contradiction.

Case $a, b \in W$, $c = 0$: Then we have $R'ab0$, $b \in X'$, and $0 \notin Y'$, which gives $R'abo^*$, $b \in X$, and $o^* \notin Y$, which entail that $a \notin X \to Y$, a contradiction.

Case $a = b = 0$, $c \in W$: Then we have $R'00c$, $0 \in X'$, and $c \notin Y'$, which gives $R'ooc$, $o \in X$, and $c \notin Y$, which entail that $o \notin X \to Y$, and thus that $a = 0 \notin CL^{\leq'}(X \to Y)$, a contradiction.

The remaining cases are similar.

For $(cq1)$, we must show that $PropFun'$ is well defined, contains constant φ_N, and that it is closed under the required operations. That each element of $PropFun'$ is a function of type $U^\omega \longrightarrow Prop'$ is given by the definition of $PropFun'$, and so the set is well-defined. To show that the set is closed under the required operations, we again proceed by cases. For example, we show that for an arbitrary elements of $\varphi^\circ, \psi^\circ \in PropFun'$, which are each either φ' or φ'' or ψ' or ψ'' for $\varphi, \psi \in PropFun$, there is a function $\Sigma \in PropFun'$ such that $\Sigma = \varphi^\circ \to \psi^\circ$.

Case $\neg$: We show that, for $\varphi^\circ \in PropFun'$ with $\varphi \in PropFun$ its source, either $\neg(\varphi^\circ) = (\neg\varphi)'$ or $\neg(\varphi^\circ) = (\neg\varphi)''$.

To make matters clear, let $\neg'$ and $\neg$ indicate the evaluation of $\neg$ in the normalized and original models, respectively. By the definitions, we can infer that

$$\neg'(\varphi^\circ) = \begin{cases} (\neg\varphi)' & \text{if } \forall f \in U^\omega,\ [\neg'(\varphi^\circ)]f = CL^{\leq'}((\neg\varphi)f); \\ (\neg\varphi)'' & \text{if } \exists f \in U^\omega,\ [\neg'(\varphi^\circ)]f = (\neg\varphi)f \cup \{0\}. \end{cases}$$

Of course, $[\neg'(\varphi^\circ)]f = \neg'(\varphi^\circ f) \in Prop'$. Depending on the choice of $^\circ$, this is either $\neg'(CL^{\leq'}(\varphi f))$ or $\neg'(\varphi f \cup \{0\})$. But $CL^{\leq'}(\varphi f)$ is either φf or $\varphi f \cup \{0\}$, so this is equivalent to either $\neg'(\varphi f)$ or $\neg'(\varphi f \cup \{0\})$. Both alternatives reduce to $\neg'(\varphi f)$. But by previous observations, this value is either $CL^{\leq'}(\varphi f)$ or is $\varphi f \cup \{0\}$. As f

was chosen arbitrarily, either for *all* f, $[\neg'(\varphi^\circ)]f = CL^{\leq'}((\neg\varphi)f)$ or for *some* f, $[\neg'(\varphi^\circ)]f = (\neg\varphi)f \cup \{0\}$, i.e., either $\neg(\varphi^\circ) = (\neg\varphi)'$ or $\neg(\varphi^\circ) = (\neg\varphi)''$.

Case $\cap$: We show that, for $\varphi^\circ, \psi^\circ \in PropFun'$ with $\varphi, \psi \in PropFun$ their respective sources, either $(\varphi)^\circ \cap (\psi)^\circ = (\varphi \cap \psi)'$ or $(\varphi)^\circ \cap (\psi)^\circ = (\varphi \cap \psi)''$.

We follow the proof strategy of the case for lifted negation, using $\cap'$ and $\cap$ to indicate evaluation of $\cap$ in the normalized and original models, respectively. Observe that for an arbitrary f, $[(\varphi^\circ) \cap' (\psi^\circ)]f = (\varphi^\circ f) \cap' (\psi^\circ f)$. Given a choice of (possibly distinct) decorations of φ° and ψ°, $\varphi^\circ f$ is either $CL^{\leq'}(\varphi f)$ or $\varphi f \cup \{0\}$ (and mutatis mutandis for $\psi^\circ f$). Trivial set theory entails two possible values for $(\varphi^\circ f) \cap' (\psi^\circ f)$: either $CL^{\leq'}(\varphi f) \cap CL^{\leq'}(\psi f) = CL^{\leq'}(\varphi f \cap \psi f)$ or $(\varphi f \cap \psi f) \cup \{0\}$. Consequently, either all f fall into the former category or some f falls into the latter category, i.e., either $(\varphi)^\circ \cap (\psi)^\circ = (\varphi \cap \psi)'$ or $(\varphi)^\circ \cap (\psi)^\circ = (\varphi \cap \psi)''$.

Case $\square$: We show that, for $\varphi^\circ \in PropFun'$ with $\varphi \in PropFun$ its source, either $\square(\varphi^\circ) = (\square\varphi)'$ or $\square(\varphi^\circ) = (\square\varphi)''$. For an arbitrary $f \in U^\omega$, we have that $[\square'(\varphi^\circ)]f = \square'(\varphi^\circ f)$. This is either $\square'(CL^{\leq'}(\varphi f))$ or $\square'(\varphi f \cup \{0\})$. As in earlier cases, this means that it is either $\square'(\varphi f)$ or $\square'(\varphi f \cup \{0\})$.

For $\square'(\varphi f)$, where $o^* \notin \varphi f$ by definition, we have mutually exhaustive subcases:
$o^* \in \square\varphi f$:   This case is impossible. By $(c10)$, $S_\square o^* o^*$. The contradiction is immediate. Thus for all cases we have $o^* \notin \varphi f$ and $o^* \notin \square\varphi f$.

$o \notin \square\varphi f$:   We show that $0 \notin \square'\varphi f$. By supposition, there is a $b \in W$ such that $S_\square ob$ and $b \notin \varphi f$. Then $S'_\square 0b$ and $b \notin \varphi f$. So $0 \notin \square'\varphi f$.

$o \in \varphi f$:   We show that $0 \notin \square\varphi f$. The assumption that $\square'(\varphi f) = \square'(CL^{\leq'}(\varphi f))$ implies that $(\varphi f) = (CL^{\leq'}(\varphi f))$. Thus, $0 \notin (CL^{\leq'}(\varphi f))$, because $o^* \notin (CL^{\leq'}(\varphi f))$. Thus the result follows from $S'_\square 00$.

In each of the subcases, $\square'\varphi f$ is either $CL^{\leq'}(\square\varphi f)$ or $(\square\varphi f) \cup \{0\}$, as required.

For $\square'(\varphi f \cup \{0\})$, we have $0 \in \square'(\varphi f \cup \{0\})$ iff $\forall x(S'_\square 0x \Rightarrow x \in \varphi f \vee x = 0)$.

The cases where $o^* \notin \varphi f$ are open, as there are two ways for 0 to enter $(\varphi f \cup \{0\})$. We have the following exhaustive subcases.

$o \in \square\varphi f$:   We show that $0 \in \square'(\varphi f \cup \{0\})$. Suppose that $S'_\square 0x$. If $x \neq 0$ then $S_\square ox$ and thus $x \in \varphi f$. If $x = 0$, then $x = 0$.

$o \notin \square\varphi f$:   From $o \notin \square\varphi f$, we get $\exists b \in W$ such that $S_\square ob$ and $b \notin \varphi f$. Thus $S'_\square 0b$ and $b \notin \varphi f$.

Now, since $\square'\varphi f$ is either $\square\varphi f$ or $\square\varphi f \cup \{0\}$, it is either the $CL^{\leq'}(\square\varphi f)$ or $\square\varphi f \cup \{0\}$, for every f. So either each f returns the former, or some f returns the latter. Thus, either $\square(\varphi^\circ) = (\square\varphi)'$ or $\square(\varphi^\circ) = (\square\varphi)''$.

Case $\rightarrow$: We show that, for $\varphi^\circ, \psi^\circ \in PropFun'$ with $\varphi, \psi \in PropFun$ their respective sources, either $(\varphi)^\circ \to (\psi)^\circ = (\varphi \to \psi)'$ or $(\varphi)^\circ \to (\psi)^\circ = (\varphi \to \psi)''$.

Observe that for an arbitrary f, $[(\varphi^\circ) \to' (\psi^\circ)]f = (\varphi^\circ f) \to' (\psi^\circ f)$. The right-hand side is always unique (given a disambiguation of $^\circ$ on the right-hand side) and an element of $Prop'$ by the verification of $(p1)$.

Four cases follow: $(\varphi^\circ f) \to' (\psi^\circ f)$ is either (i) $CL^{\leq'}(\varphi f) \to' CL^{\leq'}(\psi f)$, (ii) $CL^{\leq'}(\varphi f) \to' (\psi f \cup \{0\})$ (iii) $(\varphi f \cup \{0\}) \to' CL^{\leq'}(\varphi)$, or (iv) $(\varphi f \cup \{0\}) \to' (\psi f \cup \{0\})$.

For case (i), if 0 does not satisfy the membership requirements of $CL^{\leq'}(\varphi f) \to' CL^{\leq'}(\psi f)$, then there are $a, b \in W'$ such that $R'0ab$, $a \in CL^{\leq'}(\varphi f)$ and $b \notin CL^{\leq'}(\psi f)$. There are four possibilities. If $a, b \in W$, then $o \notin CL^{\leq'}(\varphi f) \to CL^{\leq'}(\psi f)$, and $CL^{\leq'}(\varphi f) \to' CL^{\leq'}(\psi f) = CL^{\leq'}((\varphi \to \psi)f)$. If $a = 0$, $b \in W$, then we have $Roob$ and a similar result. If $a \in W$ and $b = 0$, then we have $R'0a0$, which gives $Roao^*$, and so similarly $o \notin CL^{\leq'}(\varphi f) \to CL^{\leq'}(\psi f)$. If $a = b = 0$, then we have $R'000$. By the tonicity we have $R'0o^*0$, which gives $R'00o^*$ by $(q2)$. This entails that $Rooo^*$, which together with $0, o \in CL^{\leq'}(\varphi f)$ and $0, o^* \notin CL^{\leq'}(\psi f)$ results in $o \notin CL^{\leq'}(\varphi f) \to CL^{\leq'}(\psi f)$, as before.

Otherwise, if $0 \in CL^{\leq'}(\varphi f) \to' CL^{\leq'}(\psi f)$, then either $o^* \in CL^{\leq'}(\varphi f) \to' CL^{\leq'}(\psi f)$ or not. If it is, then by the definition we have that $CL^{\leq'}(\varphi f) \to' CL^{\leq'}(\psi f) = CL^{\leq'}(\varphi \to \psi f)$. If o^* is not a member, then there are elements $b, c \in W'$ such that Ro^*bc, $b \in CL^{\leq'}(\varphi f)$ and $c \notin CL^{\leq'}(\psi f)$. Going through the possibilities, we always obtain that $o^* \notin (\varphi f) \to (\psi f)$, but that o is always an element of the latter. Thus, $CL^{\leq'}(\varphi f) \to' CL^{\leq'}(\psi f) = (\varphi \to \psi f) \cup \{0\}$.

The cases for (ii), (iii), and (iv) are similar, and left to the reader.

Case $\forall_n$: We show that, for $\varphi^\circ \in PropFun'$ with $\varphi \in PropFun$ its source, either $\forall_n(\varphi^\circ) = (\forall_n \varphi)'$ or $\forall_n(\varphi^\circ) = (\forall_n \varphi)''$. Assume an arbitrary f, and we have $[\forall'_n(\varphi^\circ)]f = \forall'_n(\varphi^\circ f)$, the right-hand side of which is an element of $Prop'$. Now, each $\varphi^\circ g$ is either $CL^{\leq'}(\varphi g)$ or $\varphi g \cup \{0\}$, for each x_n variant g or f. Thus, the generalized intersection $\bigcap'_{g \sim_n f} \varphi g$ in the normalized frame is either the original $\bigcap'_{g \sim_n f} \varphi g$, or it is $(\bigcap_{g \sim_n f} \varphi g) \cup \{0\}$.

Case $0 \notin \forall'_n(\varphi^\circ f)$: This entails that $o \notin \forall_n \varphi f$. That is, for all $Z \in Prop$, $Z \subseteq \bigcap_{g \sim_n f} \varphi g$ implies $o \notin Z$. By definition, Z' must be the closure and not contain 0 for each such Z. And so $0 \notin \forall'_n \varphi^\circ f$, and so $\forall'_n \varphi^\circ f = \forall_n \varphi f$, an element of $Prop'$, as required.

Case $0 \in \forall'_n(\varphi^\circ f)$: Then there is an element $Z' \subseteq \bigcap'_{g \sim_n f} \varphi^\circ g$ which is an element of $Prop'$ and contains 0. Consider the source of Z'. If Z' is the closure of some set Z, then $o^* \in Z$ and so $o^* \in \forall_n \varphi f$. In this case, $\forall'_n(\varphi^\circ f) = CL^{\leq'}(\forall_n \varphi f)$.

On the other hand, suppose that $Z' = Z \cup \{0\}$ and that $o^* \notin Z$. This follows for any such Z' and so no element φg for $g \sim_n f$ contains o^*. Moreover, it follows that for each such source Z, $o \in Z$, as we formed Z' as $Z \cup \{0\}$. Thus, $\forall_n \varphi f$ must contain o and lack o^*. It follows that $\forall'_n(\varphi^\circ f) = \forall_n \varphi f \cup \{0\}$ in this case, as required, due to $\forall a \in W(a \in \forall'_n(\varphi^\circ f)$ iff $a \in \forall_n \varphi f)$.

For $(cq2)$, assume that $X', Y' \in Prop$ with sources X and Y, $\varphi^\circ \in PropFun'$ with source φ, $n \in \omega$, and $f \in U^\omega$. Further suppose that $X' \smallsetminus Y' \subseteq \bigcap'_{j \in J} \varphi^\circ(f[j/n])$.

There are two main cases.

Case $0 \in X' \smallsetminus Y'$: That is, $0 \in X'$ and $0 \notin Y'$, the latter entails that $Y' = CL^{\leq'}(Y)$. We have that $0 \in \bigcap'_{j \in J} \varphi^\circ(f[j/n])$. Thus $0 \in \varphi^\circ(f[j/n])$ for each $j \in U$. We want to show that there is an $Z \in Prop'$ such that $0 \in Z \subseteq \bigcap'_{j \in J} \varphi^\circ(f[j/n])$. Consider the set $\forall_n \varphi f \cup \{0\}$, which is an element of $Prop'$. Such a element of $Prop'$ is easily shown to be the largest subset of $\bigcap'_{j \in J} \varphi^\circ(f[j/n])$ which is an element of $Prop'$. Thus $0 \in \forall_n \varphi^\circ f$. It is trivial to show that $a \in W$ implies $a \in \forall_n \varphi^\circ f$ iff $a \in \forall_n \varphi f$, and so the result follows.

Case $0 \notin X' \smallsetminus Y'$: This case is straightforward. The sets $X', Y', \bigcap'_{j \in J} \varphi^\circ(f[j/n])$, and $(\forall_n \varphi^\circ)f$ are either themselves, or contain 0. $\square$

As the class of the normalized frames are thus a subset of the class of frames, we immediately obtain the following.

Lemma 3.6 (Normal Soundness). For any formula $\mathcal{A}$, if $\mathcal{A}$ is a theorem of a first-order logic L_{QR} or L_{QN}, then $\mathcal{A}$ is valid in the normalization of every frame in the subclass of $\mathfrak{F}_Q$ for L_{QR} or L_{QN}, respectively.

On the normalization of a frame we identify a standard valuation which is designed to correspond to the original.

Definition 3.7. If $\mathfrak{M} = \langle \mathfrak{F}, |-| \rangle$ is an model based on the L_Q-frame $\mathfrak{F}$, we take as *the standard normalization of model $\mathfrak{M}$ at 0* (for $0 \in N$) to be the tuple $\mathfrak{M}' = \langle \mathfrak{F}', |-|' \rangle$, where $\mathfrak{F}'$ is the normalization of $\mathfrak{F}$ (at o), and $|-|'$ is defined as follows:[18]

1. $|c|' = |c|$;

2. for all $\vec{j} \in \mathcal{U}^n$: $\quad |P^n|'(\vec{j}) = \begin{cases} CL^{\leq'}(|P^n|(\vec{j})), & \text{if } o \notin |P^n|(\vec{j}) \\ (|P^n|(\vec{j})) \cup \{0\}, & \text{if } o \in |P^n|(\vec{j}); \end{cases}$

---

[18]Note that the terminology of 'standard' here was introduced in [6], but tracks that the method remains that of normalization.

3. A propositional function $|\mathcal{A}|'$ is then given to every well-formed formula as defined for models above.

Lemma 3.8. Given a logic L_Q, and given a L_Q-model $\mathfrak{M} = \langle \mathfrak{F}, |-| \rangle$, the standard normalization $\mathfrak{M}' = \langle \mathfrak{F}', |-|' \rangle$ of $\mathfrak{M}$ is a L_Q-model.

Proof. That the underlying frame is an L_Q-frame is given by Lemma 3.5. We need to show that the valuation assigns an element of $PropFun'$ to every atomic proposition, which then entails that every formula is thus assigned an element of $PropFun'$ because $PropFun'$ is closed under the suitably defined operations. However, it is easy to see that every atomic proposition is assigned an element of $PropFun'$, and thus the normalized model is a L_Q-model. $\square$

Lemma 3.9. Let $\mathfrak{M}$ be an L_Q-model with set W. Further let $\mathfrak{M}'$ be the standard normalization of $\mathfrak{M}$ (at 0 for o). For all $a \in W$, for every formula $\mathcal{A}$ and $f \in U^\omega$, $a \in |\mathcal{A}|f$ iff $a \in |\mathcal{A}|'f$.

Proof. The proof is by induction on the complexity of $\mathcal{A}$. All cases are covered by the arguments of [6] except for the modalities. These cases modify the arguments of [27] to the level of propositional functions. We show one direction of the proof for the case where $\mathcal{A} = \Box\mathcal{B}$, and the reader is invited to compare it with the arguments of [27].

Suppose that $a \in |\Box\mathcal{B}|f = \Box|\mathcal{B}|f$. We must show that $\forall b \in W(S'_\Box ab \Rightarrow b \in |\mathcal{B}|'f)$. Either $b \in W$ or $b = 0$. For the former, we have that $b \in |\mathcal{B}|'f$ by the induction hypothesis (and spelling out $a \in \Box|\mathcal{B}|f$). For the latter, we first suppose that $S'_\Box a0$. By definition, this is only if $S_\Box ao^*$. We obtain that $o^* \in |\mathcal{B}|f$. By induction hypothesis, then $o^* \in |\mathcal{B}|'f$, and by $o^* \leq 0$ (Corollary 3.4) we have that $0 \in |\mathcal{B}|'f$ as required. $\square$

Theorem 3.10. For any formula $\mathcal{A}$ and any logic L_Q, $\mathcal{A}$ is a theorem of L_Q iff $\mathcal{A}$ is valid in every normal L_Q-frame.

Proof. The soundness proof completes the 'only if' direction. For the 'if' direction, suppose contrapositively that $\mathcal{A}$ is not a theorem of L_Q. Then we have in the canonical model with frame $\mathfrak{F} = \langle W, N, R, {}^*, S_\Box, S_\Diamond, e, U, Prop, PropFun \rangle$ with valuation $|-|$ and $o \in N$ whereby $o \notin |\mathcal{A}|f$ for some $f \in U^\omega$. For the $0 \notin W$, take the standard normalized modal $\mathfrak{M}' = \langle \mathfrak{F}', |-|' \rangle$. By Lemma 3.8, this is an L_Q-model. By Lemma 3.9, $o \notin |\mathcal{A}|'f$. By Corollary 3.4, we have $0 \leq' o$ which entails that $0 \notin |\mathcal{A}|'f$. But $0 \in N'$, and so $\mathcal{A}$ is not valid on $\mathfrak{F}'$. $\square$

Corollary 3.11. For any logic in the class L_Q, the rule γ is admissible.

Proof. The proof is as in [6], which we show in detail. Suppose that $\neg\mathcal{A}\vee\mathcal{B}$ and $\mathcal{A}$ are both theorems of L_Q. Then by the completeness theorem, Theorem 3.10, these formulas are valid on every normal model. Consider an arbitrary normal model $\mathfrak{M} = \langle W, N, R, {}^*, S_\square, S_\diamond, e, U, Prop, PropFun, |-|\rangle$ with point 0 ($0 = 0^*$). Since $0 \in N$, $0 \in |\neg\mathcal{A}\vee\mathcal{B}|f \cap |\mathcal{A}|f$ for every $f \in U^\omega$. Since $0 \in |\mathcal{A}|f$ and $0 = 0^*$, we have that $0 \notin |\neg\mathcal{A}|f$. But then given the definition of $|\neg\mathcal{A}\vee\mathcal{B}|f$, $0 \in |\mathcal{B}|f$, as required. $\square$

Remark 3.12. We have shown γ admissible in first-order extensions of the modal relevant logics that are L3 and in either L_{QR} or L_{QN} (and with the addition of left implication, fusion, and the Ackermann truth constant); that is, for certain extensions of $\mathbf{QG.C}^{g,d}$, the logic extending $\mathbf{B}$ with the principle of excluded middle (A10), and the disjunctive rules (R6), (Q1), and (Q2).

The extensions of the regular logics $\mathbf{QG.C}^{g,d}$ and $\mathbf{QG.K}^{g}$ that are shown to admit γ include themselves and the following:

1. extensions with extensional confinement (EC) (and its dual)

2. extensions by axiom and rule schemes from (B1)–(B24) and (Q3)–(Q12).

The schemes (B1)–(B24) and (Q3)–(Q12) have corresponding frame conditions with certain shapes. An in-depth description of these shapes can be found in [27].

We have thus shown γ admissibility in a class of first-order modal relevant logics. These logics are first-order extensions (with or without extensional confinement) of the modal logics with an L_R or L_N propositional base, as defined in Seki [27], with the inclusion of fusion, left implication, and the Ackermann truth constant.

4 γ-Admissibility with Barcan-like Axiom Schemes

The proofs of γ-admissibility for the 'easy' first-order modal relevant logics is fairly straightforward, combining the arguments of Seki [27] and Ferenz and Ferguson [6] (in the more general framework of Ferenz [5] used here). In this section, we consider some interesting first-order modal relevant logics not covered by the above results. That is, we consider extensions of the logics covered in the previous section by the Barcan formulas and the Buridan formulas.

What is important in considering the Barcan and Buridan formulas here is that the best characterization of these axioms over the general frame semantics with the Mares–Goldblatt interpretation of the quantifiers is their *transliteration* into the semantics. That is, the Barcan formula is exactly characterized by models satisfying (bf). Where $\varphi \in PropFun$:

(bf) $\forall_n \Box \varphi \subseteq \Box \forall_n \varphi$ $\qquad\qquad$ (cbuf) $\Box \exists_n \varphi \subseteq \exists_n \Box \varphi$

(bf$_\Diamond$) $\Diamond \exists_n \varphi \subseteq \exists_n \Diamond \varphi$ $\qquad\qquad$ (cbuf$_\Diamond$) $\forall_n \Diamond \varphi \subseteq \Diamond \forall_n \varphi$

Similarly, it can be shown that the conditions (bf$_\Diamond$), (cbuf), and (cbuf$_\Diamond$), correspond to (BF$_\Diamond$), (CBuF), and (CBuF$_\Diamond$). However non-illuminating these semantic conditions are, they are equivalent to any frame condition that similarly corresponds to the relevant axiom schemes.

Goldblatt [11] proves the characterization between (BF) and (bf) in the case of classical quantified modal logic, and Ferenz [5] showed this fact in the case of first-order modal relevant logics. No improved characterization has ever been given for these cases; there is no 'intuitive' semantic condition (using $R, S_\Box, S_\Diamond$ and U) for these formulas. Importantly, Goldblatt [11] shows that there is no tight relationship between expanding and contracting domains and the Barcan formula with the Mares–Goldblatt interpretation of the quantifiers in first-order modal classical logic.

It appears that these four formulas with unsatisfying frame conditions are not suitable for the method of normal models. In particular, there is evidence (but no proof) that their frame condition is not invariant under frame normalizations. We record the following.

Conjecture 4.1. The frame conditions (bf), (bf$_\Diamond$), (cbuf), and (cbuf$_\Diamond$) are not preserved by normalization. That is, for each of the four conditions, there is a frame $\mathfrak{F}_Q$ that satisfies the condition and a normalization of $\mathfrak{F}_Q$ at 0 for o that does not satisfy that condition.

This conjecture appears reasonable. First, the method of normal models is the weakest method used to prove γ admissible in relevant logics. It only works for logics with the excluded middle and certain disjunctive rules. Even then it only works for certain extensions by axioms and rules with a restricted shape for frame conditions. The transliterated frame condition (bf) (and the other three) are not obviously of these restricted forms. Now, while the restricted forms for frame conditions discussed in [27] for modal relevant logics work well with normalization, it does not rule out more complicated forms being preserved in normalization. However, as noted, we do not even have a 'concrete' frame condition for these formulas. Thus, it is natural to believe these four formula (schemes) are not preserved in the process of normalizing a model.

Second, the process of normalization itself breaks some routes for a frame to satisfy the condition for the Barcan formula. We do know that so-called *full* frames entail the validity of the Barcan formulas. Here a full frame is one in which *Prop* is every $\leq$-upset and *PropFun* is every function from U^ω into *Prop*. Our frame normalization, as noted in footnote 17 above, does not preserve the fullness of a frame. In fact, the normalized frames are certain not to be full.

Finally, because we are so ignorant of the frame condition for the Barcan formula, we don't know what shape it takes, whether it is first-order specifiable, or even whether it requires an even or odd number of points in W. As an example of where things can go wrong, we might have a strict linearity axiom and a transliterated frame condition that is equivalent to W being a strict linearly ordered set. If we normalize on a model where $o = o^*$, then we add a point and break the strict linear ordering, as we would have that $o \not\leq 0$ and $0 \not\leq o$, but also $0 \neq o$. So we cannot evaluate (bf), (bf$_\diamond$), (cbuf), and (cbuf$_\diamond$) by the normal models method (at this point in time).

5   Summary and Conclusion

We have shown that γ is admissible in a class of first-order modal relevant logics. The logics are all based on L3 logics, in the terminology of [27], which include the principle of excluded middle and a handful of disjunctive rules. These logics include some theorems such as the Converse Barcan Formula; however, the frame conditions corresponding to the Barcan formula (and some similar formulas) are seemingly not preserved under the normalization of a frame. We conjecture that the normal models method is entirely inappropriate for these axioms.

This work naturally leads in several directions. First, Seki [27] proves γ admissibility for modal relevant logics extending (those described here with) certain Sahlqvist formulas. It is natural to extend the current work to cover first-order extensions of those logics (provided an L3 base). Furthermore, additional results ought to obtain for L1 logics by dropping fusion and left implication and the tonicity condition for the middle and right-most argument place of the ternary relation. That is, provided soundness and completeness for the first-order modal logics without these connectives.

Next, it is natural to pursue other methods for proving γ-admissible in both first-order and first-order modal relevant logics. For example, one might turn to metavaluations or a first-order analog of the reduced models of Slaney [30] to tackle logics that lack the principle of excluded middle. Ferenz and Ferguson [7] have also applied the normal models method to variable domain **R**. Generalizing our results to variable domains, modal relevant logics will require, first, developing a semantics and proof theory, for it is not clear how the variable domains will interact with both the binary $S_\Box$ and the ternary R relations and their interactions.

References

[1] Wilhelm Ackermann. Begründung einer strengen Implikation. *Journal of Symbolic Logic*, 21:113–128, 1956.

[2] Ruth Barcan Marcus. A functional calculus of first order based on strict implication. *Journal of Symbolic Logic*, 11:1–16, 1946.

[3] Ross Brady. The non-triviality of extensional dialectical set theory. In G. Priest, R. Routley, and J. Norman, editors, *Paraconsistent Logics: Essays in the Inconsistent*. Philosophia Verlag, Munich, 1983.

[4] Brian Chellas. *Modal Logic*. Cambridge University Press, Cambridge, 1980.

[5] Nicholas Ferenz. Quantified modal relevant logics. *Review of Symbolic Logic*, 16:210–240, 2023.

[6] Nicholas Ferenz and Thomas M. Ferguson. γ-admissibility in first-order relevant logics: Proof using normal models in the Mares–Goldblatt setting. *Review of Symbolic Logic*, 2023. To appear.

[7] Nicholas Ferenz and Thomas M. Ferguson. γ-admissibility for Mares' varying domain **QR**. In I. Sedlár, editor, *The Logica Yearbook 2023*. College Publications, London, 2024. To appear.

[8] Kit Fine. Incompleteness for quantified relevant logics. In J. Norman and R. Sylvan, editors, *Directions in Relevant Logic*, pages 205–225. Kluwer Academic Publishers, Dordrecht, 1989. (Reprinted in A. R. Anderson, N. D. Belnap and J. M. Dunn, *Entailment*, vol. II, §52, Princeton University Press, Princeton, NJ, 1992).

[9] Harvey Friedman and Robert K. Meyer. Whither relevant arithmetic? *Journal of Symbolic Logic*, 57(3):824–831, 1992.

[10] James W. Garson. Quantification in modal logic. In D. Gabbay and F. Guenthner, editors, *Handbook of Philosophical Logic*, volume 3, pages 267–323. Kluwer, Amsterdam, 2nd edition, 2001.

[11] Robert Goldblatt. *Quantifiers, Propositions and Identity*. Cambridge University Press, Cambridge, 2011.

[12] Robert Goldblatt and Edwin Mares. A general semantics for quantified modal logic. *Advances in Modal Logic*, 6:227–246, 2006.

[13] Lloyd Humberstone. *The Connectives*. MIT Press, Cambridge, 2011.

[14] Saul Kripke. A proof of gamma. In K. Bimbó, editor, *Relevance Logics and other Tools for Reasoning. Essays in Honor of J. Michael Dunn*, volume 46 of *Tributes*, pages 261–265. College Publications, London, 2022.

[15] Edwin Mares and Robert Goldblatt. An alternative semantics for quantified relevant logic. *Journal of Symbolic Logic*, 71:163–187, 2006.

[16] Edwin Mares and Robert K. Meyer. The admissibility of γ in **R4**. *Notre Dame Journal of Formal Logic*, 33:197–206, 1992.

[17] Robert K. Meyer. Metacompleteness. *Notre Dame Journal of Formal Logic*, 17:501–516, 1976.

[18] Robert K. Meyer and J. Michael Dunn. E, R, and γ. *Journal of Symbolic Logic*, 34:460–474, 1969.

[19] Robert K. Meyer, J. Michael Dunn, and Hugues Leblanc. Completeness of relevant quantification theories. *Notre Dame Journal of Formal logic*, 15:97–121, 1974.

[20] Robert K. Meyer, Steve Giambrone, and Ross Brady. Where gamma fails. *Studia Logica*, 43:247–256, 1984.

[21] Pavel Pudlák. The lengths of proofs. In S. R. Buss, editor, *Handbook of Proof Theory*, pages 547–637. North Holland, Amsterdam, 1998.

[22] Richard Routley and Robert K. Meyer. The semantics of entailment — II. *Journal of Philosophical logic*, 1:53–73, 1972.

[23] Richard Routley and Robert K. Meyer. The semantics of entailment. In H. Leblanc, editor, *Truth, Syntax, and Modality*, pages 199–243. North-Holland, Amsterdam, 1973.

[24] Richard Routley, Valerie Plumwood, Robert K. Meyer, and Ross T. Brady. *Relevant Logics and their Rivals: Part 1. The Basic Philosophical and Semantical Theory*. Ridgewood, Atascadero, CA, 1982.

[25] Krister Segerberg. *An Essay in Classical Modal Logic*. PhD thesis, University of Uppsala, 1971.

[26] Takahiro Seki. A Sahlqvist theorem for relevant modal logics. *Studia Logica*, 73:383–411, 2003.

[27] Takahiro Seki. The γ-admissibility of relevant modal logics I — The method of normal models. *Studia Logica*, 97:199–231, 2011.

[28] Takahiro Seki. The γ-admissibility of relevant modal logics II — The method using metavaluations. *Studia Logica*, 97:351–383, 2011.

[29] Takahiro Seki. An algebraic proof of the admissibility of γ in relevant modal logics. *Studia Logica*, 100:1149–1174, 2012.

[30] John K. Slaney. Reduced models for relevant logics without WI. *Notre Dame Journal of Formal Logic*, 28(3):395–407, 1987.

[31] Timothy Smiley. Relative necessity. *Journal of Symbolic Logic*, 28:113–134, 1973.

[32] Andrew Tedder and Nicholas Ferenz. Neighbourhood semantics for quantified relevant logics. *Journal of Philosophical Logic*, 51:457–484, 2022.

[33] Alasdair Urquhart. The story of γ. In K. Bimbó, editor, *J. Michael Dunn on Information Based Logics*, volume 8 of *Outstanding Contributions to Logic*, pages 93–106. Springer, Switzerland, 2016.

[34] Yale Weiss. Cut and gamma I: Propositional and constant domain **R**. *Review of Symbolic Logic*, 13:887–909, 2020.

 Received 12 January 2025

SOME POLYADIC MODAL BOOLEAN ALGEBRAS ARE GAGGLES

NICHOLAS FERENZ*
Center for Philosophy, University of Lisbon
Lisbon, Portugal
<nicholas.ferenz@gmail.com>

ANDREW TEDDER[†]
Department of Philosophy I, Ruhr University Bochum
Bochum, Germany
<ajtedder.at@gmail.com>

Abstract

J. Michael Dunn [4] introduced gaggle theory, which develops a close correspondence between certain kinds of algebras (called gaggles) and relational semantics. The gaggle framework has been applied to many propositional logics with much success. This paper is a foray into first-order logic via gaggle-theory. We show that monadic and polyadic Boolean algebras, introduced by Halmos [10] as algebraic structures to study first-order logic, are *Boolean gaggles* (in the sense of [3]), and that their extensions with certain kinds of operations are multi-gaggles. We show these algebraic structures to have representations in terms of, and embeddings in, relational semantics, using the gaggle-theoretic framework. That is, we employ relations to model all the operations, including those representing the quantifiers and variable substitutions. This project provides a foundation to explore first-order extensions of a wide range of logics.

2020 *Mathematics Subject Classification.* Primary: 03B45, Secondary: 03G15.

---

*NF acknowledges funding from by RVO 67985807 and that this work is also financed by national funds through FCT — Fundação para a Ciência e a Tecnologia, I.P., under the scope of UIDB/00310/2020 project, identified as DOI 10.54499/UIDB/00310/2020.

[†]AT was funded by the DFG project TE 1611/1-1.

1   Introduction

The primary goal of this paper is to make headway in bringing first-order (modal) logic under the gaggle-theoretic framework. Gaggle Theory, which was introduced by Dunn [4], presents a tight link between certain algebraic structures (called gaggles) and relational semantics via representation theorems. In this paper, we begin with the simple case of algebraic structures for first-order (modal) classical logic, namely the (monadic and) polyadic Boolean algebras developed by Halmos [10].[1] We choose Halmos's polyadic algebras due to the fact that cylindric algebras, as investigated in [11, 12], include identity built in. This introduces complications beyond those involved in the quantifiers themselves, which are our focus. The main upshot, from our point of view, is that (1) monadic and polyadic Boolean algebras with certain operations are gaggles, and thus (2) these algebras can be represented and embedded into relational semantics *using the gaggle-theoretic machinery*. The fact that monadic and polyadic algebras can be modeled using relational semantics is not new, and in fact falls out of topological duality results: namely those concerning Stone spaces with binary relations (see, e.g., Halmos [8] for a detailed presentation of the dual space for monadic and polyadic Boolean algebras). Our novel results are to bring these results into conversation with existing work on gaggle-theory as a broad framework.

Halmos [10, 8, 9] introduced both monadic and polyadic Boolean algebras in order to develop an algebraic foundation to study first-order (classical) logic. In these settings, we model quantifier(s) as operation(s) on algebras appropriate for the underlying propositional logic. The operations in question represent *quantifier prefixes*, in the sense that they build in not just the quantifier expression itself, but also the variables over which it quantifiers over. In the monadic case, a single operation is added, reflecting the fact that we can only quantify over one variable. In the polyadic case, we add an operation for each subset of the domain (usually infinite), and additional operations to track the behaviour of *variable substitutions*. This algebraic approach to first-order logic does not presuppose the inclusion of the relation of identity, unlike the cylindric algebras of [11, 12]. While monadic and polyadic Boolean algebras have several representation theorems (see, e.g., Halmos [10]), we show that (some of) these algebras have additional representations via gaggle theory.

Gaggle theory links an algebra with certain properties (a gaggle) with a straightforward representation on relational semantics. It represents a generalization of the Jónsson–Tarski 'Boolean Algebras with Operators' [13] in that the properties required for such an operation to be represented in a relational semantics are fairly

---

[1]More generally, the gaggle machinery on Boolean algebras can be seen to generalize [13, 14] by replacing the condition of *normality* with a weaker condition.

weak, and that the underlying structure can be generalized to distributive lattices, semi-lattices, and more [3].

Our aim in this paper is to identify several classes of monadic and polyadic Boolean algebras with operations—monadic and polyadic modal Boolean algebras —and generalize the tools of gaggle theory to obtain representation results. In short, we aim to represent first-order logic with purely relational semantics (not making recourse to the various non-relational machinery that's been introduced to model quantifiers in, e.g., [6, 15]).

We start in the next section, §2, by introducing gaggles and fixing notation and terminology to apply in later sections. Then, in §3, we introduce monadic modal Boolean algebras, a general algebraic treatment of extensions by classical logic with modals and quantifiers. We extend this to polyadic algebras in §4, particularly introducing the representation of substitutions by *transformations*, and show a representation result for a wide class of quantified modal classical logics. In §5 we show a general embedding result, indicating that polyadic modal Boolean algebras embed into the target class of relational structures in a natural way. We close in §6 with some remarks on future directions.

2 Gaggles

Here we rehearse the concept of a Boolean gaggle. We follow Bimbó and Dunn [3] in our presentation (including choices of notation), though we note some terminology from other sources such as Dunn and Hardegree [5]. We will conventionally use the same notation for both an algebra and its carrier set: e.g., $\mathbb{A} = \langle \mathbb{A}, \dots \rangle$.

Distribution Types Let $\mathbb{A} = \langle \mathbb{A}; \wedge, \vee, \circledast^z_{i \in I} \rangle$ be a distributive lattice with operations $\circledast^z_{i \in I}$, where z is a positive integer indicating the arity of an operation. An operation $\circledast^z_i \in \circledast^z_{i \in I}$ *distributes into* $\wedge$ *or* $\vee$ when (d1) or (d2), respectively, holds.

$$(\text{d1}) \quad \forall^z_{k=1} k. \exists \mathbb{X} \in \{\wedge, \vee\}. \circledast^z_i (\vec{a}, [b_1 \mathbb{X} b_2]_k) = \circledast^z_i(\vec{a}, [b_1]_k) \wedge \circledast^z_i(\vec{a}, [b_2]_k)$$

$$(\text{d2}) \quad \forall^z_{k=1} k. \exists \mathbb{X} \in \{\wedge, \vee\}. \circledast^z_i (\vec{a}, [b_1 \mathbb{X} b_2]_k) = \circledast^z_i(\vec{a}, [b_1]_k) \vee \circledast^z_i(\vec{a}, [b_2]_k)$$

If an operation distributes into either $\wedge$ of $\vee$ then we say it *distributes*. The *distribution type* of an operation that distributes is written as one of the following:

$$\circledast^z_i : \mathbb{X}, \dots, \mathbb{X} \longrightarrow \mathbb{X}$$

where the final $\mathbb{X}$ is the operation that $\circledast^z_i$ distributes into, and each $\mathbb{X}$ on the left of the arrow is either $\vee$ or $\wedge$, as given by the corresponding positions in (d1) and (d2).

Contraposed Pair Let $\mathbb{A} = \langle A; \wedge, \vee, \otimes_{j \in J}, \circledast_1, \circledast_2 \rangle$ be a distributive lattice with operations $\otimes_{j \in J}, \circledast_1, \circledast_2$ where $\circledast_1, \circledast_2$ have the same arity. Operations $\circledast_1, \circledast_2$ are a *contraposed pair* if they satisfy one of (c1)–(c3):

(c1) $\circledast_1 : \overrightarrow{\bowtie}, [\vee]_i \longrightarrow \vee$ and $\circledast_2 : \overrightarrow{\bowtie}, [\wedge]_i \longrightarrow \wedge$

(c2) $\circledast_1 : \overrightarrow{\bowtie}, [\wedge]_i \longrightarrow \vee$ and $\circledast_2 : \overrightarrow{\bowtie}, [\wedge]_i \longrightarrow \vee$

(c3) $\circledast_1 : \overrightarrow{\bowtie}, [\vee]_i \longrightarrow \wedge$ and $\circledast_2 : \overrightarrow{\bowtie}, [\vee]_i \longrightarrow \wedge$

where, for each $j \neq i$, $[\bowtie]_j$ is the same in the distribution types of $\circledast_1$ and $\circledast_2$.

Abstract Residuation & Colligation Let $\mathbb{A} = \langle A; \wedge, \vee, \otimes_{j \in J}, \circledast_1, \circledast_2 \rangle$ be a distributive lattice with operations $\otimes_{j \in J}$ and contraposed pair $\circledast_1, \circledast_2$. The pair $\circledast_1$ and $\circledast_2$ are abstract residuals if one of (r1)–(r3) hold.

(r1) $\circledast_1(\overrightarrow{a}, [b]_i) \leq c$ iff $b \leq \circledast_2(\overrightarrow{a}, [c]_i)$

(r2) $\circledast_1(\overrightarrow{a}, [b]_i) \leq c$ iff $\circledast_2(\overrightarrow{a}, [c]_i) \leq b$

(r3) $b \leq \circledast_1(\overrightarrow{a}, [c]_i)$ iff $c \leq \circledast_2(\overrightarrow{a}, [b]_i)$

The transitive closure of abstract residuation is *colligation*. Two operations $\circledast_1$ and $\circledast_2$ are *colligated* when there is a chain of operations with $\circledast_1$ and $\circledast_2$ at the endpoints such that each operation in the chain is abstractly residuated with its immediate neighbors.

Families For a distributive lattice with operations $\mathbb{A} = \langle A; \wedge, \vee, \otimes_{j \in J}, \circledast_{i \in I} \rangle$ with each $\circledast_{i \in I}$ (with $I \neq \emptyset$) being a z-ary operation, we say that the set of operations $\circledast_{i \in I}$ is a *family of operations* when all operations in the set are colligated with each other, and each has a distribution type. We say that $\circledast_{i \in I}$ is a *complete family of operations* when $\circledast_{i \in I}$ is a family containing $z + 1$ operations.

In [5], the notion of a *founded family* is used. The definition is equivalent to a family as defined here, except that one of the operations in a founded family is selected as the "head" of that family. We elide this detail here.

0-1 Operations Let $\mathbb{A} = \langle A; \wedge, \vee, \neg, \circledast \rangle$ be a Boolean algebra with an operation $\circledast$ with a distribution type and arity n. $\circledast$ is a *0-1 operation* when for all i $(1 \leq i \leq n)$, (a) holds if $\circledast$ distributes into $\vee$ and (b) holds if $\circledast$ distributes into $\wedge$:

(a) If $\circledast : \overrightarrow{\bowtie}, [\vee]_i \longrightarrow \vee$ then $\circledast(\overrightarrow{a}, [0]_i) = 0$, and
if $\circledast : \overrightarrow{\bowtie}, [\wedge]_i \longrightarrow \vee$ then $\circledast(\overrightarrow{a}, [1]_i) = 0$.

(b) If $\circledast : \overrightarrow{\mathbb{W}}, [\wedge]_i \longrightarrow \wedge$ then $\circledast(\overrightarrow{a}, [1]_i) = 1,$ and
    if $\circledast : \overrightarrow{\mathbb{W}}, [\vee]_i \longrightarrow \wedge$ then $\circledast(\overrightarrow{a}, [0]_i) = 1.$

This definition fixes the behaviour of the operations with respect to the bounds, which, since these are also meets and joins (of the whole algebra), can be seen as a generalisation of the requirement that the operations in question have distribution types.

Boolean Gaggles An algebra $\mathbb{A} = \langle \mathbb{A}; \wedge, \vee, \neg, \circledast_{i \in I} \rangle$, where $\langle \mathbb{A}; \wedge, \vee, \neg \rangle$ is a Boolean algebra, is a *Boolean Gaggle* (BG) if $\circledast_{i \in I}$ is a family of 0-1 operations and a *multi-BG* if $\circledast_{i \in I}$ draws from multiple families of operations.

In [3, p. 40], some examples are presented for how unary modalities appear in this framework. The trick is to add, for each family of operations, a single accessibility relation with properties determined by the distribution types of the operations in the family. For example, one might introduce the binary relation R for $\Diamond$ as primitive. Then we can model 8 two-member families of unary modalities (given interdefinability in the Boolean setting) using the binary relation R, its negation $\overline{R}$, and the permutations of these relations $R^{1 \rightleftarrows 2}$ and $\overline{R}^{1 \rightleftarrows 2}$, where $R^{1 \rightleftarrows 2} \alpha \beta$ iff $R \beta \alpha$. Operations on $\wp(W)$ can then be defined using these relations, as in page 40 of [3]. We give two examples using their terminology. Where $A \in \wp(W)$:

(fa1) $\beta \in \Diamond A$ iff $\exists \alpha (R \alpha \beta \wedge \alpha \in A)$;

(fa3) $\beta \in \Box A$ iff $\forall \alpha (\overline{R} \alpha \beta \vee \alpha \in A)$.

Moreover, in general $R^{n \rightleftarrows m}$ permutes the arguments in the nth and mth places in R.

3 Monadic Modal Boolean Algebras

For now, we focus on the monadic case, as it serves as an excellent proof of concept. In a sense, we dualize the presentation of Halmos, giving the algebras in terms of $\forall$ and the top 1. Although the adjective 'modal' used here is particular to unary modal operations that satisfy the particular equations we require, we recognize the term 'modal operation' can be interpreted to have less requirements. The requirements we set out here are the minimal for our gaggle-theoretic representation.

Monadic Modal Boolean Algebras A monadic (multi-)modal Boolean algebra is a tuple $\langle \mathbb{A}; \vee, \wedge, \neg, 1, \forall, \circledast^1_{i \in I} \rangle$, where

(a) $\langle \mathbb{A}; \vee, \wedge, \neg, 1 \rangle$ is a Boolean algebra;

(b) $\forall$ is a *quantifier* on $\mathbb{A}$, i.e., it is a mapping from $\mathbb{A}$ into $\mathbb{A}$ s.t., when $p, q \in \mathbb{A}$:

$$\text{(1) } \forall 1 = 1, \qquad \text{(2) } \forall p \leq p, \qquad \text{(3) } \forall(p \vee \forall q) = (\forall p \vee \forall q).$$

(c) $\langle \mathbb{A}; \vee, \wedge, \neg, 1, \circledast^1_{i \in I} \rangle$ is a Boolean algebra with unary modal operations $\circledast^1_{i \in I}$. That is, each $\circledast_i \in \circledast^1_{i \in I}$ is a mapping from $\mathbb{A}$ into $\mathbb{A}$ that has a distribution type and is 0-1.

The resulting structures are, not too surprisingly, multi-gaggles, recorded in the following.

Fact 1. Every monadic modal Boolean algebra is a multi-gaggle.

Proof. The underlying algebra is a Boolean algebra, and the requirements for each modal operation in $\circledast^1_{i \in I}$ are built into the definition. It remains to show only that the quantifier is a 0-1 operation. It is straightforward to show that $\forall(p \wedge q) = \forall p \wedge \forall q$, which means that $\forall$ has the distribution type $\wedge \longrightarrow \wedge$. Moreover, since $\forall 1 = 1$, it follows that $\forall$ respects bounds. Therefore every monadic modal Boolean algebra is a multi-gaggle. $\qquad\square$

Note that the operation $\exists$—definable via negation (i.e., $\exists p = \neg \forall \neg p$)—is the other member of the family of 0-1 operations containing $\forall$. This follows from the **S5**-ish properties of the quantifier, as the $\square$ and $\diamond$ operations in normal modal logics weaker than **S5** tend not to be in the same family. The operation $\exists$ has type $\vee \longrightarrow \vee$ and the appropriate bound respecting behaviour (for note $\exists \neg 1 = \neg \forall 1 = \neg 1$).

Having shown that the relevant algebraic structure are (multi-)gaggles, we can go on to prove a simple representation theorem with respect to relational semantics via some interesting augmentation of the usual gaggle-theory presentation.

3.1 Gaggle-theoretic Representation of Monadic Modal Boolean Algebras

The usual gaggle-theoretic representation via relational semantics is applicable in this setting. In the polyadic case, we'll have to model the interaction between quantifiers (and substitution operations). In the monadic case, the situation is much simplified. The single universal quantifier needs only a single relation that can be thought of as modeling 'differing on a single variable'. Something like the standard relational interpretation can be given for $\overline{R}_\forall$, and suggests the reading that $\overline{R}_\forall \alpha \beta$ holds when β and α differ on the assignment of the atomic formulas, each of which has a single variable. Thus we define the following relational structures as models. In the framework below, however, we define the negation $R_\forall$ as primitive.

MMBA Frames An MMBA frame with families of unary operations $\circledast^1_{i \in I}$ is a tuple $\mathfrak{F} = \langle W; \{R_i\}_{i \in I} \rangle$ where W is a non-empty set, each $R_i \subseteq W^2$, and $\overline{R}_\forall^{1 \rightleftarrows 2}$ is an equivalence relation on W.

The operations on the frame are defined as in [3, Definition 1.4.3]. Notably, there are 8 families of pairs of operations definable for unary modal operations, which are items (fa1)–(fa8) of [3, p. 40]. We take these to be present in the background, as our focus is on the operation for $\forall$. This operation is $\square$-like. In fact, as per (fa3), if $B \in \wp(W)$, then $\forall B$ is defined by the following:[2]

$$\forall \alpha \in W (R_\forall \alpha\beta \vee \alpha \in B) \quad \Leftrightarrow \quad b \in \forall B$$

MMBA Models An MMBA-model for $\mathbb{A}$ is a tuple $\mathfrak{M} = \langle \mathfrak{M}, v \rangle$ where $\mathfrak{F}$ is an MMBA-frame and $v : \mathbb{A} \longrightarrow \wp(W)$ such that

(v1) $v(a \wedge b) = v(a) \cap v(b)$ $\qquad$ (v4$_i$) $v(\circledast_i a) = \circledast_i(v(a))$ (for each $i \in I$)

(v2) $v(a \vee b) = v(a) \cup v(b)$ $\qquad$ (v5) $\quad v(\forall a) = \forall(v(a))$

(v3) $v(\neg a) = W \setminus v(a)$

Truth, Validity We say that an equation $\tau_1 = \tau_2$ is *true in* $\mathfrak{M}$ when $v(\tau_1) = v(\tau_2)$, and *valid on* $\mathfrak{F}$ if it is true in every $\mathfrak{M}$ based on $\mathfrak{F}$.

3.2 Representation & Embedding

The representation theorem that follows is essentially a soundness theorem: every model makes at least all the right identity statements true.

Theorem 1 (Representation of MMBAs). Let $\mathbb{A} = \langle \mathbb{A}; \vee, \wedge, \neg, 1, \forall, \circledast^1_{i \in I} \rangle$ be an MMBA. If $\mathfrak{M}$ is a model for $\mathbb{A}$ on frame $\mathfrak{F}$ (as in the definition above), then the equations, which characterise $\mathbb{A}$ as an MMBA, are true in $\mathfrak{M}$. That is, there is an algebra of sets on $\mathfrak{F}$ into which $\mathbb{A}$ can be homomorphically mapped.

Proof. The proof follows that of [3, Theorem 1.4.10]. The main differences are due to the extra properties of the quantifier over a 0–1 operation. Essentially, we have an **S5** modal operation, and the representation follows from the representation of (multi-modal extensions of) **S5**, which is covered in [3, §1.2]. $\qquad \square$

---

[2]The reader is to note the ordering of the argument places in $R_\forall ab$ is the reverse of typical. This is in keeping with gaggle-theoretic standards, where the operation in question always takes the final argument place. The existential quantifier, $\exists$, would be given by $R^{1 \rightleftarrows 2}$, which reverses the first and second argument place: the notation typically used (well, the notation typical for $\square$, anyways).

The representation theorem is just a representation theorem for an **S5** logic with additional modalities. This relation between monadic Boolean logic and **S5** is well-known, and the details of which are expounded well in many place, e.g., Mints [16].

Canonical Model for an MMBA Fix an MMBA $\mathbb{A} = \langle \mathbb{A}; \vee, \wedge, \neg, 1, \forall, \circledast^1_{i \in I} \rangle$. We define the canonical frame $\langle W^c, \{R^c_i\}_{i \in I/f}, R^c_\forall \rangle$ by setting W^c to be the set of ultrafilters on $\mathbb{A}$, the set $\{R^c_i\}_{i \in I/f}$ to consist of a single canonical relation for each family of operations in I, and $R^c_\forall$ to be defined, similar to a $\square : \wedge \longrightarrow \wedge$ relation in [3, Definition 1.4.11], as follows:

$$R^c_\forall \alpha \beta \quad \text{iff} \quad \exists a \in \mathbb{A}(a \notin \alpha \wedge \forall a \in \beta).$$

Note that this definition for $R_\forall$ differs from what is normally employed. In fact, it is the negation of the usual clause. If we had started with $\exists$ primitive, the existence predicate would be given a canonical relation as we would expect (but for $\exists$ instead of $\forall$, in analogy with the modalities). What we have, via [3, Definition 1.4.2], is that $R_\exists = \overline{R_\forall} = \overline{R_\forall}^{1 \rightleftarrows 2}$. Combined with the definition for $\forall$ on $\wp(W)$, this gives us the desired result.

Over and above what Bimbó and Dunn call *pure* gaggles [3, p. 27] (i.e., those which satisfy all and only those equations mandated by gaggle-hood), due to $\forall$ having extra properties and $R^c_\forall$ subsequently having more requirements, we have to show the canonical frame is a frame. For this it is sufficient to show that $\overline{R}^c_\forall$ is an equivalence relation. First, it cannot hold that $R^c_\forall \alpha \alpha$, for any α, as $\forall \alpha \leq \alpha$. Symmetry and Transitivity are straightforward to verify, and are left to the reader.

The following lemma is key to the embedding and is stated and proved in [3, Lemma 1.4.15], excepting the new cases for $\forall$. For its statement, we need notation for the canonical relation defined except allowing for each argument place except the final one to be a filter (rather than an ultrafilter). We use R', with subscripts, to denote these relations. In particular, $R'_\circledast$ is the relaxation of $\overline{R}^c_\circledast$ to have filters in the first arguments place when $\circledast$ distributes into $\wedge$, and of $R^c_\circledast$ to have filters in the first arguments place when $\circledast$ distributes into $\vee$.

Lemma 1 (Boolean Squeeze Lemma). Let $\mathbb{A} = \langle \mathbb{A}; \vee, \wedge, \neg, 1, \forall, \circledast^1_{i \in I} \rangle$ be an MMBA. Suppose $R'_\circledast(\alpha_1, \ldots, \alpha_n, \beta)$ for non-empty, proper filters $\alpha_1, \ldots, \alpha_n$ and ultrafilter β, where $R_\circledast$ is the canonical relation for operation $\circledast$. Then there are ultrafilters $\alpha'_1, \ldots, \alpha'_n$ extending the filters $\alpha_1, \ldots, \alpha_n$, respectively, such that $R^c_\circledast(\alpha'_1, \ldots, \alpha'_n, \beta)$ or $\overline{R}^c_\circledast(\alpha'_1, \ldots, \alpha'_n, \beta)$ as appropriate.

Proof. The proof for most cases is given by Bimbó and Dunn in [3]. The remaining case is that of the relation $R^c_\forall$, for which the argument is essentially the same as that for $\square$ in [3, Ch. 1]. $\qquad\square$

Theorem 2 (Embedding Theorem)**.** Let $\mathbb{A} = \langle \mathbb{A}; \vee, \wedge, \neg, 1, \forall, \circledast^1_{i \in I} \rangle$ be an MMBA. There is a model $\mathfrak{M}$ on a frame $\mathfrak{F}$ for $\mathbb{A}$ for which the algebra of sets on $\mathfrak{F}$ is isomorphic to $\mathbb{A}$. That is, there is a valuation v_c that is an injective homomorphism from $\mathbb{A}$ into the algebra of sets on $\mathfrak{F}$: $v_c(a) = v_c(b)$ in $\mathfrak{M}$ iff $a = b$ in $\mathbb{A}$.

Proof. On the canonical frame, we define the canonical valuation as follows:

$$v_c(a) = \{\alpha \colon \alpha \text{ is an ultrafilter and } a \in \alpha\}$$

What remains to be verified is that v_c has the desired properties to be an isomorphism, which is straightforward and left to the reader (compare the proof of [3, Thm. 1.4.16] for inspiration). $\qquad\square$

The embedding and representation of monadic first-order logic into relational semantics is perhaps not too unexpected, given the close relationship between monadic first order logics and **S5**-ish modal logics, as in [16]. The novelty here is primarily due to the use of the gaggle-theoretic machinery, showing how monadic first order logics plug into the usual framework, and setting the stage for similar work in the polyadic case. First and foremost, the above can be generalized to any operations that are 0-1 operations. Thus, this will work for monadic Boolean algebras with any set of 0-1 operations. To see this, note that the above proofs only dealt with the 'new' case of the monadic quantifier, and the remaining cases can be dealt with the usual gaggle-theoretic arguments.

4 Polyadic Modal Boolean Algebras

First, we rehearse Halmos's definition of polyadic Boolean algebras.

Polyadic Boolean Algebras A *Polyadic Boolean Algebra* is a tuple $\langle \mathbb{A}, \vee, \wedge, \neg, 1, \mathcal{I}, S, \forall \rangle$ where $\mathbb{A}$ is a Boolean algebra, I is a set, S is a mapping that associates a Boolean endomorphism $S(\tau)$ of $\mathbb{A}$ with every transformation τ on $\mathcal{I}$ (i.e., $\tau \colon I \longrightarrow I$), and $\forall$ is a mapping that associates a quantifier $\forall(J)$ on $\mathbb{A}$ for every subset J of $\mathcal{I}$, and furthermore:

(P1) $\forall(\emptyset)$ is the identity mapping on $\mathbb{A}$;

(P2) $\forall(J \cup K) = \forall(J)\forall(K)$ for all $J, K \subseteq \mathcal{I}$;[3]

---

[3]The operation of juxtaposing operations (with an arity of 1) is taken to be application from right to left. That is, for $p \in \mathbb{A}$, $\forall(J)\forall(K)p$ is the result of first applying $\forall(K)$ to p, then applying $\forall(J)$ to the result.

(P3) $S(\delta)$ is the identity mapping on $\mathbb{A}$, where δ is the identity transformation;

(P4) $S(\sigma\tau) = S(\sigma)S(\tau)$ for all transformations σ, τ;

(P5) $S(\sigma)\forall(J) = S(\tau)\forall(J)$ when $\tau = \sigma$ outside J;

(P6) $\forall(J)S(\tau) = S(\tau)\forall(\tau^{-1}J)$ whenever $\tau^{-1}J$ is well defined.

The defining identities above are the identities given by Halmos, but with $\exists$ replaced with $\forall$. The main job of the identities given is in the tracking of free and bound variables, and the same identities will work with $\forall$-quantifiers.

For any set $J \subseteq \mathcal{I}$, we say that $a \in \mathbb{A}$ is *independent of J* when $\forall(J)a = a$; and that *J supports a* when a is independent of $\mathcal{I} \setminus J$. We will denote by x/y the transformation that substitutes x for y, and otherwise leaves $\mathcal{I}$ alone. These definitions are as in [10]. Although we do not need the assumption, because it involves a tighter correlation with the standard non-algebraic presentations of first-order logic, we assume that we are dealing with *finite $\mathcal{I}$-algebras*, that is, each $a \in \mathbb{A}$ has finite support.

PBAs with 0-1 operations A *polyadic Boolean algebra with 0-1 operations* (or a *PMBA*) is tuple $\langle \mathbb{A}, \vee, \wedge, \neg, 1, \circledast^{z}_{i\in I}, \mathcal{I}, S, \forall \rangle$ where

1. $\langle \mathbb{A}, \vee, \wedge, \neg, 1, \mathcal{I}, S, \forall \rangle$ is a finite polyadic Boolean I-algebra

2. $\circledast^{z}_{i\in I}$ is a set of 0-1 operations on $\mathbb{A}$ (drawn from, potentially, multiple families)

$$(\text{P7}_i) \quad S(\tau) \circledast^{z}_{i} (a_1, \ldots, a_z) = \circledast^{z}_{i}(S(\tau)a_1, \ldots, S(\tau)a_z) \qquad\qquad (\circledast_i \in \circledast_{i\in I})$$

The (P7_i) conditions essentially encode the fact that a substitution does not substitute modalities, but rather only terms.

The next result is a straightforward of the related result concerning monadic modal Boolean algebras.

Fact 2. Every polyadic modal Boolean algebra is a multi-gaggle.

Proof. The proof consists in showing that each $S(\tau)$ and $\forall(J)$ are 0-1 operations. For the $\forall(J)$s this is guaranteed by the fact that each is a quantifier. For the $S(\tau)$s, this is guaranteed by the constraint that each be a Boolean endomorphism so that, e.g., $S(\tau)1 = S(\tau)(a \vee \neg a) = S(\tau)a \vee \neg S(\tau)a = 1$, and $S(\tau)0 = 0$ can be shown similarly. That they have distribution types follows from the fact that the former is a Boolean endomorphism and that the latter is a quantifier. In particular, $S(\tau) : \wedge \longleftarrow \wedge, \vee \longrightarrow \vee$: that is, substitutions have two distribution types. We also have that substitutions respect both bounds, as required. $\qquad\square$

PMBA Frames A PMBA frame for PMBA $\langle \mathbb{A}, \vee, \wedge, \neg, 1, \circledast_{i \in I}^{z}, \mathcal{I}, S, \forall \rangle$ is a tuple $\mathfrak{F} = \langle W; \{R_i\}_{i \in I}, \{R_{S(\tau)}\}_{\tau \in I^I}, \{R_{\forall(J)}\}_{J \subseteq I} \rangle$ where W is a non-empty set, each relation $R \subseteq W^2$, regardless of decoration, and the following conditions are satisfied:

(c1) $\overline{R}_{\forall(J)}^{1 \rightleftarrows 2}$ is an equivalence relation on W, for each $J \subseteq \mathcal{I}$;[4]

(c2) for every τ of $\mathbb{A}$ and $\beta \in W$, $\exists \alpha \in W(R_{S(\tau)}\alpha\beta)$;   (seriality)

(c3) $R_{S(\tau)}\alpha\beta$ and $R_{S(\tau)}\gamma\beta$ imply $\alpha = \gamma$;   (functionality)

(cP1) $\overline{R}_{\forall(\emptyset)}\alpha\beta$ iff $\alpha = \beta$;

(cP2) $\overline{R}_{\forall(J \cup K)}\alpha\beta$ iff $\exists\gamma(\overline{R}_{\forall(K)}\alpha\gamma \;\&\; \overline{R}_{\forall(J)}\gamma\beta)$;

(cP3) $R_{S(\delta)}\alpha\beta$ iff $\alpha = \beta$;

(cP4) $R_{S(\sigma\tau)}\alpha\beta$ iff $\exists\gamma(R_{S(\tau)}\alpha\gamma \;\&\; R_{S(\sigma)}\gamma\beta)$;

(cP5) $R_{S(\sigma)}\beta\gamma$ implies $\exists\alpha(R_{S(\tau)}\alpha\gamma \;\&\; \overline{R}_{\forall(J)}\alpha\beta)$, when $\tau = \sigma$ outside J;

(cP6) $\exists\alpha(R_{S(\tau)}\gamma\alpha \;\&\; \overline{R}_{\forall(J)}\alpha\beta)$ iff $\exists\alpha(R_{S(\tau)}\alpha\beta \;\&\; \overline{R}_{\forall(\tau^{-1}J)}\alpha\gamma)$, whenever $\tau^{-1}J$ is well defined;

(cP7$_i$) $R_{S(\tau)}\alpha\beta$ and $R_i(\gamma_1, \ldots, \gamma_z, \alpha)$ (where R_i is for the z-ary operation $\circledast_i$) implies $R_i(\gamma_1', \ldots, \gamma_z', \beta)$ and $R_{S(\tau)}\gamma_1\gamma_1'$ through $R_{S(\tau)}\gamma_z\gamma_z'$,   (for each $i \in I$).

As the naming suggests, (cP1) is a condition for the validity of (P1), and so forth. In condition (cP2), the ordering of the relation does not matter, as $\cup$ is commutative. In condition (cP4), however, the order is crucial, as composition of transformations is not, in general, commutative.

The universal quantifier is, again, boxy, and so it will correspond to an operation on $\wp(W)$ like (fa3) in [3, p. 40]. As $S(\tau)$ has distribution types $\wedge \longrightarrow \wedge$ and $\vee \longrightarrow \vee$, and we have chosen the diamondy condition (fa1) as primitive, we obtain the following definitions of the operations on $\wp(W)$:

$$\forall\alpha \in W(R_{\forall(J)}\alpha\beta \vee \alpha \in B) \;\;\Leftrightarrow\;\; \beta \in \forall(J)B$$
$$\exists\alpha \in W(R_{S(\tau)}\alpha\beta \wedge \alpha \in B) \;\;\Leftrightarrow\;\; \beta \in S(\tau)B$$

Using these, we obtain the following definition of models.

---

[4]We hereby drop the superscript from '$\overline{R}$' from the notation to avoid unnecessary clutter, since we require this relation to be an equivalence relation, and thus symmetric.

1195

PMBA Models A PMBA-model for $\mathbb{A}$ is a tuple $\mathfrak{M} = \langle \mathfrak{M}, v \rangle$ where $\mathfrak{F}$ is a PMBA-frame and $v : \mathbb{A} \longrightarrow \wp(W)$ such that

(v1) $v(a \wedge b) = v(a) \cap v(b)$

(v2) $v(a \vee b) = v(a) \cup v(b)$

(v3) $v(\neg a) = W \setminus v(a)$

(v4$_i$) $v(\circledast_i a) = \circledast_i(v(a))$

(v5$_J$) $v(\forall(J)a) = \forall(J)(v(a))$

(v6$_\tau$) $v(S(\tau)a) = S(\tau)(v(a))$

When $J \subseteq I$, τ is a transformation in $\mathbb{A}$, and $\circledast_i \in \circledast_{i \in \mathcal{I}}$. Here, we define *truth* and *validity* as in §3.1 above.

4.1 Representation

Theorem 3 (Representation of PMBAs). Fix a PMBA $\mathbb{A} = \langle \mathbb{A}; \vee, \wedge, \neg, 1, \forall, \circledast_{i \in I}^z, \mathcal{I}, S, \forall \rangle$. If $\mathfrak{M}$ is a model for $\mathbb{A}$ with frame $\mathfrak{F}$, then the equations characterising $\mathbb{A}$ as a PMBA are true in $\mathfrak{M}$. That is, there is an algebra of sets on $\mathfrak{F}$ into which $\mathbb{A}$ can be homomorphically mapped.

Proof. The arguments of [3, Theorem 1.4.10] can be used to show the modal operations defined on $\wp(W)$, the quantifiers, and the substitution operations have the correct distribution type and they are all 0-1 operations. In our structures, however, there are additional identities that must be shown to hold. Furthermore, we must show that each $S(\tau)$ is a Boolean endomorphism and that each quantifier is a quantifier.

We can show that every putative quantifier is, in fact, a quantifier by arguments similar to those of the monadic case, and details are left to the reader. That is, the identities in (b) in the definition of MMBAs (see, p. 5) are satisfied.

Let's consider the $S(\tau)$ operations. Using functionality and seriality, we can straightforwardly derive that $\exists \alpha \in W(R_{S(\tau)}\alpha\beta \wedge \alpha \in B)$ iff $\forall \alpha \in W(\overline{R}_{S(\tau)}\alpha\beta \vee \alpha \in B)$, making concrete the link between the operations on the frame for the two distribution types $\vee \longrightarrow \vee$ and $\wedge \longrightarrow \wedge$. That $S(\tau)$ in $\mathfrak{M}$ is a Boolean endomorphism is immediate from condition (v6$_\tau$).

We now show that (P1)–(P7) hold.

For (P1), we have already shown that $\forall(\emptyset)X \leq X$, as we have a quantifier. To show the converse, let $\alpha \in X$. Then by (cP1) we have that $\overline{R}_{\forall(\emptyset)}\alpha\alpha$ With $\alpha \in X$ and α being the only point accessible from α, we have that $\alpha \in \forall(\emptyset)X$.

For (P2), for the left-to-right direction suppose that $\alpha \in \forall(J \cup K)X$, but also $\beta \notin \forall(J)\forall(K)X$. The former means that $\forall \gamma \in W(\overline{R}_{\forall(J \cup K)}\gamma\beta \Rightarrow \gamma \in X)$. The latter entails that there is an $x \in W$ such that $\overline{R}_{\forall(J)}x\beta$, $x \notin \forall(K)X$, $\overline{R}_{\forall(K)}\alpha x$, and $\alpha \notin X$. By condition (cP2), we obtain $\overline{R}_{\forall(J \cup K)}\alpha\beta$ from the last two relation statements, and thus $\alpha \in X$, a contradiction. The right-to-left direction is similar.

For (P3), the case is similar to (P1).

For (P4), suppose for the left-to-right direction that $\beta \in S(\sigma\tau)X$. It follows that $\exists\alpha(R_{S(\sigma\tau)}\alpha\beta \ \& \ \alpha \in X)$. By condition (cP4) we obtain $\exists x \in W(R_{S(\sigma)}x\beta \ \& \ R_{S(\tau)}\alpha x)$. From the latter with $\alpha \in X$ we obtain that $x \in S(\tau)X$. With $R_{S(\sigma)}x\beta$ this entails that $\beta \in S(\sigma)S(\tau)X$, as required. The other direction is similar and left to the reader.

For (P5), the two directions are symmetrical, so let's assume that $\gamma \in S(\sigma)\forall(J)X$. Then there is a unique β where $R_{S(\sigma)}\beta\gamma$ and $\beta \in \forall(J)X$. By condition (cP5), it follows that $\exists\alpha(R_{S(\tau)}\alpha\gamma \ \& \ \overline{R}_{\forall(J)}\alpha\beta)$. From the former conjunct, we know that α is unique. From the latter, and that the relation is an equivalence relation, that $\alpha \in \forall(J)X$. This gives that $\gamma \in S(\tau)\forall(J)X$, as required.

For (P6), for the left-to-right direction assume that $\beta \in \forall(J)S(\tau)X$ and that $\tau^{-1}J$ is well defined. Suppose that $\beta \notin S(\tau)\forall(\tau^{-1}J)X$. It follows by (c2) and (c3) that there is a unique α such that $R_{S(\tau)}\alpha\beta$ and $\alpha \notin \forall(\tau^{-1}J)X$. The latter entails that there is an $x \in W$ where $\overline{R}_{\forall(\tau^{-1}J)}x\alpha$ and $x \notin X$. From (cP6), $\overline{R}_{\forall(\tau^{-1}J)}x\alpha$, and $R_{S(\tau)}\alpha\beta$ we obtain that there is a γ where $R_{S(\tau)}x\gamma$ and $R_{\forall(J)}\gamma\beta$. From these relations with γ and our original assumption we then get that $\gamma \in S(\tau)X$, and thus also $x \in X$, a contradiction.

For the other direction, assume first that $\beta \in S(\tau)\forall(\tau^{-1}J)X$ and that $\tau^{-1}J$ is well defined. Further assume that $\beta \notin \forall(J)S(\tau)X$. From the latter, we obtain that there is a x such that $\overline{R}_{\forall(J)}x\beta$ and $x \notin S(\tau)X$. This gives that there is a unique α such that $R_{S(\tau)}\alpha x$ where $\alpha \notin X$. With (cP6), $R_{S(\tau)}\alpha x$, and $\overline{R}_{\forall(J)}x\beta$, we obtain that there is a γ such that $R_{S(\tau)}\gamma\beta$ and $\overline{R}_{\forall(\tau^{-1}J)}\alpha\gamma$. With our first assumption, the former entails that $\gamma \in \forall(\tau^{-1}J)X$. With the latter, we obtain that $\alpha \in X$, a contradiction.

For (P7), we need to show that the each substitution commutes with the modal operations. Note that, for any $\oplus_i^z \in \circledast_{i\in I}$ and $X_1,\ldots,X_z \in \wp(W)$ we have that $S(\tau)(\oplus_i^z(X_1,\ldots,X_z)) = \{\beta \in W : \exists\gamma \in W(R_{S(\tau)}\gamma\beta \ \& \ \gamma \in \oplus_i^z(X_1,\ldots,X_z)\}$.

Now, we have two cases for general cases for $\gamma \in \oplus^z(X_1,\ldots,X_z)$.

1. Suppose that $\oplus_i^z : \mathbb{X},\ldots,\mathbb{X} \longrightarrow \vee$. Then $\gamma \in \oplus_i^z(X_1,\ldots,X_z)$ holds just in case $\exists\alpha_1,\ldots,\alpha_z(R_i(\alpha_1,\ldots,\alpha_z,\gamma) \ \& \ \alpha_1 \pitchfork X_1 \ \& \ \cdots \ \& \ \alpha_z \pitchfork X_z)$, where $\alpha \pitchfork X$ is either $\alpha \in X$ or $\alpha \notin X$, as appropriate as in the proof of [3, Lem. 1.4.9]. By condition (cP7$_i$), we have that $\exists\alpha'_1,\ldots,\alpha'_z$ such that $R_{S(\tau)}\alpha_1\alpha'_1,\ldots,R_{S(\tau)}\alpha_z\alpha'_z$ and $R_i(\alpha'_1,\ldots,\alpha'_z,\beta)$. It is easy to show that $\alpha'_i \in S(\tau)X_i$, for each $1 \leq i \leq z$. Thus, we have that $\beta \in \oplus_i^z(S(\tau)X_1,\ldots,S(\tau)X_n)$, as required.

2. Suppose that $\gamma : \mathbb{X},\ldots,\mathbb{X} \longrightarrow \wedge$. The argument proceeds similarly. $\qquad\square$

5   Embedding

The biggest departure from—rather, generalization of—the gaggle-framework is in the definition of canonical relations for the quantifiers and substitution operations. For the quantifiers, e.g., a point in the situation should, via a relation $\overline{R}_{\forall(J)}$, see all other points that differ *in the evaluation of the 'variables' in J*.

The idea, intuitively, is that when evaluating a quantified expression, we pay attention only to different interpretations of the quantified expression which depend on how we interpret those variables actually quantified. What we do in the relational semantics is, in effect, to situate these varying interpretations arising from different assignments to the variables in different points of the frame. The relations associated with the quantifiers, then, track the similarities between points in which the evaluation of formulas differs only up to the point that the variables of some quantifier prefix are interpreted differently (though, strictly speaking, the relational semantics do not interpret variables, talking this way is a useful fiction). In the concrete example $\forall x \forall y \Box P(x, y)$, we consider a quantifier $\forall(\{x, y\})$ in the PMBA and a relation $\overline{R}_{\forall(\{x,y\})}$ in the associated frame which associates a pair of worlds α, β just in case the value of $\Box P(x, y)$ at α differs from its interpretation at β up to the point that α, β assign x and y differently. Thinking in these terms allows us to place the machinery used to interpret quantifiers "on the same level" as that machinery used to interpret the modal operations.

We will, in keeping with the gaggle-theoretic framework, define R^c, rather than the equivalence class of its negation.

Canonical Frame Let $\mathbb{A} = \langle \mathbb{A}; \vee, \wedge, \neg, 1, \forall, \circledast^1_{i \in I}, \mathcal{I}, S, \forall \rangle$ be a PMBA. The canonical frame for $\mathbb{A}$ is a tuple $\mathfrak{F}^c = \langle W^c; \{R^c_i\}_{i \in I}, \{R^c_{S(\tau)}\}_{S(\tau) \in S}, \{R^c_{\forall(J)}\}_{J \subseteq I} \rangle$ where:

1. W^c is the set of ultrafilters on $\mathbb{A}$.

2. The set $\{R^c_i\}_{i \in I/f}$ consists of a single canonical relation for each family of operations in I (as in the monadic case).

3. For every $S(\tau) \in S$, due to having two distribution types, define

$$R^c_{S(\tau)}\alpha\beta \text{ iff both } \forall a \in \mathbb{A}(a \notin \alpha \vee S(\tau)a \in \beta)$$

$$\text{and } \neg \exists a \in \mathbb{A}(a \notin \alpha \ \& \ S(\tau)a \in \beta)$$

4. For every $\forall(J) \in \forall$, define

$$R^c_{\forall(J)}\alpha\beta \text{ iff } \exists a \in \mathbb{A}(a \notin \alpha \ \& \ \forall(J)a \in \beta)$$

Note that we obtain for the relations for substitutions the following, by combining the conjuncts and performing a little logic:

$$R^c_{S(\tau)}\alpha\beta \text{ iff } \forall a \in \mathbb{A}(a \in \alpha \Leftrightarrow S(\tau)a \in \beta).$$

In the following squeeze lemma, the relations for substitutions do not conform nicely to the gaggle-framework. That is, when we define a theory to bear the relation to an ultrafilter, the theory is already the unique ultrafilter that bears that relation. We note this more carefully in the proof.

The condition (cP5) and the corresponding equality (p5) play a crucial role in making this all work. For example, if $\forall(J)a \in \alpha$ then each substitution instance of a is in α (and any other $\overline{R}^c_{\forall(J)}$ related element). Suppose that $\forall(J)a \in \gamma$ but $S(x/y)p \notin \gamma$. By the latter we have $R^c_{S(x/y)}\beta\gamma$ where $p \notin \beta$. But then (cP5) gives us that $R^c_{S(x/y)}\alpha\beta$ and $\overline{R}^c_{\forall(J)}\alpha\beta$. We clearly obtain that $p \in \beta$, a contradiction.

We remind the reader that $R'_\circledast$ is the relaxation of $\overline{R}^c_\circledast$ to have filters in the first arguments place when $\circledast$ distributes into $\wedge$, and of $R^c_\circledast$ to have filters in the first arguments place when $\circledast$ distributes into $\vee$.

Lemma 2 (Polyadic Squeeze). Let $\mathbb{A} = \langle \mathbb{A}; \vee, \wedge, \neg, 1, \forall, \circledast^1_{i \in I}, \mathcal{I}, S, \forall \rangle$ be a PMBA. Suppose $R'_\circledast(\alpha_1, \ldots, \alpha_n, \beta)$ for non-empty, proper filters $\alpha_1, \ldots, \alpha_n$ and ultrafilter β, where $R_\circledast$ is the canonical relation for operation $\circledast$. Then there are ultrafilters $\alpha'_1, \ldots, \alpha'_n$ extending filters $\alpha_1, \ldots, \alpha_n$, respectively, such that we have $R^c_\circledast(\alpha'_1, \ldots, \alpha'_n, \beta)$ or $\overline{R}^c_\circledast(\alpha'_1, \ldots, \alpha'_n, \beta)$ as appropriate.

Proof. For the cases for $\circledast \in \circledast_{i \in I}$, the arguments come directly from [3]. For the relations determined by substitutions—Boolean endomorphisms—we show the stronger claim: $R'_{S(\tau)}\alpha\beta$ iff $R^c_{S(\tau)}\alpha\beta$. This means that any theory α that bears the relation $R'_{S(\tau)}$ to β is (1) unique and (2) an ultrafilter on $\mathbb{A}$. The proof follows straightforwardly by setting $\alpha = \{a \in \mathbb{A} : S(\tau)a \in \beta\}$.

For the relations $R'_{\forall(J)}$, the arguments are the same as in the monadic case. $\square$

Lemma 3 (Canonical Frame is a Frame). The canonical frame is a frame.

Proof. We show that (c1)–(c3), (cP1)–(cP7$_i$) hold. For (c1), we can use the arguments from the monadic case, for each quantifier's relation. Conditions (c2) and (c3) follow immediately from the proof of Lemma 2, where the unique α such that $R_{S(\tau)}\alpha\beta$ is defined as $\{a \in \mathbb{A} : S(\tau)a \in \beta\}$. Condition (cP1) follows straightforwardly because $\forall(\emptyset)p = p$. For (cP3), the proof is similar to the case of (cP1).

For the left-to-right direction of (cP2), let $\overline{R}^c_{\forall(J \cup K)}\alpha\beta$. Then this means, by the definition of the canonical relations, plus (p2), $\neg\exists a \in \mathbb{A}(a \notin \alpha \,\&\, \forall(J \cup K)a =$

$\forall(J)\forall(K)a \in \beta$). We must construct a γ with the right properties. First, consider the theory $\gamma' = \{a \in \mathbb{A} : \forall(J) \in \beta\}$. Clearly $\overline{R}'_{\forall(J)}\gamma'\beta$, by definition. To see that $\overline{R}'_{\forall(K)}\alpha\gamma'$, note that if $\forall(K)b \in \gamma'$, then $\forall(J)\forall(K)b = \forall(J \cup K)b \in \beta$, which gives that $b \in \alpha$, as required. Thus, we use the squeeze lemma to obtain a γ extending γ' such that $\overline{R}^c_{\forall(K)}\alpha\gamma$.[5] Clearly, now we also have $\overline{R}^c_{\forall(J)}\gamma\beta$, as γ is an ultrafilter and extends γ'.

The right-to-left direction is straightforward and left to the reader.

As in (cP2), we give only the interesting left-to-right direction. Assume that $R^c_{S(\sigma\tau)}\alpha\beta$. This is, applying (P4), $\forall a \in \mathbb{A}(a \in \alpha \Leftrightarrow S(\sigma)S(\tau)a \in \beta)$. There is a unique γ such that $R^c_{S(\sigma)}\gamma\beta$. We show that $R^c_{S(\tau)}\alpha\gamma$. Suppose that $a \in \alpha$. This is if and only if $S(\sigma)S(\tau)a \in \beta$, itself if and only if $S(\tau)a \in \gamma$.

For (cP5), first assume that $\tau = \sigma$ outside J. For any γ, by the Squeeze Lemma there is a unique α and unique β such that $R^c_{S(\sigma)}\beta\gamma$ and $R^c_{S(\tau)}\alpha\gamma$. We show that $\overline{R}_{\forall(J)}\alpha\beta$. Assume for reduction, symmetrically, that $\forall(J)a \in \alpha$, but $a \notin \beta$. From the former, we have that $S(\tau)\forall(J)a \in \gamma$, which by (cP5) gives $S(\sigma)\forall(J)a \in \gamma$. This latter fact entails that $\forall(J)a \in \beta$, which itself gives $a \in \beta$, a contradiction.

For (cP6), let $R^c_{S(\tau)}\gamma\alpha$ and $\overline{R}^c_{\forall(J)}\alpha\beta$. Consider the unique x where $R^c_{S(\tau)}x\beta$. This means $a \in x$ iff $S(\tau)a \in \beta$. We need to show that $\overline{R}_{\forall(\tau^{-1}J)}x\gamma$. For reductio, suppose that $\forall(\tau^{-1}J))b \in \gamma$ but $b \notin x$, for some $b \in \mathbb{A}$. Then $S(\tau)\forall(\tau^{-1}J)b \in \alpha$. By (P6), $\forall(J)S(\tau)b \in \alpha$. Then by $\overline{R}^c_{\forall(J)}\alpha\beta$, we have $S(\tau)b \in \beta$. This then entails, but an 'iff' above, that $b \in \beta$, which results in the contradiction.

For the left-to-right direction of (cP7$_i$), we show the case where $\circledast^z_i$ distributes into $\vee$ (and hence that R'_o is the relaxation of R^c_i and not its negation). Assume that $R^c_{S(\tau)}\alpha\beta$ and that $R^c_i(\gamma_1, \ldots, \gamma_z, \alpha)$. The former entails that $a \in \alpha$ iff $S(\tau)a \in \beta$, and the latter that $\forall a_1, \ldots, a_z(a_1 \pitchfork \gamma_1 \& \cdots \& a_z \pitchfork \gamma_z \Rightarrow \circledast^z_i(a_1, \ldots, a_z) \in \alpha)$.[6] Note that $\circledast^z_i(a_1, \ldots, a_z) \in \alpha$, by (P7$_i$) and a previous 'iff', just in case $\circledast^z_i(S(\tau)a_1, \ldots, S(\tau)a_z) \in \beta$.

Let $\gamma''_j = \{S(\tau)a : a \in \gamma_j\}$ for $1 \leq j \leq z$. It follows that $R'_i(\gamma''_1, \ldots, \gamma''_z, \beta)$: this is because (i) $S(\tau)a_j \pitchfork \gamma''_j$ iff $a_j \pitchfork \gamma_j$, and (ii) $\circledast^z_i(a_1, \ldots, a_z) \in \alpha$) just in case $\circledast^z_i(S(\tau)a_1, \ldots, S(\tau)a_z) \in \beta$. Using the Squeeze Lemma, we then obtain each $\gamma'_j \supseteq \gamma''_j$. It remains to be shown that $R_{S(\tau)}\gamma_j\gamma'_j$. Suppose that $a \in \gamma$, then by definition $S(\tau)a \in \gamma'' \subseteq \gamma'$. For the other direction, suppose $a \notin \gamma$. Then $\neg a \in \gamma$, and so $S(\tau)\neg a \in \gamma''$, which implies that $\neg S(\tau)a \in \gamma'' \subseteq \gamma'$. This entails $S(\tau)a \notin \gamma'$, as required.

---

[5]For this to follow using the Squeeze Lemma, we must be careful that this is really $\overline{R}^{c,1\rightleftarrows 2}_{\forall(K)}\alpha\gamma$ and $\overline{R}'^{1\rightleftarrows 2}_{\forall(K)}\alpha\gamma$.

[6]Note that $\pitchfork$ stands for whichever of $\in$ or $\notin$ is appropriate. Conventions for the use of this symbol are discussed in more detail in [3].

For the right-to-left direction of (cP7$_i$), we again only show the case for $\circledast_i^z$ distributing into $\vee$. Suppose that $R_i^c(\gamma_1', \ldots, \gamma_z', \beta)$ and that $R_{S(\tau)}\gamma_j\gamma_j'$, for $1 \le j \le z$. Let α be the unique ultrafilter such that $R_{S(\tau)}\alpha\beta$. We show that $R_i^c(\gamma_1, \ldots, \gamma_z, \alpha)$.

Suppose that $a_1, \ldots, a_z$ are such that $a_1 \pitchfork \gamma_1$ & $\cdots$ & $a_z \pitchfork \gamma_z$ as required for R_i^c. Then we obtain $S(\tau)a_1 \pitchfork \gamma_1'$ & $\cdots$ & $S(\tau)a_z \pitchfork \gamma_z'$ due to each $R_{S(\tau)}\gamma_j\gamma_j'$. It follows that $\circledast_i^z(S(\tau)_1, \ldots, S(\tau)a_z) \in \beta$, because $R_i^c(\gamma_1', \ldots, \gamma_z', \beta)$. By (P7$_i$), we have that $S(\tau) \circledast_i^z (a_1, \ldots, a_z) \in \beta$, which implies that $\circledast_i^z(a_1, \ldots, a_z) \in \alpha$, as required. $\qquad\square$

Theorem 4 (Embedding). Let $\mathbb{A} = \langle \mathbb{A}; \vee, \wedge, \neg, 1, \forall, \circledast_{i\in I}^1, \mathcal{I}, S, \forall \rangle$ be a PMBA. There is a model $\mathfrak{M}$ on a frame $\mathfrak{F}$ for $\mathbb{A}$ such that the algebra of sets on $\mathfrak{F}$ is isomorphic to $\mathbb{A}$. That is, there is a valuation v_c that is an injective homomorphism from $\mathbb{A}$ into the algebra of sets on $\mathfrak{F}$: $v_c(a) = v_c(b)$ in $\mathfrak{M}$ iff $a = b$ in $\mathbb{A}$.

Proof. We use the canonical frame, which we have shown to be a frame. On this, we add the valuation defined as follows:

$$v_c(a) = \{\alpha : a \in \alpha \ \& \ \alpha \text{ is an ultrafilter on } \mathbb{A}\}.$$

It is to be shown that v_c is a valuation, as in the definition of PMBA-models (on p. 12). The cases for $\wedge, \vee, \neg$, and $\circledast_i$ are covered by the arguments of [3, Theorem 1.4.16]. The remaining cases follow similar arguments:

For a quantifier $\forall(J)$, for the left-to-right direction, assume that $\beta \in v_c(\forall(J)a)$. Then $\forall(J)a \in \beta$. For any α such that $\overline{R}_{\forall(J)}^c\alpha\beta$, we have that $a \in \alpha$. So, in particular, $\overline{R}_{\forall(J)}^c\alpha\beta$ implies $\alpha v_c(a)$, as required. For the right-to-left direction, assume that $\beta \notin v_c(\forall(J)a)$. Then $\forall(J)a \notin \beta$. Take the principal upset $(\neg a)^\uparrow$ on $\mathbb{A}$ generated by $\neg a$. Clearly $R_{\forall(J)}'(\neg a)^\uparrow\beta$ and $a \notin (\neg a)^\uparrow$. By the Squeeze Lemma, there is an α extending $(\neg a)^\uparrow$ such that $a \notin \alpha$ (and hence $\alpha \notin v_c(a)$) where $\overline{R}_{\forall(J)}^c\alpha\beta$. These last facts show that $\beta \notin \forall(J)v_c(a)$, as required.

For the left-to-right direction for $S(\tau)$, assume that $\beta \in v_c(S(\tau)a)$. Then $S(\tau)a \in \beta$. Consider the ultrafilter $\alpha = \{a \in \mathbb{A} : S(\tau)a \in \beta\}$. Clearly this is the unique α such that $R_{S(\tau)}\alpha\beta$. Moreover, we have that $a \in \alpha$, and so $\alpha \in v_c(a)$. For the other direction, assume that $\beta \notin v_c(S(\tau)a)$. Then $S(\tau)a \notin \beta$. Let α be the unique ultrafilter such that $R_{S(\tau)}\alpha\beta$. Then $a \notin \alpha$, and so $\alpha \notin v_c(a)$, as required. $\qquad\square$

6 Conclusion

In summary, we have shown that monadic and polyadic Boolean algebras are gaggles, and that some of their extensions with z-ary 0-1 operations are also gaggles. We have also used the tools of gaggle theory to construct relational semantics for these (multi-)gaggles. This lays the foundation for future work in employing the gaggle framework generally to first-order logics extending propositional logics that enjoy gaggle presentation. We point to future work.

Gaggles come in many shapes and forms: we have gaggles on distributive lattices, non-distributive lattices, semi-lattices, as well as partial gaggles and dual gaggles [2]. These structures are ripe for the development of generalized monadic and polyadic variants. Indeed, many non-classical logics that fall into these categories have been given monadic or polyadic extensions: see, e.g., monadic **BL** [7], polyadic algebras over some non-classical logics [17]. We aim to abstract to the more general settings described in the first sentence of the paragraph, with particular novel specifications to, e.g., polyadic Ackermann groupoids though polyadic De Morgan monoids for first-order relevant logics.

There is a general correspondence between gaggles and display calculi that was developed in Restall [18]. With the present work, there is the suggestion that Restall's recipe will give display calculi for first-order logics. It is our hope to develop this suggestion into a complete examination into the relation of polyadic gaggles and the display calculi methods for first-order logics developed by Balco et al. [1]. In [1], idea of *properly displaying* quantifiers is developed. Restall's methods ought to produce properly displayed quantifiers. Similarly, there may be an interesting comparison available with Wansing's work on display calculi for first-order modal logics [19].

References

[1] Samuel Balco, Giuseppe Greco, Alexander Kurz, Andrew Moshier, Alessandra Palmigiano, and Apostolos Tzimoulis. First order logic properly displayed, 2021. arXiv:2105.06877.

[2] Katalin Bimbó. Dual gaggle semantics for entailment. *Notre Dame Journal of Formal Logic*, 50:23–41, 2009.

[3] Katalin Bimbó and J. Michael Dunn. *Generalized Galois Logics: Relational Semantics of Nonclassical Logical Calculi*, volume 188 of *CSLI Lecture Notes*. CSLI Publications, Stanford, CA, 2008.

[4] J. Michael Dunn. Gaggle theory: An abstraction of Galois connections and residuation, with applications to negation, implication, and various logical operators. In J. van Eijck, editor, *Logics in AI: European Workshop JELIA '90*, number 478 in Lecture Notes in Computer Science, pages 31–51. Springer, Berlin, 1991.

[5] J. Michael Dunn and Gary M. Hardegree. *Algebraic Methods in Philosophical Logic*, volume 41 of *Oxford Logic Guides*. Oxford University Press, Oxford, 2001.

[6] James Garson. Quantification in modal logic. In D. Gabbay and F. Guenthner, editors, *Handbook of Philosophical Logic*, volume 3, pages 267–323. Kluwer, Amsterdam, 2nd edition, 2001.

[7] Revaz Grigolia. Monadic BL-algebras. *Georgian Mathematical Journal*, 13: 267–276, 2006.

[8] Paul R. Halmos. Algebraic logic, I: Monadic Boolean algebras. *Composition Mathematica*, 12:217–249, 1955. (Reprinted in [10]).

[9] Paul R. Halmos. Algebraic logic, II: Homogeneous locally finite polyadic Boolean algebras of infinite degree. *Fundamenta Mathematicae*, 43:256–325, 1956. (Reprinted in [10]).

[10] Paul R. Halmos. *Algebraic Logic*. Chelsea, New York, NY, 1962.

[11] Leon Henkin, James Donald Monk, and Alfred Tarski. *Cylindric Algebras. Part I*. North-Holland, Amsterdam, 1971.

[12] Leon Henkin, James Donald Monk, and Alfred Tarski. *Cylindric Algebras. Part II*. North-Holland, Amsterdam, 1985.

[13] Bjarni Jónsson and Alfred Tarski. Boolean algebras with operators. I. *American Journal of Mathematics*, 73:891–939, 1951.

[14] Bjarni Jónsson and Alfred Tarski. Boolean algebras with operators. II. *American Journal of Mathematics*, 74:127–162, 1952.

[15] Edwin Mares and Robert Goldblatt. A general semantics for quantified modal logic. *Advances in Modal Logic*, 6:227–246, 2006.

[16] Grigori Mints. *A Short Introduction to Modal Logic*. CSLI Publications, 1992.

[17] Don Pigozzi and Antonino Salibra. Polyadic algebras of nonclassical logics. In *Algebraic Methods in Logic and Computer Science, vol. 28*, pages 51–66. Banach Center Publications, Warszawa, 1993.

[18] Greg Restall. Display logic and gaggle theory. *Reports on Mathematical Logic*, 29:133–146, 1995.

[19] Heinrich Wansing. *Displaying Modal Logic*. Springer, Dordrecht, 1998.

Received 16 December 2024

The Constructive Logic of Paradox: Paraconsistency on the Woodruff Plan

Rohan French
University of California, Davis
`<rfrench@ucdavis.edu>`

Abstract

In this paper we investigate the constructive companion of the logic of paradox. After presenting the logic semantically, and examining how it relates to both intuitionistic logic and the logic of paradox, we prove completeness for two semantically conservative extensions of the constructive logic of paradox—a non-paraconsistent one with a 'true-only' predicate, and a non-constructive one with a De Morgan negation.

2020 *Mathematics Subject Classification.* Primary: 03B45, Secondary: 03B80.

1 Introduction

In his paper "On Constructive Nonsense Logic" Woodruff states his goal as the following, where D is the logic of nonsense from [10]:

> Our goal will be to produce a calculus which stands to D as intuitionistic does to classical logic, and at the same time stands to intuitionistic logic as D does to classical logic. [14, p. 195]

The logic that Woodruff goes on to characterize combines features of the semantics of intuitionistic logic—its frames being pre-orders—with features of 'dual valuation' semantics for logics of nonsense.[1] While the semantics for the logic we will investigate

---

I'd like to thank Katalin Bimbó, Ellie Ripley, Shawn Standefer, Andrzej Indrzejczak, Nils Kurbis, Hitoshi Omori, and a referee for this journal for helpful discussions on various aspects of the present paper as well as audiences at the LLAL GIS (II) Workshop at Tohoku University, and the 2nd Third Workshop at the University of Alberta.

[1]Arenhart & Omori argue in [1] that because of this dual valuational semantics, among many other reasons, Woodruff's logic is not really a logic of nonsense. The interested reader should consult [6, p. 358–362] for a discussion of Woodruff's and some neighboring systems of constructive nonclassical logic.

below exhibits similar features (being built on pre-orders, but combined with the 'dual valuation' treatment of many-valued logics implicit in [4]), what we take as our starting point for the present paper is the idea, implicit in the above quote, of understanding the constructivisation of a logic L as involving it being appropriately similar to both L and to IL. We can think of this as a 'relational' dimension of similarity between logics, alluded to by the claim that DI should be appropriately related to both D and IL, to which we might plausibly wish to add a more 'intrinsic' dimension of similarity—in the above case it seems plausible that the constructive nonsense logic DI ought to be both constructive and a logic of nonsense.

In this paper, we set out to apply what we will call 'The Woodruff Plan' for building constructive nonclassical logics in a new setting, paraconsistent logic. In particular, we are interested in characterizing and understanding the constructive analogue of the three-valued logic of paradox LP of [9].[2]

To do this we will need to get clear on how to understand the following two claims:

- L stands to LP as IL stands to CL.

- L stands to IL as LP stands to CL.

We will give a relatively coarse grained syntactic notion of 'standing to' here, understanding the claim that L to stand to LP as IL stands to CL as requiring that there be some collection of sequents S such that the extension of L by S results in LP and the extension of IL by S results in CL. We will then combine this relational notion of 'standing to' with the restriction that our logic L must be constructive/paraconsistent to get our notion of a *constructive companion* and a *paraconsistent companion*.

Definition 1. S^c is a *constructive companion* of S iff (i) S^c has the disjunction property, and (ii) there is an S-valid sequent s which, when added to IL gives CL, and which when added to S^c gives S.

Definition 2. S^p is a *paraconsistent companion* of S iff (i) in S^p there is no nonempty finite set of sentences Γ such that $[\Gamma \succ\ ]$ is valid, and (ii) the result of adding explosion (i.e., $[\Gamma, A, \neg A \succ \Delta]$) to S^p gives S.

Note that here we are understanding paraconsistency in a language neutral way requiring that for a logic to be paraconsistent there can be no nonempty finite set

---

[2]Our main interest in this logic, although not on display otherwise in the present paper, arises out of an interest in the constructive versions of the non-transitive logic ST of [4], it being well known that this logic is in some sense translationally equivalent to LP, as pointed out in [2] and [5].

of formulas which behave like $\{A, \neg A\}$ do in classical logic.[3]

Our goal in the present paper will be to investigate a particular logic cLP, which we will show is a constructive companion of LP and a paraconsistent companion of IL. For the sake of simplicity (and ease of comparison with LP) we will be considering only languages with the primitive connectives $\land$, $\lor$ and $\neg$. The omission of the conditional is purely for ease of exposition and its addition would not repair any of the difficulties which we point out along the way.

The plan for the remainder of the paper is as follows. In §2 we introduce the constructive logic of paradox model-theoretically and show that it is a constructive companion of LP and a paraconsistent companion of IL. We will then look at two conservative extensions of this logic for which we will prove soundness and completeness results, in §3 by the addition of a 'just true' operator, and in §4 by the addition of a De Morgan negation.

Some preliminaries before we begin. Our language $\mathcal{L}$ consists of formulas constructed out of a countable stock $p_1, \ldots, p_n, \ldots$, of propositional variables, the first three of which we will usually abbreviate as p, q, r, using the connectives $\land$, $\lor$ and $\neg$. Capital Roman letters are schematic for formulas of our language; capital Greek letters (that are not also Roman letters) are schematic for (typically finite) multisets of formulas. Throughout we work in the SET-SET framework, with logics being identified with generalized consequence relations $\vdash \,\subseteq\, \wp(\mathcal{L}) \times \wp(\mathcal{L})$. Accordingly, throughout, a *sequent* is an ordered pair $\langle \Gamma, \Delta \rangle$ of finite multisets of formulas which we will write as $[\Gamma \succ \Delta]$.

2 Semantics for the Constructive Logic of Paradox

Our models are a particular kind of intuitionistic Kripke models with a pair of valuations, investigated initially in [8], and more recently in [7]. Formally speaking, a model is a tuple $\langle W, \leq, V_s, V_t \rangle$ where W is a nonempty set, $\leq$ is a reflexive and transitive relation of accessibility on W, and V_s and V_t are functions from propositional atoms to subsets of W which satisfy the following three conditions for all $w, v \in W$ and propositional variables p_i:

- PERSISTENCE-S If $w \in V_s(p_i)$ and $w \leq v$, then $v \in V_s(p_i)$.

- PERSISTENCE-T If $w \in V_t(p_i)$ and $w \leq v$, then $v \in V_t(p_i)$.

- CONTAINMENT $V_s(p_i) \subseteq V_t(p_i)$.

---

[3]That is to say, we require that a paraconsistent logic not satisfy the principle $sEQC$ of 'set-based explosion', the strongest explosion principle surveyed in [3].

Truth at a point in a model is defined as follows.

1. $\mathcal{M}, w \Vdash_s p_i$ iff $w \in V_s(p_i)$

2. $\mathcal{M}, w \Vdash_t p_i$ iff $w \in V_t(p_i)$

3. $\mathcal{M}, w \Vdash_s A \wedge B$ iff $\mathcal{M}, w \Vdash_s A$ and $\mathcal{M}, w \Vdash_s B$

4. $\mathcal{M}, w \Vdash_t A \wedge B$ iff $\mathcal{M}, w \Vdash_t A$ and $\mathcal{M}, w \Vdash_t B$

5. $\mathcal{M}, w \Vdash_s A \vee B$ iff $\mathcal{M}, w \Vdash_s A$ or $\mathcal{M}, w \Vdash_s B$

6. $\mathcal{M}, w \Vdash_t A \vee B$ iff $\mathcal{M}, w \Vdash_t A$ or $\mathcal{M}, w \Vdash_t B$

7. $\mathcal{M}, w \Vdash_s \neg A$ iff for all $w \leq w'$ we have $\mathcal{M}, w' \not\Vdash_t A$

8. $\mathcal{M}, w \Vdash_t \neg A$ iff for all $w \leq w'$ we have $\mathcal{M}, w' \not\Vdash_s A$

We will say that a sequent $[\Gamma \succ \Delta]$ *holds in a model* $\mathcal{M}$ iff for all worlds w we have either $\mathcal{M}, w \not\Vdash_t \gamma$ for some $\gamma \in \Gamma$, or $\mathcal{M}, w \Vdash_t \delta$ for some $\delta \in \Delta$. A sequent $[\Gamma \succ \Delta]$ is *valid on a frame* $\langle W, \leq \rangle$ iff it holds in every model $\langle W, \leq, V_s, V_t \rangle$ on that frame. Finally, a sequent is *valid in a class of models* M iff it holds in every $\mathcal{M} \in \mathsf{M}$, and is *valid on a class of frames* F iff it is valid on every frame $\mathfrak{F} \in \mathsf{F}$. Finally, let $\Gamma \vdash_{\mathsf{cLP}} \Delta$ iff $[\Gamma \succ \Delta]$ is valid on all frames.

Just like with the Kripke semantics for intuitionistic logic, our semantic clauses are such that by stipulating the conditions PERSISTENCE-T, PERSISTENCE-S, and CONTAINMENT for propositional variables we are able to show that these conditions apply to all formulas in the language.

Lemma 1 (Persistence). *For all models $\mathcal{M} = \langle W, \leq, V_s, V_t \rangle$, formulas A and points $w, v \in W$ we have:*

1. *If $\mathcal{M}, w \Vdash_s A$ and $w \leq v$, then $\mathcal{M}, v \Vdash_s A$.*

2. *If $\mathcal{M}, w \Vdash_t A$ and $w \leq v$, then $\mathcal{M}, v \Vdash_t A$.*

Lemma 2 (Containment). *For all models $\mathcal{M} = \langle W, \leq, V_s, V_t \rangle$, formulas A and points $w, v \in W$ we have that if $\mathcal{M}, w \Vdash_s A$, then $\mathcal{M}, w \Vdash_t A$.*

Moreover, witnessing the fact that cLP is constructive, we can show that it satisfies the Disjunction property.

Lemma 3 (Disjunction Property). *If $\vdash_{\mathsf{cLP}} A \vee B$, then either $\vdash_{\mathsf{cLP}} A$ or $\vdash_{\mathsf{cLP}} B$.*

Proof. This follows by the standard model-theoretic argument. If both sequents are invalid then there are models $\mathcal{M}_A$ and $\mathcal{M}_B$ where they fail to hold. Construct a new model $\mathcal{M}$ by taking the disjoint union of these two models and adding a new point w from which every point in each model is accessible. By T-PERSISTENCE we cannot have either $\mathcal{M}, w \Vdash_t A$ or $\mathcal{M}, w \Vdash_t B$, and so we have $\mathcal{M}, w \nVdash_t A \vee B$, as desired. $\qquad\square$

What we will show in the next two subsections is that this logic is indeed both a constructive counterpart of LP, and a paraconsistent counterpart of IL.

2.1 Constructive Counterpart

It is very easy to show that, just as classical logic is determined by the class of all intuitionistic Kripke frames where the accessibility relation is the identity relation, so too is the logic of paradox determined by the class of all frames of the above kind. Moreover, these frames are precisely those frames which validate the law of excluded middle.

Theorem 1. *All instances of the law of excluded middle are valid on a frame iff its accessibility relation is the identity relation.*

Proof. Let $\langle W, = \rangle$ be a frame, and suppose $\mathcal{M} = \langle W, =, V_s, V_t \rangle$ is a model on that frame. Suppose that $\mathcal{M}, w \nVdash_t A$. Then by CONTAINMENT we have $\mathcal{M}, w \nVdash_s A$. So at all worlds v such that $w = v$ we have $\mathcal{M}, v \nVdash_s A$, and so $\mathcal{M}, w \Vdash_t \neg A$, as desired.

Let $\langle W, \leq \rangle$ be a frame where $w \leq v$ and $w \neq v$. Let $V_s(p) = V_t(p) = \{u : v \leq u\}$. As $w \leq v$ and $v \Vdash_s p$ it follows that $w \nVdash_t \neg p$. It is clear that $w \notin V_t(p)$ and so $w \nVdash_t p$ and so the instance $[\succ p \vee \neg p]$ fails to be valid on $\langle W, \leq \rangle$. $\qquad\square$

To show that the logic determined by the class Id of frames whose accessibility relation is the identity relation is LP it will be helpful to specify LP-validity and counterexamples in the following way.

Definition 3. A *Strong-Kleene valuation* is a function $v : \mathcal{L} \to \{1, \frac{1}{2}, 0\}$ where:

- $v(\neg A) = 1 - v(A)$

- $v(A \wedge B) = \min(v(A), v(B))$

- $v(A \vee B) = \max(v(A), v(B))$

A Strong Kleene valuation v is an LP-*counterexample* to a sequent $[\Gamma \succ \Delta]$ iff $v[\Gamma] \subseteq \{1, \frac{1}{2}\}$ and $v[\Delta] = 0$. A sequent is LP-*valid* iff it has no LP-counterexamples.

The following results are analogues of the familiar results connecting up the endpoints of intuitionistic Kripke frames with boolean valuations.

Theorem 2. *Suppose that $\langle W, \leq, V_s, V_t \rangle$ is a model where $\leq$ is identity. Consider the following trivaluation v_w*

$$
v_w(A) = \begin{cases} 1 & \text{if } w \Vdash_s A \\ \frac{1}{2} & \text{if } w \Vdash_t A \text{ and } w \not\Vdash_s A \\ 0 & \text{if } w \not\Vdash_t A \end{cases}
$$

Then v_w is a Strong Kleene valuation such that $v_w(A) \in \{1, \frac{1}{2}\}$ iff $\mathcal{M}, w \Vdash_t A$.

Proof. The only difficulty in the proof is verifying that v_w is a Strong-Kleene valuation. We show the case of negation, the others being similar.

$v_w(B) = 1$: $v_w(B) = 1$ iff $w \Vdash_s B$ iff $w \not\Vdash_t \neg B$ (as $\leq$ is identity) iff $v_w(\neg B) = 0$, as desired.

$v_w(B) = \frac{1}{2}$: $v_w(B) = \frac{1}{2}$ iff $w \Vdash_t A$ and $w \not\Vdash_s A$ iff $w \not\Vdash_s \neg A$ and $w \Vdash_t \neg A$ (as $\leq$ is identity) iff $v_w(\neg B) = \frac{1}{2}$, as desired.

$v_w(B) = 0$: $v_w(B) = 0$ iff $w \not\Vdash_t B$ iff $w \Vdash_s \neg B$ (as $\leq$ is identity) iff $v_w(\neg B) = 1$, as desired. $\square$

Theorem 3. *Suppose that v is a Strong Kleene valuation. Let $\mathcal{M}_v = \langle \{w\}, =, V_s^v, V_t^v \rangle$ be the model where $w \in V_s^v(p_i)$ iff $v(p_i) = 1$ and $w \in V_t^v(p_i)$ iff $v(p_i) \in \{1, \frac{1}{2}\}$. Then $v(A) \in \{1, \frac{1}{2}\}$ iff $\mathcal{M}_v, w \Vdash_t A$.*

Proof. We prove something stronger, namely, that the following equivalences hold:

1. $v(B) = 1$ iff $\mathcal{M}_v, w \Vdash_s B$;

2. $v(B) = \frac{1}{2}$ iff $\mathcal{M}_v, w \not\Vdash_s B$ and $\mathcal{M}_v, w \Vdash_t B$;

3. $v(B) = 0$ iff $\mathcal{M}_v, w \not\Vdash_t B$.

We proceed by induction on the complexity of A. The basis case is trivial. Suppose then, that for all formulas of complexity less than n the above three equivalences hold. We treat the case in the induction step for conjunction.

$v(B \wedge C) = 1$: $v(B \wedge C) = 1$ iff $v(B) = 1$ and $v(C) = 1$ iff by the induction hypothesis $\mathcal{M}_v, w \Vdash_s B$ and $\mathcal{M}_v, w \Vdash_s C$ iff $\mathcal{M}_v, w \Vdash_s B \wedge C$, as desired.

$v(B \wedge C) = \frac{1}{2}$: $v(B \wedge C) = \frac{1}{2}$ iff either (i) $v(B) = \frac{1}{2}$ or $v(C) = \frac{1}{2}$ and (ii) $v(B) \in \{1, \frac{1}{2}\}$ and (iii) $v(C) \in \{1, \frac{1}{2}\}$. Suppose wlog that $v(B) \in \frac{1}{2}$. Then by the induction hypothesis, $\mathcal{M}_v, w \Vdash_s B$ and so $\mathcal{M}_v, w \Vdash_s B \wedge C$. Further, from (iii) by the induction hypothesis, we have $\mathcal{M}_v, w \Vdash_t C$ and so by (i) and the induction hypothesis $\mathcal{M}_v, w \Vdash_t B \wedge C$.

$v(B \wedge C) = 0$: $v(B \wedge C) = 0$ iff $v(B) = 0$ or $v(C) = 0$ iff by the induction hypothesis $\mathcal{M}_v, w \nVdash_t B$ or $\mathcal{M}_v, w \nVdash_t C$ iff $\mathcal{M}_v, w \nVdash_t B \wedge C$, as desired. $\qquad\square$

We can use this result to easily show that cLP is paraconsistent in the strong sense of there being no set of formulas Γ such that $[\Gamma \succ\;]$.

Lemma 4 (Strong Paraconsistency). *For no set of formulas Γ, do we have $\Gamma \vdash_{\mathsf{cLP}}$.*

Proof. The valuation v_* where $v_*(p_i) = \frac{1}{2}$ for all propositional variables p_i is such that $v_*(A) = \frac{1}{2}$ for all formulas A. So by Theorem 3, it follows that $\mathcal{M}_v, w \Vdash_t A$ for all formulas A, and so for all formulas $C \in \Gamma$. $\qquad\square$

Theorem 4. *A sequent $[\Gamma \succ \Delta]$ is valid in all models on frames in Id iff $[\Gamma \succ \Delta]$ is LP valid.*

Proof. Suppose that $[\Gamma \succ \Delta]$ is not LP-valid. Then there is some Strong Kleene valuation v such that $v[\Gamma] \subseteq \{1, \frac{1}{2}\}$ and $v[\Delta] = 0$. By Theorem 3, it follows that $\mathcal{M}_v, w \Vdash_t C$ for all $C \in \Gamma$ and $\mathcal{M}_v, w \nVdash_t D$ for any $D \in \Delta$, and so $[\Gamma \succ \Delta]$ is not valid on Id.

Suppose then, that $[\Gamma \succ \Delta]$ is not valid on Id. Then there is a model $\mathcal{M} = \langle W, \leq, V_s, V_t \rangle$ in Id where at some $w \in W$, we have $\mathcal{M}, w \Vdash_t C$ for all $C \in \Gamma$ and $\mathcal{M}, w \nVdash_t D$ for all $D \in \Delta$. By Theorem 2 it follows that $v_w[\Gamma] \subseteq \{1, \frac{1}{2}\}$, and $v_w[\Delta] = 0$ and so $[\Gamma \succ \Delta]$ is invalid in LP. $\qquad\square$

From the above we can see that cLP is a constructive companion of LP, as the logic determined by the class of frames which validate the law of excluded middle is LP.

2.2 Paraconsistent Counterpart

Unlike the case for the law of excluded middle, there is no class of *frames* which validates explosion. In other words the logic we get by extending cLP by all instances of $[\Gamma, A, \neg A \succ \Delta]$ is Kripke-incomplete.[4]

---

[4]A similar result can be seen to hold for the condition X^+ in [14, p. 203] that meaningfulness is effectively decidable—there the proof essentially relies on the fact that for every frame there is a model on that frame which does not have a strong valuation.

Theorem 5. *Let $\langle W, \leq \rangle$ be a frame. Then there are V_s and V_t such that some instance of $[\Gamma, A, \neg A \succ \Delta]$ fails to hold in $\langle W, \leq, V_s, V_t \rangle$.*

Proof. In particular, we will show that $[p, \neg p \succ q]$ fails to hold in $\langle W, \leq, V_s, V_t \rangle$. Let w be a world in W. Then let $V_t(p) = \{u : w \leq u\}$ (so $w \Vdash_t p$), $V_s(p) = \emptyset$ (so in particular, for all $w \leq v$ we have $v \nVdash_s p$ and so $w \Vdash_t \neg p$), and let $V_t(q) = V_s(q) = \emptyset$ (and so $w \nVdash_t q$). Then the result follows. $\square$

So there is no condition on *frames* which we can impose which will result in all models on frames which meet that condition validating explosion. But we can impose a condition directly on *models*. In particular, we will show that all instances of explosion hold in a model exactly when that model meets a condition we call *delayed converse containment* or DCC.

Theorem 6. *If all instances of $[\Gamma, A, \neg A \succ \Delta]$ hold in $\mathcal{M}$, then $\mathcal{M}$ satisfies*

(DCC) $$\forall w \forall p_i (w \in V_t(p_i) \Rightarrow \exists v \geq w.v \in V_s(p_i)).$$

Proof. We prove the contrapositive. Suppose that the above condition is false, that is we have a model $\mathcal{M}$ where for some $w \in W$ we have:

$$w \in V_t(p_i) \text{ and } \forall v \geq w.v \notin V_s(p_i).$$

From this it follows directly that we have $w \Vdash_t p$, and as for all $w \leq v$ we have $v \nVdash_s p$ also $w \Vdash_t \neg p$, and so it is not the case that $[p, \neg p \succ\ ]$ holds in $\mathcal{M}$, and so not all instances of $[\Gamma, A, \neg A \succ \Delta]$ hold in $\mathcal{M}$. $\square$

As with containment, we can also show that models which satisfy delayed converse containment for propositional variables also satisfy it for all formulas.

Theorem 7. *Suppose that $\mathcal{M}$ satisfies (DCC). Then for all formulas A and worlds w we have:*

$$w \Vdash_t A \Rightarrow \exists v \geq w.v \Vdash_s A.$$

Proof. By induction on the complexity of A. Basis is trivial.

Conjunction For the conjunction case, suppose that $w \Vdash_t A \wedge B$. Then $w \Vdash_t A$ and $w \Vdash_t B$. So there are worlds u, v such that $w \leq u$ and $w \leq v$ where $u \Vdash_s A$ and $v \Vdash_s B$. If $u \leq v$ or $v \leq u$ then at the later one we will have $A \wedge B$ s-verified. If not then note by persistence that at u we have $u \Vdash_t B$ and so (again by IH) $x \Vdash_s B$ for some $u \leq x$, and so $x \Vdash_s A \wedge B$.

Disjunction For disjunction suppose that $w \Vdash_t A \vee B$. Then wlog suppose $w \Vdash_t A$. Then by IH we have that $w \leq u$ and $u \Vdash_s A$ and so $u \Vdash_s A \vee B$.

Negation Suppose $w \Vdash_t \neg A$. Then we have for all $u \geq w$ that $u \nVdash_s A$. Now let v be an arbitrary world where $w \leq v$. If $v \Vdash_t A$ then by the IH we must have some $v \leq x$ such that $x \Vdash_s A$. By the transitivity of accessibility relation, this means that we must have $w \leq x$ and so $x \nVdash_s A$, giving us a contradiction. So it follows that all the worlds $u \geq w$ are such that $u \nVdash_t A$ and so $w \Vdash_s \neg A$. $\square$

One important consequence of this condition is that we can recover the usual intuitionistic truth condition for negation.

Proposition 1. *If $\mathcal{M}$ satisfies (DCC), then we have*

$$w \Vdash_t \neg A \;\Leftrightarrow\; \forall u \geq w.u \nVdash_t A.$$

Proof. Suppose that $\mathcal{M}$ satisfies condition (DCC). Then by Theorem 7, we have that for all formulas A and worlds w, if $w \Vdash_t A$ then there is a v such that $v \geq w$ and $v \Vdash_s A$, and so contraposing this we have

$$\forall v \geq w.v \nVdash_s A \;\Rightarrow\; w \nVdash_t A$$

Now suppose that $w \Vdash_t \neg A$. Then by our truth conditions, it follows that for all $v \geq w$ we have $v \nVdash_s A$. Now suppose that u is a point such that $u \geq w$. Then by the transitivity of $\leq$, we have that all $u' \geq u$ are such that $u' \nVdash_s A$, and so by the above inset condition $u \nVdash_t A$. But u was arbitrary, so we have that $\forall u \geq w.u \nVdash_t A$, as desired.

For the other direction, suppose that we have that for all $u \geq w$, $u \nVdash_t A$. Then by Lemma 2, we have $u \nVdash_s A$, and so by our truth conditions $w \Vdash_t \neg A$, as desired. $\square$

Corollary 5. *If $\mathcal{M}$ satisfies (DCC), then all instances of $[\Gamma, A, \neg A \succ \Delta]$ hold in $\mathcal{M}$.*

Theorem 8 (Paraconsistent Companion)**.** *A sequent $[\Gamma \succ \Delta]$ is valid in the class of all models satisfying DCC iff $\Gamma \vdash_{\mathsf{IL}} \Delta$.*

Proof. Given Proposition 1, it follows that any model $\langle W, \leq, V_s, V_t \rangle$ which satisfies DCC is modally equivalent, in the sense of validating the same formulas, (w.r.t. t-satisfaction) to the intuitionistic Kripke model $\langle W, \leq, V_t \rangle$. As a result if a sequent fails to hold in a given DCC model, it also fails to hold in the corresponding intuitionistic Kripke model. For the converse, if a sequent fails in some intuitionistic Kripke model $\langle W, \leq, V \rangle$ the aforementioned result tells us that the same sequent fails to hold in the model $\langle W, \leq, V, V \rangle$. $\square$

From the above we can see that cLP is a paraconsistent companion of IL, understanding the result of extending cLP by some sequent s as involving taking the logic determined by the class of models which validate s.

2.3 Characterizing cLP

It is a routine exercise to see that the following sequents are valid and invalid according to the above semantics.

VALID	INVALID
$[A \succ \neg\neg A]$	$[\neg\neg A \succ A]$
$[\neg A \vee \neg B \succ \neg(A \wedge B)]$	$[\neg(A \wedge B) \succ \neg A \vee \neg B]$
$[\neg A \wedge \neg B \succ \neg(A \vee B)]$	$[A, \neg A \succ B]$
$[\neg(A \vee B) \succ \neg A \wedge \neg B]$	$[\succ A \vee \neg A]$
$[\succ \neg(A \wedge \neg A)]$	
$[\neg\neg\neg A \succ \neg A]$	

Looking at this one might reasonably wonder whether cLP is just the intersection of the generalized consequence relations for LP and IL. As it turns out, this is not the case, as the sequent $[p, \neg p \succ q, \neg q]$ is valid in both of these logics, but is not valid in cLP. To see this consider the model $\langle \{0, 1\}, \leq, V_s, V_t \rangle$ where $V_t(p) = \{0, 1\}$, $V_s(p) = \emptyset$, $V_t(q) = V_s(q) = \{1\}$ and $\leq$ is the reflexive closure of $0 < 1$. In this model, $0 \Vdash_t p$, and as $i \nVdash_s p$ for $i \in \{0, 1\}$ we have $0 \Vdash_t \neg p$. But as we have $1 \Vdash_s q$ we have $0 \nVdash_t \neg q$ and also $0 \nVdash_t q$, so the sequent fails to hold at 0.

So cLP is not the intersection of the logic of paradox and intuitionistic logic, although as we have seen above, it is strongly related to them. Unfortunately, we have been unable, in this signature, to give a proof-theoretic characterization of this logic. We can get an indirect grip on cLP proof-theoretically though, through the (conservative) addition of additional connectives, as we will see in the following sections. Each of these extensions repair the expressive deficit we face in proving completeness in distinct ways. Unfortunately, each comes at a cost of either constructivity or paraconsistency.

3 cLP with a Strict Truth Operator: Constructive, but not Paraconsistent

In this section we will consider a signature similar to that considered by Woodruff in [13] (and [14]). The base signature used therein includes not just the connectives

$\{\wedge, \vee, \neg, \rightarrow\}$ but also two additional operators T and $*$ with essentially the following semantic clauses [13, p. 94f]:

$$\mathcal{M}, w \Vdash_s \mathsf{T}A \ \text{ iff } \ \mathcal{M}, w \Vdash_s A \qquad\qquad \mathcal{M}, w \Vdash_t \mathsf{T}A \ \text{ iff } \ \mathcal{M}, w \Vdash_s A$$

$$\mathcal{M}, w \Vdash_s *A \ \text{ iff } \ \mathcal{M}, w \Vdash_t A \qquad\qquad \mathcal{M}, w \Vdash_t *A \ \text{ iff } \ \mathcal{M}, w \Vdash_t A$$

Woodruff refers to these two operators as the 'unconditional' and 'hedged' truth operators. Given its behavior in the present semantics, we will prefer to refer to T as the 'strict truth' operator—roughly speaking the 'unconditional' truth-operator reports that the sentence following it is true, while the 'hedged' truth operator merely reports that the sentence following is not false.

Let $\mathcal{L}^{\mathsf{T}}$ be the language, which results from extending $\mathcal{L}$ by the addition of the strict truth operator T. We will say that a sequent $[\Gamma \succ \Delta]$ of $\mathcal{L}^{\mathsf{T}}$ is valid in $\mathsf{cLP}^{\mathsf{T}}$ iff it is valid on all frames, sometimes writing this as $\Gamma \vdash_{\mathsf{cLP}^{\mathsf{T}}} \Delta$. As it happens extending our language with just this operator allows us to give a sound and complete sequent calculus for the logic $\mathsf{cLP}^{\mathsf{T}}$. The proof system $\mathcal{SCLP}^{\mathsf{T}}$ consists of all the rules and basic sequents in Figure 1.

It is a routine matter to show that $\mathcal{SCLP}^{\mathsf{T}}$ is sound w.r.t. our semantics enriched with the clauses for the T-operator.

Theorem 9 (Soundness for $\mathcal{SCLP}^{\mathsf{T}}$). *If a sequent $[\Gamma \succ \Delta]$ is derivable in $\mathcal{SCLP}^{\mathsf{T}}$, then it is valid.*

Proof. By induction on the length of derivations. For the basis case it is easy to see that $[Id]$ is valid. For the case of $[\mathsf{T}ax]$ suppose that $\mathcal{M} = \langle W, \leq, V_s, V_t \rangle$ is a model and for some $w \in W$ we have $\mathcal{M}, w \Vdash_t \mathsf{T}A$. Then it follows that $\mathcal{M}, w \Vdash_s A$ and so by Lemma 2, that $\mathcal{M}, w \Vdash_t A$, as desired.

In the induction step, we treat $[\neg R]$, the other cases either being routine or sufficiently similar. We show that if the conclusion sequent fails to hold in a model then so does the premise sequent. Suppose that $[\Gamma \succ \neg A]$ fails to hold at $w \in W$. Then we have $\mathcal{M}, w \Vdash_t C$ for all $C \in \Gamma$ and $\mathcal{M}, w \not\Vdash_t \neg A$. So there is some $w \leq v$ such that $\mathcal{M}, v \Vdash_s A$ and so $\mathcal{M}, w \Vdash_t \mathsf{T}A$ as desired. $\square$

For completeness, we will use a relatively straightforward canonical model proof, the worlds in our canonical model being suitably saturated pairs of sets of formulas.

Definition 4 (T-Saturation). Say that a pair of sets of formulas (l, r) is T-*saturated* iff $[l \succ r]$ is unprovable in $\mathcal{SCLP}^{\mathsf{T}}$ and we have the following:

   1. If $A_1 \wedge A_2 \in l$, then $A_1 \in l$ and $A_2 \in l$.

INITIAL SEQUENTS

$$\frac{}{[\Gamma, A \succ A, \Delta]} \; [Id] \qquad \frac{}{[\Gamma, \mathsf{T}A \succ \Delta, A]} \; [Tax]$$

STRUCTURAL RULES

$$\frac{[\Gamma \succ \Delta, A] \quad [A, \Gamma' \succ \Delta']}{[\Gamma, \Gamma' \succ \Delta, \Delta']} \; [Cut]$$

$$\frac{[\Gamma \succ \Delta]}{[A, \Gamma \succ \Delta]} \; [KL] \qquad \frac{[\Gamma \succ \Delta]}{[\Gamma \succ \Delta, A]} \; [KR] \qquad \frac{[A, A, \Gamma \succ \Delta]}{[A, \Gamma \succ \Delta]} \; [WL] \qquad \frac{[\Gamma \succ \Delta, A, A]}{[\Gamma \succ \Delta, A]} \; [WR]$$

OPERATIONAL RULES

$$\frac{[\Gamma, A, B \succ \Delta]}{[\Gamma, A \wedge B \succ \Delta]} \; [\wedge L] \qquad \frac{[\Gamma \succ \Delta, A] \quad [\Gamma' \succ \Delta', B]}{[\Gamma, \Gamma' \succ \Delta, \Delta', A \wedge B]} \; [\wedge R]$$

$$\frac{[\Gamma, A \succ \Delta] \quad [\Gamma', B \succ \Delta']}{[\Gamma, \Gamma', A \vee B \succ \Delta, \Delta']} \; [\vee L] \qquad \frac{[\Gamma \succ \Delta, A, B]}{[\Gamma \succ \Delta, A \vee B]} \; [\vee R]$$

T-INTERACTION RULES

$$\frac{[\Gamma \succ \Delta, \mathsf{T}A] \quad [\Gamma \succ \Delta, \mathsf{T}B]}{[\Gamma \succ \Delta, \mathsf{T}(A \wedge B)]} \; [\wedge RT] \qquad \frac{[\Gamma, \mathsf{T}A, \mathsf{T}B \succ \Delta]}{[\Gamma, \mathsf{T}(A \wedge B) \succ \Delta]} \; [\wedge LT]$$

$$\frac{[\Gamma, \mathsf{T}A \succ \Delta] \quad [\Gamma, \mathsf{T}B \succ \Delta]}{[\Gamma, \mathsf{T}(A \vee B) \succ \Delta]} \; [\vee LT] \qquad \frac{[\Gamma \succ \Delta, \mathsf{T}A, \mathsf{T}B]}{[\Gamma \succ \Delta, \mathsf{T}(A \vee B)]} \; [\vee RT]$$

$$\frac{[\Gamma \succ \Delta, \mathsf{T}A]}{[\Gamma \succ \Delta, \mathsf{T}\mathsf{T}A]} \; [TTR] \qquad \frac{[\Gamma, \mathsf{T}A \succ \Delta]}{[\Gamma, \mathsf{T}\mathsf{T}A \succ \Delta]} \; [TTL]$$

NEGATION RULES

$$\frac{[\Gamma, \mathsf{T}A \succ \;]}{[\Gamma \succ \neg A]} \; [\neg R] \qquad \frac{[\Gamma, A \succ \;]}{[\Gamma \succ \mathsf{T}\neg A]} \; [\neg RT] \qquad \frac{[\Gamma \succ \Delta, A]}{[\Gamma, \mathsf{T}\neg A \succ \Delta]} \; [\neg LT] \qquad \frac{[\Gamma \succ \Delta, \mathsf{T}A]}{[\Gamma, \neg A \succ \Delta]} \; [\neg L]$$

Figure 1: The proof system $\mathcal{SCLP}^{\mathsf{T}}$ for the Constructive Logic of Paradox with Strict Truth

2. If $A_1 \wedge A_2 \in r$, then $A_i \in r$ for some $i \in \{1, 2\}$.

3. If $A_1 \vee A_2 \in l$, then $A_i \in l$ for some $i \in \{1, 2\}$.

4. If $A_1 \vee A_2 \in r$, then $A_1 \in r$ and $A_2 \in r$.

5. If $\mathsf{T}(A_1 \wedge A_2) \in l$, then $\mathsf{T}(A_1) \in l$ and $\mathsf{T}(A_2) \in l$.

6. If $\mathsf{T}(A_1 \wedge A_2) \in r$, then $\mathsf{T}(A_i) \in r$ for some $i \in \{1, 2\}$.

7. If $\mathsf{T}(A_1 \vee A_2) \in l$, then $\mathsf{T}(A_i) \in l$ for some $i \in \{1, 2\}$.

8. If $\mathsf{T}(A_1 \vee A_2) \in r$, then $\mathsf{T}(A_1) \in r$ and $\mathsf{T}(A_2) \in r$.

9. If $\neg A \in l$, then $\mathsf{T}(A) \in r$.

10. If $\mathsf{T}(\neg A) \in l$ then $A \in r$.

11. If $\mathsf{T}(A) \in l$, then $A \in l$.

12. If $\mathsf{TT}(A) \in l$, then $\mathsf{T}(A) \in l$.

13. If $\mathsf{TT}(A) \in r$, then $\mathsf{T}(A) \in r$.

Definition 5 (T-complexity). Given a formula A, let the T-complexity $tc(A)$ be defined as follows.

1. $tc(p) = 0$

2. $tc(\mathsf{T}p) = 0$

3. $tc(\neg A) = 1 + tc(A)$

4. $tc(\mathsf{T}\neg A) = 1 + tc(A)$

5. $tc(\mathsf{TT}A) = 1 + tc(A)$

6. $tc(A \wedge B) = tc(\mathsf{T}(A \wedge B)) = 1 + tc(A) + tc(B)$

7. $tc(A \vee B) = tc(\mathsf{T}(A \vee B)) = 1 + tc(A) + tc(B)$

Lemma 6. *Any unprovable sequent $[\Gamma \succ \Delta]$ can be expanded to a T-saturated pair (l, r) s.t. $\Gamma \subseteq l$ and $\Delta \subseteq r$.*

Definition 6 (T-Canonical Model). The T-canonical model $\mathcal{M}^{\mathsf{T}} = \langle W, \leq, V_s, V_t \rangle$ is defined by letting W be the set of all T-saturated pairs (l, r), $(l, r) \leq (l', r')$ iff $l \subseteq l'$, and by letting $(l, r) \in V_t(p_i)$ iff $p_i \in l$ and $(l, r) \in V_s(p_i)$ iff $\mathsf{T}(p_i) \in l$.

It is easy to see that a T-canonical model is indeed a model, the only non-obvious condition being that it satisfies CONTAINMENT, but this follows directly from Definition 4.11.

Theorem 10. *For any* T*-saturated pair* $(l, r) \in W$ *in* $\mathcal{M}^{\mathsf{T}}$ *we have:*

- *If* $A \in l$, *then* $(l, r) \Vdash_t A$.

- *If* $\mathsf{T}(A) \in l$, *then* $(l, r) \Vdash_s A$.

- *If* $A \in r$, *then* $(l, r) \nVdash_t A$.

- *If* $\mathsf{T}(A) \in r$, *then* $(l, r) \nVdash_s A$.

Proof. By induction on the T-complexity of A. For the basis case, it follows from the definition of the T-canonical model that if $p_i \in l$ then $(l, r) \in V_t(p_i)$ and so $(l, r) \Vdash_t p_i$, and likewise for $\mathsf{T}p_i \in l$. Suppose then, that $p_i \in r$. As (l, r) is a T-saturated pair it follows that $p_i \notin l$, as otherwise $[l \succ r]$ would be derivable, and so $(l, r) \notin V_t(p_i)$. The case when $\mathsf{T}p_i \in r$ is similar.

For the induction step, we treat the cases where $A = B \wedge C$, $A = \mathsf{TT}B$, and $A = \neg B$. We begin with the cases where $A = B \wedge C$.

$B \wedge C \in l$: If $B \wedge C \in l$, then by Definition 4.1 we have $B \in l$ and $C \in l$. So by the induction hypothesis, we have $(l, r) \Vdash_t B$ and $(l, r) \Vdash_t C$ and so $(l, r) \Vdash_t B \wedge C$.

$\mathsf{T}(B \wedge C) \in l$: If $\mathsf{T}(B \wedge C) \in l$, then by Definition 4.5, we have $\mathsf{T}B \in l$ and $\mathsf{T}C \in l$. So by the induction hypothesis, we have $(l, r) \Vdash_s B$ and $(l, r) \Vdash_s C$ and so $(l, r) \Vdash_s B \wedge C$.

$B \wedge C \in r$: If $B \wedge C \in r$, then by Definition 4.2, we must have either $B \in r$ or $C \in r$. So by the induction hypothesis, we have $(l, r) \nVdash_t B$ or $(l, r) \nVdash_t C$, and so either way we have $(l, r) \nVdash_t B \wedge C$.

$\mathsf{T}(B \wedge C) \in r$: If $\mathsf{T}(B \wedge C) \in r$, then by Definition 4.6, we must have either $\mathsf{T}B \in r$ or $\mathsf{T}C \in r$. So by the induction hypothesis, we have either $(l, r) \nVdash_s B$ or $(l, r) \nVdash_s C$, and so either way we have $(l, r) \nVdash_s B \wedge C$.

Consider now, the case where $A = \mathsf{TT}B$.

$\mathsf{TT}B \in l$: If $\mathsf{TT}B \in l$ then by Definition 4.12, we have $\mathsf{T}B \in l$ and so by the induction hypothesis, we have $(l, r) \Vdash_s B$ and so by our truth conditions, we have $(l, r) \Vdash_s \mathsf{T}(B)$ and so $(l, r) \Vdash_t \mathsf{TT}(B)$ as desired.

$\mathsf{TT}B \in r$: If $\mathsf{TT}B \in r$ then by Definition 4.13, we have $\mathsf{T}B \in r$ and so by the induction hypothesis, we have $(l,r) \not\Vdash_s B$ and so by our truth conditions we have $(l,r) \not\Vdash_s \mathsf{T}(B)$ and so $(l,r) \not\Vdash_t \mathsf{TT}(B)$ as desired.

$\mathsf{TTT}B \in l$: If $\mathsf{TTT}B \in l$ then by Definition 4.12, we have $\mathsf{TT}B \in l$ and so by Definition 4.12 again we have $\mathsf{T}B \in l$. By the induction hypothesis, it follows that $(l,r) \Vdash_s B$ and so $(l,r) \Vdash_s \mathsf{TT}B$ by our truth conditions for T.

$\mathsf{TTT}B \in r$: If $\mathsf{TTT}B \in r$ then by Definition 4.13, we have $\mathsf{TT}B \in r$ and so by Definition 4.13 again we have $\mathsf{T}B \in r$. By the induction hypothesis, it follows that $(l,r) \not\Vdash_s B$ and so $(l,r) \not\Vdash_s \mathsf{TT}B$, by our truth conditions for T.

Consider now, the case where $A = \neg B$.

$\neg B \in l$: If $\neg B \in l$ then for every accessible (l',r'), we must also have $\neg B \in l$ and so by Definition 4.9 $\mathsf{T}(B) \in r$. So by the induction hypothesis, for all such (l',r') we have $(l',r') \not\Vdash_s B$, and so $(l,r) \Vdash_t \neg B$.

$\mathsf{T}\neg B \in l$: If $\mathsf{T}\neg B \in l$ then we have $\mathsf{T}\neg B \in l'$ for all accessible (l',r'). So by Definition 4.10 we have $B \in r'$ and so by the induction hypothesis, $(l',r') \not\Vdash_t B$, and so $(l,r) \Vdash_s \neg B$.

$\neg B \in r$: We show that $[l, \mathsf{T}B \succ]$ is unprovable. If not then by $[\neg R]$ and $[KR]$ we would have $[l \succ r]$ provable, contradicting T-saturation. So we extend $(l \cup \{\mathsf{T}(B)\}, \emptyset)$ to a T-saturated pair (l',r') using Lemma 6. Then we have $(l,r) \leq (l',r')$. As $\mathsf{T}(B) \in l'$ we have $(l',r') \Vdash_s B$, an so $(l,r) \not\Vdash_t \neg B$.

$\mathsf{T}\neg B \in r$: We show $[l, B \succ]$ is unprovable. If not then by $[\neg RT]$ we would have $[l \succ \mathsf{T}\neg B]$ provable, and so by weakening $[l \succ r]$ provable, contradicting T-saturation. So we extend $(l \cup \{B\}, \emptyset)$ to a T-saturated pair (l',r') using Lemma 6. Then we have $(l,r) \leq (l',r')$, and as $B \in l$ by the induction hypothesis, $(l',r') \Vdash_t B$, and so $(l,r) \not\Vdash_s \neg B$. $\qquad \square$

Given this we can prove completeness.

Theorem 11 (Completeness for $\mathcal{L}_\mathsf{T}$). *If the $\mathcal{L}_\mathsf{T}$-sequent $[\Gamma \succ \Delta]$ is valid in the class of all models, then $[\Gamma \succ \Delta]$ is derivable in $\mathcal{SCLP}^\mathsf{T}$.*

Proof. We prove the contrapositive. Suppose that $[\Gamma \succ \Delta]$ is not derivable in $\mathcal{SCLP}^\mathsf{T}$. Then by lemma 6, there is a T-saturated pair (l,r) where $\Gamma \subseteq l$ and $\Delta \subseteq r$. So by Theorem 10, we have that in the T-canonical model $\mathcal{M}^\mathsf{T}$ that $\mathcal{M}^\mathsf{T}, (l,r) \Vdash_t C$ for all $C \in \Gamma$, and that $\mathcal{M}^\mathsf{T}, (l,r) \not\Vdash_t D$ for all $D \in \Delta$, and so the sequent is invalid. $\qquad \square$

Interestingly, we have bought completeness here at the cost of (strong) paraconsistency. The essential problem here is that using T we can define an intuitionistic negation as $\mathsf{T}\neg A$. As explosion is valid for intuitionistic negation our logic fails to be strongly paraconsistent.

Lemma 7. $\vdash_{\mathsf{cLP}\mathsf{T}}$ *is not paraconsistent.*

Proof. To see this note that the sequent $[A, \mathsf{T}\neg A \succ]$ is derivable, as shown below:

$$\frac{\dfrac{}{[A \succ A]} \; [Id]}{[A, \mathsf{T}\neg A \succ]} \; [\neg LT]$$

So by Theorem 11, we have $A, \mathsf{T}\neg A \vdash_{\mathsf{cLP}\mathsf{T}} \emptyset$. $\qquad\square$

It is easy to show that the disjunction property still holds for $\vdash_{\mathsf{cLP}\mathsf{T}}$, though, using the same proof as was given in §2. So as we can see here, we can buy completeness for cLP, but at the cost of it no longer being paraconsistent in the extended language.

4 cLP with a De Morgan Negation: Paraconsistent, but not Constructive

In this section, we will consider the addition of a De Morgan negation '$\sim$' to cLP with the following clauses:

- $\mathcal{M}, w \Vdash_s \sim A$ iff $\mathcal{M}, w \not\Vdash_t A$.

- $\mathcal{M}, w \Vdash_t \sim A$ iff $\mathcal{M}, w \not\Vdash_s A$.

As before we will let $\mathcal{L}^\sim$ be the language which results from the addition of $\sim$ to $\mathcal{L}$, and say that a sequent $[\Gamma \succ \Delta]$ from the language $\mathcal{L}^\sim$ is valid in $\mathsf{cLP}^\sim$ iff it is valid on all frames, sometimes writing this as $\Gamma \vdash_{\mathsf{cLP}^\sim} \Delta$. The proof system $\mathcal{SCLP}^\sim$ for this logic consists of the rules and basic sequents in Figure 2.

One curious feature of this extension is that it is no longer the case that all formulas are persistent. In particular, formulas which contain the De Morgan negation are not persistent. This is reflected in the $[\sim\neg L]$ and $[\neg R]$ rules of $\mathcal{SCLP}^\sim$. Consider, for example the rule $[\neg R]$, where $\overrightarrow{A_i}$ is a metavariable over sequences of formulas $A_1, \ldots, A_n$ for some n.

$$\frac{[\overrightarrow{\neg A_i}, \overrightarrow{p_i} \succ \overrightarrow{\sim\neg A_i}, \overrightarrow{\sim p_i}, \sim A]}{[\overrightarrow{\neg A_i}, \overrightarrow{p_i} \succ \overrightarrow{\sim\neg A_i}, \overrightarrow{\sim p_i}, \neg A]} \; [\neg R]$$

INITIAL SEQUENTS

$$\frac{}{[\Gamma, A \succ A, \Delta]} \; [Id] \qquad \frac{}{[\Gamma \succ \Delta, A, \sim A]} \; [\sim LEM]$$

STRUCTURAL RULES

$$\frac{[\Gamma \succ \Delta, A] \quad [A, \Gamma' \succ \Delta']}{[\Gamma, \Gamma' \succ \Delta, \Delta']} \; [Cut]$$

$$\frac{[\Gamma \succ \Delta]}{[A, \Gamma \succ \Delta]} \; [KL] \qquad \frac{[\Gamma \succ \Delta]}{[\Gamma \succ \Delta, A]} \; [KR] \qquad \frac{[A, A, \Gamma \succ \Delta]}{[A, \Gamma \succ \Delta]} \; [WL] \qquad \frac{[\Gamma \succ \Delta, A, A]}{[\Gamma \succ \Delta, A]} \; [WR]$$

OPERATIONAL RULES

$$\frac{[\Gamma, A, B \succ \Delta]}{[\Gamma, A \wedge B \succ \Delta]} \; [\wedge L] \qquad \frac{[\Gamma \succ \Delta, A] \quad [\Gamma' \succ \Delta', B]}{[\Gamma, \Gamma' \succ \Delta, \Delta', A \wedge B]} \; [\wedge R]$$

$$\frac{[\Gamma, A \succ \Delta] \quad [\Gamma', B \succ \Delta']}{[\Gamma, \Gamma', A \vee B \succ \Delta, \Delta']} \; [\vee L] \qquad \frac{[\Gamma \succ \Delta, A, B]}{[\Gamma \succ \Delta, A \vee B]} \; [\vee R]$$

DE MORGAN RULES

$$\frac{[\Gamma \succ \Delta, \sim A] \quad [\Gamma \succ \Delta, \sim B]}{[\Gamma \succ \Delta, \sim(A \vee B)]} \; [\sim \vee R] \qquad \frac{[\Gamma, \sim A, \sim B \succ \Delta]}{[\Gamma, \sim(A \vee B) \succ \Delta]} \; [\sim \vee L]$$

$$\frac{[\Gamma, \sim A \succ \Delta] \quad [\Gamma, \sim B \succ \Delta]}{[\Gamma, \sim(A \wedge B) \succ \Delta]} \; [\sim \wedge L] \qquad \frac{[\Gamma \succ \Delta, \sim A, \sim B]}{[\Gamma \succ \Delta, \sim(A \wedge B)]} \; [\sim \wedge R]$$

$$\frac{[\Gamma \succ A, \Delta]}{[\Gamma, \sim\sim A \succ \Delta]} \; [\sim\sim L] \qquad \frac{[\Gamma \succ \Delta, A]}{[\Gamma \succ \Delta, \sim\sim A]} \; [\sim\sim R]$$

NEGATION RULES

$$\frac{[A, \overrightarrow{\neg A_i}, \overrightarrow{p_i} \succ \overrightarrow{\sim\neg A_i}, \overrightarrow{\sim p_i}]}{[\sim\neg A, \overrightarrow{\neg A_i}, \overrightarrow{p_i} \succ \overrightarrow{\sim\neg A_i}, \overrightarrow{\sim p_i}]} \; [\sim\neg L] \qquad \frac{[\overrightarrow{\neg A_i}, \overrightarrow{p_i} \succ \overrightarrow{\sim\neg A_i}, \overrightarrow{\sim p_i}, \sim A]}{[\overrightarrow{\neg A_i}, \overrightarrow{p_i} \succ \overrightarrow{\sim\neg A_i}, \overrightarrow{\sim p_i}, \neg A]} \; [\neg R]$$

$$\frac{[\Gamma \succ A, \Delta]}{[\Gamma \succ \sim\neg A, \Delta]} \; [\sim\neg R] \qquad \frac{[\Gamma, \sim A \succ \Delta]}{[\Gamma, \neg A \succ \Delta]} \; [\neg L]$$

Figure 2: The Proof System $\mathcal{SCLP}^{\sim}$ for The Constructive Logic of Paradox with De Morgan Negation.

The non-principal formulas of this rule which occur on the left hand side of the sequent (i.e., formulas of the form $\neg A$ or p_i) are all formulas which are T-PERSISTENT, while those on the right hand side (i.e., formulas of the form $\sim\neg A$ and $\sim p_i$) are T-ANTI-PERSISTENT, where A is X-ANTI-PERSISTENT iff if $w \nVdash_x A$ and $w \leq v$, then $v \nVdash_x A$.

As before soundness is a relatively routine matter.

Theorem 12 (Soundness for $\mathcal{SCLP}^\sim$). *If a sequent $[\Gamma \succ \Delta]$ is derivable in $\mathcal{SCLP}^\sim$ then it is valid in $\mathsf{cLP}^\sim$.*

Proof. By induction on the length of derivations. For the basis case it is easy to see that $[Id]$ is valid. For the case of $[LEM\sim]$, suppose that for some model $\mathcal{M} = \langle W, \leq, V_s, V_t \rangle$ and $w \in W$ we have $\mathcal{M}, w \Vdash_t C$ for all $C \in \Gamma$ and $\mathcal{M}, w \nVdash_t \sim A$. Then we have $\mathcal{M}, w \Vdash_s A$ and so by Lemma 2, $\mathcal{M}, w \Vdash_t A$, as desired.

In the induction step, we treat $[\neg R]$, the other cases either being routine or similar. We show that if the conclusion sequent fails to hold in a model, then so does the premise sequent. Suppose, then, that we have a model $\mathcal{M} = \langle W, \leq, V_s, V_t \rangle$ and a $w \in W$ such that (i) $\mathcal{M}, w \Vdash_t \overrightarrow{\neg A_i}$, (ii) $\mathcal{M}, w \Vdash_t \overrightarrow{p_i}$, (iii) $\mathcal{M}, w \nVdash_t \overrightarrow{\sim\neg A_i}$, (iv) $\mathcal{M}, w \nVdash_t \overrightarrow{\sim p_i}$, and (v) $\mathcal{M}, w \nVdash_t \neg A$. From (v) it follows that for some $v \geq w$ we have $\mathcal{M}, v \Vdash_s A$, from which it follows that $\mathcal{M}, v \nVdash_t \sim A$. But as the formulas in (i) and (ii) are T-PERSISTENT they also hold at v, and as the formulas in (iii) and (iv) are T-ANTI-PERSISTENT they also fail at v and so the premise sequent fails to hold at v, and so fails to hold in $\mathcal{M}$. $\square$

To prove completeness in this setting we will draw on some techniques (particularly concerning how to treat rules containing non-persistent formulas) used in [12] to prove completeness for the logic **C+J**, which combines classical and intuitionistic logic.

Definition 7. Say that a pair of sets of formulas (l, r) is *d-saturated* iff $[l \succ r]$ is unprovable and we have the following:

1. If $A_1 \wedge A_2 \in l$ then $A_1 \in l$ and $A_2 \in l$.

2. If $A_1 \wedge A_2 \in r$ then $A_i \in r$ for some $i \in \{1, 2\}$.

3. If $\sim(A_1 \wedge A_2) \in l$ then $\sim A_i \in l$ for some $i \in \{1, 2\}$.

4. If $\sim(A_1 \wedge A_2) \in r$ then $\sim A_1 \in r$ and $\sim A_2 \in r$.

5. If $A_1 \vee A_2 \in l$ then $A_i \in l$ for some $i \in \{1, 2\}$.

6. If $A_1 \vee A_2 \in r$ then $A_1 \in r$ and $A_2 \in l$.

7. If $\sim(A_1 \vee A_2) \in l$ then $\sim A_1 \in l$ and $\sim A_2 \in l$.

8. If $\sim(A_1 \vee A_2) \in r$ then $\sim A_i \in r$ for some $i \in \{1, 2\}$.

9. If $\sim\sim A \in l$ then $A \in l$.

10. If $\sim\sim A \in r$ then $A \in r$.

11. If $\sim A \in r$, then $A \in l$.

12. If $A \in r$, then $\sim A \in l$.

13. If $\neg A \in l$ then $\sim A \in l$.

14. If $\sim\neg A \in r$ then $A \in r$.

Lemma 8. *Any unprovable sequent $[\Gamma \succ \Delta]$ can be expanded to a d-saturated pair (l, r) s.t. $\Gamma \subseteq l$ and $\Delta \subseteq r$.*

Definition 8. The *d*-canonical model $\mathcal{M}^d = \langle W, \leq, V_t, V_s \rangle$ is defined by letting W be the set of all saturated pairs (l, r) of formulas, and letting $(l, r) \leq (l', r')$ whenever we have all of the following conditions met:

1. If $p \in l$ then $p \in l'$.

2. If $\sim p \in r$ then $\sim p \in r'$.

3. If $\neg A \in l$ then $\neg A \in l'$.

4. If $\sim\neg A \in r$ then $\sim\neg A \in r'$.

Finally, let $(l, r) \in V_t(p_i)$ iff $p_i \in l$ and $(l, r) \in V_s(p_i)$ iff $\sim p_i \in r$.

Again to see that the *d*-canonical model is a model the only non-obvious condition which we need to check is that it satisfies CONTAINMENT. In this case suppose $(l, r) \in V_s(p_i)$, and so $\sim p_i \in r$. Then by Definition 7.11, it follows that $p_i \in l$ and, and so $(l, r) \in V_t(p_i)$, as desired.

Definition 9. Given a formula A, let the *De Morgan Complexity $dmc(A)$* be defined as follows.

1. $dmc(p) = 0$

2. $dmc(\sim p) = 0$

3. $dmc(\neg A) = 1 + dmc(A)$

4. $dmc(\sim\neg A) = 1 + dmc(A)$

5. $dmc(\sim\sim A) = 1 + dmc(A)$

6. $dmc(A \wedge B) = dmc(\sim(A \wedge B)) = 1 + dmc(A) + dmc(B)$

7. $dmc(A \vee B) = dmc(\sim(A \vee B)) = 1 + dmc(A) + dmc(B)$

Note that it follows from our definition of De Morgan complexity that $dmc(\sim A)$ is $dmc(\neg A) - 1$. We will appeal to this fact in a few places in the following theorem.

Theorem 13. *For any d-saturated pair $(l, r) \in W$ in $\mathcal{M}^d$ we have:*

1. *If $A \in l$ then $(l, r) \Vdash_t A$.*

2. *If $\sim A \in l$ then $(l, r) \not\Vdash_s A$.*

3. *If $A \in r$ then $(l, r) \not\Vdash_t A$.*

4. *If $\sim A \in r$ then $(l, r) \Vdash_s A$.*

Proof. By induction on $dmc(A)$. For the basis case, cases 1 and 4 are covered by the definition of V_s and V_t. For case 2, if $\sim p_i \in l$ then as (l, r) is d-saturated it follows that $\sim p_i \notin r$, and so $(l, r) \notin V_s(p_i)$, and thus $(l, r) \not\Vdash_s p_i$. 3 follows by similar reasoning.

The only remaining cases of interest are those for negation, and those involving De Morgan negation. Here we will treat those for negation and for the De Morgan negation of negations. We begin with **negation**.

1. Suppose $\neg A \in l$. Consider any point (l', r') such that $(l, r) \leq (l', r')$. Then by definition $\neg A \in l'$. So by Definition 7.13, we have that $\sim A \in l'$. As $dmc(\sim A) = dmc(\neg A) - 1$, we can appeal to induction hypothesis 2, allowing us to conclude that $(l', r') \not\Vdash_s A$. But (l', r') was arbitrary, and so we have $(l, r) \Vdash_t \neg A$.

2. Suppose that $\sim\neg A \in l$. We show that $[\Gamma' \succ \Delta'] =$

$$[\{p_i : p_i \in l\} \cup \{\neg A_i : \neg A_i \in l\} \cup \{A\} \succ \{\sim\neg A_i : \sim\neg A_i \in r\} \cup \{\sim p_i : \sim p_i \in r\}]$$

   is unprovable. If not then by $[\sim\neg R]$ and $[K]$ we would have (l, r) provable contradicting saturation. So extend $[\Gamma' \succ \Delta']$ to a saturated set $(l', r') \in W$. By definition $(l, r) \leq (l', r')$. As $A \in l$ by 1 and the hypothesis of induction, we have $(l', r') \Vdash_t A$. So we have $(l, r) \not\Vdash_s \neg A$ as desired.

3. Suppose that $\neg A \in r$. We show that $[\Gamma' \succ \Delta'] =$

$$[\{p_i : p_i \in l\} \cup \{\neg A_i : \neg A_i \in l\} \succ \{\sim\neg A_i : \sim\neg A_i \in r\} \cup \{\sim p_i : \sim p_i \in r\} \cup \{\sim A\}]$$

is unprovable. If not then $[\Gamma' \succ \Delta']$ is provable, and so by $[\neg R]$ and $[K]$ we have $[l \succ r]$ provable, contradicting saturation! So $[\Gamma' \succ \Delta']$ is unprovable and so we can extend it to a saturated set $(l', r') \in W$. By construction, $(l, r) \le (l', r')$ and $\sim A \in r'$ and so, as $dmc(\sim A) = dmc(\neg A) - 1$, by 4 and the inductive hypothesis, we have $(l', r') \Vdash_s A$, and so $(l, r) \nVdash_t \neg A$.

4. Suppose $\sim\neg A \in r$. Then consider a point (l', r') where $(l, r) \le (l', r')$. By definition we have $\sim\neg A \in r'$ and so by Definition 7.14 $A \in r'$. So by 3 and the inductive hypothesis, we have $(l', r') \nVdash_t A$. So it follows that $(l, r) \Vdash_s \neg A$.

Now for the **De Morgan negation of negation**

1. Suppose $\sim\neg A \in l$. Then we can proceed as in case 2 for negation to conclude that $(l, r) \nVdash_s \neg A$, and thus $(l, r) \Vdash_t \sim\neg A$, as desired.

2. Suppose $\sim\sim\neg A \in l$. Then by Definition 7.9, we have $\neg A \in l$, and so by case 1 for negation above $(l, r) \Vdash_t \neg A$ and so $(l, r) \nVdash_s \sim\neg A$, as desired.

3. Suppose that $\sim\neg A \in r$. Then by case 4 for negation, we have $(l, r) \Vdash_s \neg A$, and so $(l, r) \nVdash_t \sim\neg A$, as desired.

4. Suppose that $\sim\sim\neg A \in r$. Then by Definition 7.10, we have $\neg A \in r$. So by case 3 for negation above we have $(l, r) \nVdash_t \neg A$, and so $(l, r) \Vdash_s \sim\neg A$, as desired. $\square$

Theorem 14 (Completeness for $\mathcal{SCLP}^\sim$)**.** *If a sequent is valid in* $\mathsf{cLP}^\sim$*, then it is derivable in* $\mathcal{SCLP}^\sim$*.*

Again, we have bought our completeness at a cost, in this case the cost of constructivity.

Lemma 9. $\vdash_{\mathsf{cLP}^\sim}$ *is not constructive.*

Proof. In this case, we do not have the disjunction property, as we have $\vdash_{\mathsf{cLP}^\sim} p \vee \sim p$ but neither $\vdash_{\mathsf{cLP}^\sim} p$ nor $\vdash_{\mathsf{cLP}^\sim} \sim p$. $\square$

It is easy to show though that $\vdash_{\mathsf{cLP}^\sim}$ is still paraconsistent, again using essentially the same proof as we used in §2 but conflating the distinction between the negation of cLP and our De Morgan negation. So we have bought completeness for cLP , but this time at the cost of it not longer being constructive in the extended language.

5   Conclusion

In this paper, we introduced a logic cLP, the constructive logic of paradox, and showed that it bore strong affinities to both intuitionistic logic and the logic of paradox. We have been unable to prove completeness for this logic in the standard signature, although we were able to prove completeness for two different extensions of our base language. In the first case, we showed that we could prove completeness if we enriched our language by the addition of a 'strict truth' operator, moving to a signature more similar to those used by Woodruff in [13, 14], although at the cost of paraconsistency of the logic in the extended language. In the second case, we showed that we could prove completeness if we enriched the language by the addition of a De Morgan negation, although this time at the cost of the constructivity of the logic in the extended language.

What should we make of this seeming incompleteness of cLP in the base language? Incompleteness has its source in the semantics for a logic being too expressive for the deductive apparatus of a logic. One result which is very suggestive that this is what is going on in the present setting is the following.

Proposition 2. *The logics* cLP$^\mathsf{T}$ *and* cLP$^\sim$ *are both conservative extensions of* cLP.

Proof. This result follows from the fact that models for cLP$^\mathsf{T}$ and cLP$^\sim$ are just models for cLP with the addition of new truth-in-a-model clauses for the new operators. $\square$

Our inability to prove completeness in the standard signature, combined with the fact that we are able to prove it in a semantically-conservative extension of our language, suggests that our semantics is too expressive for us to be able to treat it in the standard signature. This thought is reinforced by the fact that the semantics for non-transitive constructive logics seem to need to resort to either partial semantic clauses (as is done in [11]), or labelled proof-systems (as is done in [7]). We leave further consideration of this issue for future work.

References

[1] Jonas R. B. Arenhart and Hitoshi Omori. On Woodruff's constructive nonsense logic. *Studia Logica*, January 2024. doi: 10.1007/s11225-023-10092-z.

[2] Eduardo Barrio, Lucas Rosenblatt, and Diego Tajer. The logics of strict-tolerant logic. *Journal of Philosophical Logic*, 44(5):551–571, 2015.

[3] Sankha S. Basu and Sayantan Roy. Generalized explosion principles. *Unpublished Manuscript*, 2024.

[4] Pablo Cobreros, Paul Egré, Ellie Ripley, and Robert van Rooij. Tolerant, classical, strict. *Journal of Philosophical Logic*, 41:347–385, 2012.

[5] Bogdan Dicher and Francesco Paoli. ST, LP, and tolerant metainferences. In C. Başkent and T. M. Ferguson, editors, *Graham Priest on Dialetheism and Paraconsistency*, pages 383–407. Springer, 2019.

[6] Thomas Macaulay Ferguson. A computational interpretation of conceptivism. *Journal of Applied Non-Classical Logics*, 24(4):333–367, 2014.

[7] Rohan French. Invitation to constructive nonreflexive and nontransitive logics. *Unpublished Manuscript*, 2024.

[8] Jean-Yves Girard. *Three-Valued Logic and Cut-Elimination: The Actual Meaning of Takeuti's Conjecture*. Państwowe Wydawn. Naukowe, Warszawa, 1976.

[9] Graham Priest. The logic of paradox. *Journal of Philosophical Logic*, 8:219–241, 1979.

[10] Krister Segerberg. A contribution to nonsense-logics. *Theoria: A Swedish Journal of Philosophy and Psychology*, 31(3):199–217, 1965.

[11] Morgan Thomas. A Kripke-style semantics for paradox-tolerant, nontransitive intuitionistic logic. *Unpublished Manuscript*, 2014.

[12] Masanobu Toyooka and Katsuhiko Sano. Combining intuitionistic and classical propositional logic: Gentzenization and Craig interpolation. *Studia Logica*, January 2024. doi: 10.1007/s11225-023-10067-0.

[13] Peter W. Woodruff. *Foundations of Three-Valued Logic*. PhD thesis, University of Pittsburgh, Pittsburgh, 1969.

[14] Peter W. Woodruff. On constructive nonsense logic. In S. Halldén, editor, *Modality, Morality and Other Problems of Sense and Nonsense: Essays Dedicated to Sören Halldén*, pages 192–205. GWK Gleerup Bokforlag, 1973.

Received 24 December 2025

State Semantics for Substructural Logics

Edwin Mares

Te Herenga Waka – Victoria University of Wellington
Wellington, New Zealand
<edwin.mares@vuw.ac.nz>

Abstract

This paper presents semantics for a wide range of substructural logics and shows soundness and completeness for these systems. The logics characterised by the semantics include both systems that do not contain distribution, such as Full Lambek calculus with negation, Multiplicative–Additive Linear Logic, and Lattice-**R**, as well as those that are distributive, such as the mainstream relevant logics and **R**-Mingle. The semantics is an indexical semantics, in which truth is relativised to points.

2020 *Mathematics Subject Classification.* Primary: 03B47, Secondary: 03B35.

1 Introduction

This essay presents a general semantics for a wide range of substructural logics. It is an indexical or pointed semantics; it relativises the truth of formulae to points. In its treatment of implication, it borrows heavily from the work of Alasdair Urquhart [24] and Kit Fine [10]. It uses an application operator, $\circ$, to give a truth condition for implication. Where a and b are points in the frame, $a \circ b$ is the result of applying a to b and is itself a point in the frame. An implication $A \to B$ is satisfied by a if and only if for all b that satisfies A, then B is satisfied by $a \circ b$.

The treatment of negation is borrowed from Garrett Birkoff's treatment of lattice complementation [6] and Robert Goldblatt's semantics for orthologic [12]. There is an incompatibility relation, $\perp$, on points. A negation $\neg A$ is true at a point, a, if and only if all points at which A holds are incompatible with a. In the present paper, for each point a there is a "minimal" point $-a$ which is incompatible with a. The minus operator in the present semantics can be seen as a dual of the star operator from the

I am grateful to Katalin Bimbó and an anonymous referee for extremely useful corrections and comments and to Rob Goldblatt for saving me from making a very embarrassing mistake.

Routley–Meyer semantics for relevant logics [22, 21] and similar conditions placed on it make similar theses valid. For example, the condition that $--a = a$ forces the double negation equivalence, $\neg\neg A \leftrightarrow A$, which has previously been difficult to satisfy in semantics for non-distributive logics such as linear logic (see, e.g., [1]).

The points themselves are called *states*. "State" here is a place-holder for other more specific notions. For example, in the semantics for the relevant logic **R**, a state might be taken to be a piece of information, in the sense of Urquhart [24].[1] Then the expression, '$a \vDash A$', in the semantical theory can be taken to mean that a contains A as information. In my book [16], I use a similar semantics in which points are interpreted as theories (in the sense of scientific theories) to characterise the logic **E** of relevant entailment.[2] In that book, '$a \vDash A$' means that the theory a postulates A. To treat other logics, perhaps, one might read states as properties of worlds. If we do so, then '$a \vDash A$' can be understood as saying that A is true in any world that instantiates a.

There is some similarity between the present semantics and some variants of truthmaker semantics, especially Mark Jago's semantics for relevant logic [15]. I am, however, quite wary of claiming that the present semantics can be interpreted as a truthmaker semantics. The problem is that the semantics does not satisfy the so-called "disjunction thesis" (see [19]). The disjunction thesis states that the truthmaker for each disjunction is one of its disjuncts. Intuitively, the potential truthmakers for the proposition that either Susan is at home or she is walking Nova are that Susan is at home and that Susan is walking Nova.

Instead, in my semantics disjunction is treated by means of an intersection operator on states, $\sqcap$. I borrow the idea of using an intersection operator from Lloyd Humberstone [14], Kosta Došen [8], Hiroakira Ono [18] and Ross Brady [7].[3] The state $a \sqcap b$ satisfies whatever formulae are satisfied by both a and b. Where $a \vDash A$ and $b \vDash B$, $a \sqcap b \vDash A \vee B$.[4]

The novelty of the present semantics is how generally it can be applied.[5] The base

---

[1]Urquhart [24] only gives a semantics for the implication fragment of **R** (which can easily be extended to treat conjunction as well).

[2]My interpretation develops the notion of a theory as a point in the semantics for relevant logic as presented in Fine [10].

[3]Mathematically similar means of representing non-distributing disjunction can also be found in Urquhart's representation of lattices [25] and Katalin Bimbó has dualised semantics for relevant logics in Bimbó [3] (for a more complete discussion, see [2]).

[4]Jago [15] uses a forcing-like mechanism to deal with disjunction and so, strictly speaking, he rejects the disjunction thesis in his truthmaker semantics. Perhaps, then, I can reject the disjunction thesis and still call my model theory a truthmaker semantics after all.

[5]The theory of residuated lattices [11] is a general semantics for *positive* substructural logics. It does not include a treatment of De Morgan negation, unlike the present semantics. It would be

system, **LB** for "lattice-**B**", is a very simple logic — it is a fragment of Full Lambek Logic (**FL**). A list of standard axioms and corresponding semantic postulates is given to extend **LB** to other systems. The logics treated include systems in which the law of distribution does not hold and those in which distribution does hold. Among those in which distribution does not hold are negation extensions of **FL** and **BCK**, Multiplicative–Additive Linear Logic (**MALL**), and Lattice-**R** (**LR**). Among those in which distribution does hold are the mainstream relevant logics (**B**, **DJ**, **DK**, **RW**, **R**, and **E**) and **R**-Mingle (**RM**). All of these logics are formulated as having negation, implication, disjunction, conjunction, and fusion. Fusion is required, even though relevant and substructural logicians sometimes reject it, because it is needed for the present paper's strategy of proving completeness. I also explain how to extend the appropriate systems with backwards implication ($\leftarrow$).

The plan of the paper is as follows. In section 2, I develop a state-based model theory for the logic, **LB**. In section 3, an axiomatisation of **LB** is given and in section 4, soundness is proven. In section 5, the changes to the semantics that enable the modelling of distributive logics are given. These changes include a change to the frame and a change to the satisfaction condition for disjunction. In section 6, frame conditions that correspond to standard implication and negation axioms (and the law of excluded middle) are given. In various combinations, adding these axioms and/or distribution to **LB** produces logics such as **MALL**, **LR**, **BCK$^\neg$**, **B**, **DJ**, **DK**, **RW**, **R**, **RM**, and so on. In section 6.1, semantic conditions are explained for the addition of a second implication connective ($\leftarrow$) the addition of which produces **FL$^\neg$** and some of its extensions.

In section 7, completeness is proven. In sections 7.1 and 7.2, a canonical model is constructed. The states of the canonical model are not arbitrary theories, but rather principal theories, that is, for each formula A, the theory $[A]_L$ is in the canonical model, where $[A]_L$ is the set of all the formulae that A entails, according to the logic L. The use of principal theories, rather than all theories, enables me to prove that every proposition in the canonical model has a least element.

2 Models

I begin by producing a model theory for the base logic, **LB**. **LB** does not contain the thesis of the distribution of conjunction over disjunction. In section 5, I set out the additional constraints on models for logics that have distribution. The frame and the constraints on assignments to propositional variables work together to ensure

---

very interesting, however, to discover the exact formal relationship between state semantics and residuated lattices.

that every formula expresses a proposition that has a least state. The treatment of negation, in particular, requires this.

2.1 Frames

The underlying algebraic structure of a state frame is a *meet semilattice*, and before I define the class of frames I define this algebraic notion. A meet semilattice is a pair $\langle S, \sqcap \rangle$, such that S is a non-empty set and $\sqcap$ is a binary operator on S that satisfies the following equations. For all $a, b, c \in S$:

1. $a \sqcap a = a$ (idempotence);

2. $a \sqcap b = b \sqcap a$ (commutativity);

3. $a \sqcap (b \sqcap c) = (a \sqcap b) \sqcap c$ (associativity).

In a meet semilattice, we can define an ordering relation $\leq$ such that

$$a \leq b \text{ if and only if } a \sqcap b = a.$$

The idempotence of $\sqcap$ entails that $\leq$ is reflexive. Suppose that $a \leq b$ and $b \leq c$. By definition, $a \sqcap b = a$ and $b \sqcap c = b$. By associativity, $a = a \sqcap b = a \sqcap (b \sqcap c) = (a \sqcap b) \sqcap c = a \sqcap c$. Thus, $a \leq c$. Therefore, $\leq$ is transitive. Moreover, $\leq$ is antisymmetric. For suppose that $a \leq b$ and $b \leq a$. By commutativity, $a = a \sqcap b = b \sqcap a = b$.

I often appeal to the following fact:

Fact 1. *If $a \leq a'$ and $b \leq b'$ then $a \sqcap b \leq a' \sqcap b'$.*

Proof. Suppose that $a \leq a'$ and $b \leq b'$, i.e., $a = a \sqcap a'$ and $b = b \sqcap b'$. Then $a \sqcap b = (a \sqcap a') \sqcap (b \sqcap b')$. By associativity and commutativity, $a \sqcap b = (a \sqcap b) \sqcap (a' \sqcap b')$, i.e., $a \sqcap b \leq a' \sqcap b'$. $\square$

A *state frame* is a quintuple $\langle S, \Sigma, \circ, \sqcap, \bot \rangle$ such that S is a non-empty set (of states), $\Sigma \in S$, $\circ$ and $\sqcap$ are binary operators on S, and $\bot$ is a binary relation on S such that the following conditions hold:

1. $\langle S, \sqcap \rangle$ is a meet semilattice;

2. $\Sigma \circ a = a$;

3. if $a \leq a'$ and $b \leq b'$ then $a \circ b \leq a' \circ b'$;

4. $\bot$ is symmetrical;

5. if $a \perp b$ and $b \leq b'$ then $a \perp b'$;

6. where $a^\perp$ is the set of states b such that $a \perp b$, $a^\perp$ has a unique least member under $\leq$, called "$-a$";

7. $--a = a$;

8. for all states b, c, there is a unique least state under $\leq$, a, such that $c \leq a \circ b$;

9. if $a \leq b \circ c$ and $a \leq b' \circ c$ then $a \leq (b \sqcap b') \circ c$;

10. if $a \leq b \circ c$ and $a \leq b \circ c'$ then $a \leq b \circ (c \sqcap c')$;

11. $(a \sqcap a') \circ (b \sqcap b') \leq (a \circ b) \sqcap (a' \circ b')$;

12. $(a \sqcap b)^\perp = a^\perp \cap b^\perp$.

The incompatibility relation, $\perp$, is used to treat negation. It is borrowed from Robert Goldblatt's semantics for orthologic [12]. A negation, $\neg A$, holds at a state a if A holds only at states that are incompatible with a. Michael Dunn [9] adapts the incompatibility relation to the Routley–Meyer semantics for relevant logic. Dunn shows that the Routley star a^* of a world a, can be understood as the maximal world compatible with a. In the present semantics, this idea is dualised, that is, for each state a, there is a minimal state $-a$ that is incompatible with a.

The set of states is also closed under least upper bounds or "joins", as the following lemmas and theorem show.

Lemma 2. *If $a \leq b$ then $a^\perp \subseteq b^\perp$.*

Proof. Assume that $a \leq b$. By the definition of $\leq$, $a = a \sqcap b$. Thus, $a^\perp = (a \sqcap b)^\perp$. By semantic postulate 12, $a^\perp = a^\perp \cap b^\perp$, therefore $a^\perp \subseteq b^\perp$. $\qquad\square$

Lemma 3. *If $a \leq b$, then $-b \leq -a$.*

Proof. Suppose that $a \leq b$. By lemma 2, $a^\perp \subseteq b^\perp$, and so $-a \in b^\perp$, i.e., $b \perp -a$. Therefore $-b \leq -a$. $\qquad\square$

Theorem 4. *For all states a and b, $-(-a \sqcap -b)$ is the least upper bound of $\{a, b\}$.*

Proof. $-a \sqcap -b \leq -a$ and $-a \sqcap -b \leq -b$. By lemma 3, $--a \leq -(-a \sqcap -b)$ and $--b \leq -(-a \sqcap -b)$. Since $--a = a$ and $--b = b$, $-(-a \sqcap -b)$ is an upper bound of $\{a, b\}$.

To show that it is the least upper bound, let c be such that $a \leq c$ and $b \leq c$. By lemma 3, $-c \leq -a$ and $-c \leq -b$ and so $-c \leq -a \sqcap -b$. By lemma 3 and semantic postulate 7, $-(-a \sqcap -b) \leq c$. Thus, $-(-a \sqcap -b)$ is the least upper bound of $\{a, b\}$. $\qquad\square$

Theorem 4 enables me to define a join operator on states, $\sqcup$:

$$a \sqcup b \ =_{df} \ -(-a \sqcap -b)$$

The following lemma is used to prove that De Morgan's laws are valid.

Lemma 5. *For all states a, b, c, $b \sqcap c \leq a$ if and only if $a \perp -b \sqcup -c$.*

Proof. $\Rightarrow$ Suppose that $b \sqcap c \leq a$. By lemma 3, $-a \leq -(b \sqcap c)$ and so $a \perp -(b \sqcap c)$. By the definition of $\sqcup$ and semantic postulate 7, $a \perp -b \sqcup -c$.

$\Leftarrow$ Suppose that $a \perp -b \sqcup -c$. By the symmetry of $\perp$, $-b \sqcup -c \perp a$. By semantic postulate 6, $-(-b \sqcup -c) \leq a$. By the definition of $\sqcup$, $- - (- - b \sqcap - - c) \leq a$ and therefore, by semantic postulate 7, $b \sqcap c \leq a$. $\qquad\qquad\square$

2.2 Truth

The language contains a non-empty set of propositional variables, $\mathcal{A} = \{p, q, r, \ldots\}$, and the set of well formed formulae is defined in Bakus–Naur form as follows:

$$\mathcal{A} \mid t \mid \neg B \mid (B \to B) \mid (B \wedge B) \mid (B \vee B) \mid (B \circ B)$$

I also define the equivalence $A \leftrightarrow B$ as $(A \to B) \wedge (B \to A)$.

A *state model* is a pair of a state frame and a variable assignment v. Variable assignments are constrained by the following conditions. For every propositional variable, p,

1. $v(p)$ is closed upwards under $\leq$, that is, for all states a and b, if $a \leq b$ and $a \in v(p)$, then $b \in v(p)$;

2. for all states a and b, if both a and b are in $v(p)$, then $a \sqcap b$ is in $v(p)$;

3. $v(p)$ has a unique minimal state under $\leq$.

Condition 2 is in fact redundant, but I appeal to it from time to time and so it is useful to state it. Algebraically, $v(p)$ is a principal filter on $\langle S, \sqcap \rangle$.

Each variable assignment v determines a satisfaction relation $\vDash_v$ that is recursively defined by the following truth clauses.

Truth Conditions

PV. $a \vDash_v p$ if and only if $a \in v(p)$;

$t.$ $a \vDash_v t$ if and only if $\Sigma \leq a$;

∧. $a \vDash_v A \wedge B$ if and only if $a \vDash_v A$ and $a \vDash_v B$;

∨. $a \vDash_v A \vee B$ if and only if there are states b and c such that $b \vDash_v A$, $c \vDash_v B$, and $b \sqcap c \leq a$;

∘. $a \vDash_v A \circ B$ if and only if there are states b and c such that $b \vDash A$, $c \vDash B$, and $b \circ c \leq a$;

→. $a \vDash_v A \to B$ if and only if for all states $b \vDash_v A$, $a \circ b \vDash_v B$;

¬. $a \vDash_v \neg A$ if and only if, for all states b, if $b \vDash_v A$ then $a \perp b$.

I use the notation '$[\![A]\!]_v$' to denote the set of states in which A is true (i.e., the truth set of A). Using this notation, the truth condition for negation can be rewritten as $a \vDash_v \neg A$ if and only if $[\![A]\!]_v \subseteq a^{\perp}$. In what follows, I do not discuss relationships between variable assignments, and so subscripted 'v' play no role and I drop them.

A formula A is valid on a model if and only if, for the Σ of that model, $\Sigma \vDash A$. A is valid on a class of models if and only if it is valid on each of the models in that class.

The hereditariness and semantic entailment lemmas below are crucial to the proofs that follow.

Lemma 6 (Hereditariness). *For all formulae A and all states a, b, if $a \leq b$ and $a \vDash A$, then $b \vDash A$.*

Proof. Assume $a \leq b$ and $a \vDash A$. The proof is by induction on the complexity of A.

Case 1. A is a propositional variable, p. This case follows from the definition of variable assignments and the truth condition for propositional variables.

Case 2. A is t. Suppose that $a \vDash t$. By the truth condition for t, $\Sigma \leq a$, hence $\Sigma \leq b$ and so $b \vDash t$.

Case 3. A is a conjunction, $B \wedge C$. By the inductive hypothesis, $b \vDash B$ and $b \vDash C$. By the truth condition for conjunction, $b \vDash B \wedge C$.

Case 4. A is a disjunction, $B \vee C$. By the truth condition for disjunction, there are $c, d \in S$, $c \vDash B$, $d \vDash C$, and $c \sqcap d \leq a$. By assumption, $c \sqcap d \leq b$ and so $b \vDash B \vee C$.

Case 5. A is an implication, $B \to C$. By the truth condition for implication, if $c \vDash B$, then $a \circ c \vDash C$. By semantic postulate 3, $a \circ c \leq b \circ c$ and so, by the inductive hypothesis, $b \circ c \vDash C$. Thus, by the truth condition for implication, $b \vDash B \to C$.

Case 6. A is a fusion, $B \circ C$. By the truth condition for fusion, there are $c, d \in S$ such that $c \circ d \leq a$, $c \vDash B$ and $d \vDash C$. By the transitivity of $\leq$, $c \circ d \leq b$ and so $b \vDash B \circ C$.

Case 7. A is a negation, $\neg B$. By lemma 2, $a^{\perp} \subseteq b^{\perp}$. Thus, if $[\![B]\!] \subseteq a^{\perp}$, then $[\![B]\!] \subseteq b^{\perp}$. $\square$

Lemma 7 (Semantic Entailment). $\Sigma \vDash A \to B$ *if and only if* $[\![A]\!] \subseteq [\![B]\!]$.

Proof. $\Rightarrow$ Suppose that $\Sigma \vDash A \to B$ and $a \vDash A$. By semantic postulate 2, $a = \Sigma \circ a$. By the truth condition for implication, $\Sigma \circ a \vDash B$ and so $a \vDash B$.

$\Leftarrow$ Suppose that $[\![A]\!] \subseteq [\![B]\!]$ and that $a \vDash A$. Then $a \vDash B$, but $a = \Sigma \circ a$. Thus, by the truth condition for implication, $\Sigma \vDash A \to B$. $\square$

Lemma 8. *For all formulas A, $[\![A]\!]$ contains a unique least state under $\leq$.*

Proof. By induction on the complexity of A.

Case 1. A is a propositional variable, p. Then $[\![p]\!] = v(p)$ and, by a condition on models, $v(p)$ has a unique least state.

Case 2. A is t. Σ is the unique least state in $[\![t]\!]$.

Case 3. A is a conjunction, $B \wedge C$. By the inductive hypothesis, $[\![B]\!]$ and $[\![C]\!]$ both have unique minimal points. Let us call them b and c respectively. By theorem 4, $\{b, c\}$ has a least upper bound $b \sqcup c$. By lemma 6 and the truth condition for conjunction, $b \sqcup c \vDash B \wedge C$. Since $b \sqcup c$ is a the least upper bound of $\{b, c\}$ and any other state that satisfies both B and C must be greater than both b and c, $b \sqcup c$ is the least state in $[\![B \wedge C]\!]$.

Case 4. A is a disjunction, $B \vee C$. By the inductive hypothesis, there are unique least members of $[\![B]\!]$ and $[\![C]\!]$. Let us call them b and c respectively. By the truth condition for disjunction, $b \sqcap c$ is the unique least state that satisfies $B \vee C$.

Case 5. A is a negation, $\neg B$. By the inductive hypothesis, $[\![B]\!]$ has a unique least member, b. By semantic postulate 6, $b^{\perp}$ has a unique least member, $-b$. $\perp$ is symmetric, and so $-b \perp b$. By semantic postulate 5, $[\![B]\!] \subseteq -b^{\perp}$, i.e., $-b \vDash \neg B$. To show that $-b$ is the minimal state in $[\![\neg B]\!]$, suppose that $c \vDash \neg B$. Then $c \perp b$ and so $b \perp c$, but $-b$ is the least state in $b^{\perp}$, so $-b \leq c$. Thus, $-b$ is the unique least state that satisfies $\neg B$.

Case 6. A is an implication, $B \to C$. By the inductive hypothesis, $[\![B]\!]$ has a unique minimal point, b. If $a \vDash B \to C$, then $a \circ b \vDash C$. Let c be the least member of $[\![C]\!]$. Then $c \leq a \circ b$. By semantic postulate 8, there is a least a' such that $c \leq a' \circ b$. Thus, a' is the unique least member of $[\![B \to C]\!]$.

Case 7. A is a fusion, $B \circ C$. By the inductive hypothesis, there are c and d that are $\leq$-minimal points satisfying B and C respectively. Hence, by semantic postulate 3 and the truth condition for fusion, $c \circ d$ is the unique least state in $[\![B \circ C]\!]$. $\square$

To highlight the relationship between negation and the $-$ operator in the semantics, I extract what was proven in lemma 8 case 5 as a fact of its own:

Fact 9. *If a is the unique least state in $[\![A]\!]$ then $-a$ is the unique least state in $[\![\neg A]\!]$.*

Lemma 10. *If a is the unique least state in $[\![\neg A]\!]$ then $-a$ is the unique least state in $[\![A]\!]$.*

Proof. Suppose that a is the unique least state in $[\![\neg A]\!]$. Let b be the minimal state in $[\![A]\!]$. Then $a \perp b$ and so $b \perp a$, hence $-b \leq a$. But by lemma 9, $-b$ is the least state in $[\![\neg A]\!]$ and so $-b = a$, thus $-a = --b = b$. Therefore, $-a$ is the minimal state in $[\![A]\!]$. $\square$

Lemma 11. *If $a \vDash A$ and $b \vDash A$, then $a \sqcap b \vDash A$.*

Proof. By induction on the complexity of A.

Case 1. A is a propositional variable, p. Follows by the conditions on variable assignments.

Case 2. A is t. Assume that $a \vDash t$ and $b \vDash t$. Then $\Sigma \leq a$ and $\Sigma \leq b$. Since $a \sqcap b$ is the greatest lower bound of a and b, $\Sigma \leq a \sqcap b$ and so $a \sqcap b \vDash t$.

Case 3. A is a conjunction, $B \wedge C$. Follows by the inductive hypothesis and truth condition for conjunction.

Case 4. A is a disjunction, $B \vee C$. Assume that $a \vDash B \vee C$ and $b \vDash B \vee C$. Then both $c \sqcap d \leq a$ and $c \sqcap d \leq b$, where c and d are the minimum states in $[\![B]\!]$ and $[\![C]\!]$ respectively. Since $a \sqcap b$ is the greatest lower bound of $\{a, b\}$, $c \sqcap d \leq a \sqcap b$. Therefore, by the truth condition for disjunction, $a \sqcap b \vDash B \vee C$.

Case 5. A is an implication $B \to C$. Assume that $a \vDash B \to C$ and $b \vDash B \to C$. Now suppose that c and d are the minimum states in $[\![B]\!]$ and $[\![C]\!]$ respectively. So $d \leq a \circ c$ and $d \leq b \circ c$. By semantic postulate 9, $d \leq (a \sqcap b) \circ c$, and so $a \sqcap b \vDash B \to C$.

Case 6. A is a negation, $\neg B$. Assume that $a \vDash \neg B$ and $b \vDash \neg B$. Then $[\![B]\!] \subseteq a^{\perp}$ and $[\![B]\!] \subseteq b^{\perp}$, so $[\![B]\!] \subseteq a^{\perp} \cap b^{\perp}$. By semantic postulate 12, $a^{\perp} \cap b^{\perp} = (a \sqcap b)^{\perp}$, and so $a \sqcap b \vDash \neg B$.

Case 7. A is a fusion, $B \circ C$. Assume that $a \vDash B \circ C$ and $b \vDash B \circ C$. Then there are states c, d such that $c \vDash B$ and $d \vDash C$ and $c \circ d \leq a$, and states c', d' such that $c' \vDash B$ and $d' \vDash C$ and $c' \circ d' \leq b$. By the inductive hypothesis, $c \sqcap c' \vDash B$ and $d \sqcap d' \vDash C$. $a \sqcap b$ is a greatest lower bound of a and b, so $(c \circ d) \sqcap (c' \circ d') \leq a \sqcap b$. By semantic postulate 11, $(c \sqcap c') \circ (d \sqcap d') \leq (c \circ d) \sqcap (c' \circ d')$ and so $(c \sqcap c') \circ (d \sqcap d') \leq a \sqcap b$. By the inductive hypothesis, $c \sqcap c' \vDash B$ and $d \sqcap d' \vDash C$, so by the truth condition for fusion, $a \sqcap b \vDash B \circ C$. $\square$

3   The Base Logic LB

The base logic is the logic that is sound and complete over the class of state models. I call this logic **LB** for "Lattice-**B**". It is Routley and Meyer's base relevant logic

B minus the distribution axiom.[6] **LB** is also the logic **FL**, without backwards implication (see section 6.1) but with negation. Like **B**, the implication fragment of **LB** is just the set of formulae of the form $A \to A$.

Axiom Schemata

1. $A \to A$

2. $A \to (A \vee B);$ $\qquad B \to (A \vee B)$

3. $(A \to C) \to ((A \wedge B) \to C);$ $\qquad (B \to C) \to ((A \wedge B) \to C)$

4. $((A \to B) \wedge (A \to C)) \to (A \to (B \wedge C))$

5. $((A \to C) \wedge (B \to C)) \to ((A \vee B) \to C)$

6. $(A \vee B) \leftrightarrow \neg(\neg A \wedge \neg B)$

7. $\neg\neg A \to A$

8. t

Rules

$$\frac{\vdash A \to B \quad \vdash A}{\vdash B} \qquad \frac{\vdash A \quad \vdash B}{\vdash A \wedge B} \qquad \frac{\vdash A \to B \quad \vdash C \to D}{\vdash (B \to C) \to (A \to D)}$$

$$\frac{\vdash A \to \neg B}{\vdash B \to \neg A} \qquad \frac{\vdash A \to (B \to C)}{\vdash (A \circ B) \to C} \qquad \frac{\vdash (A \circ B) \to C}{\vdash A \to (B \to C)} \qquad \frac{\vdash A}{\vdash t \to A}$$

The rules that do not explicitly concern t are modus ponens, adjunction, affixing, contraposition, importation, and exportation respectively. The final rule can be thought of as a form of necessitation, where '$t \to A$' is read as saying that A follows from the conjunction of the theorems of the logic.

The following fact collects together theorems of the logic that are used in proving metatheorems:

Fact 12. *(1)* $\vdash A \to (B \to (A \circ B));$ *(2)* $\vdash A \to \neg\neg A;$ *(3)* $\vdash ((A \vee B) \to C) \to ((A \to C) \wedge (B \to C)).$

Proof. (1) By axiom 1, $\vdash (A \circ B) \to (A \circ B)$. By exportation, $\vdash A \to (B \to (A \circ B))$.

(2) By axiom 1, $\vdash \neg A \to \neg A$. By contraposition, $\vdash A \to \neg\neg A$.

(3) By axiom 2, $\vdash A \to (A \vee B)$. By affixing, $\vdash ((A \vee B) \to C) \to (A \to C)$. Similarly, $\vdash ((A \vee B) \to C) \to (B \to C)$. So, by adjunction and axiom 4, $\vdash ((A \vee B) \to C) \to ((A \to C) \wedge (B \to C))$. $\qquad\square$

---

[6]Note that the version of **B** I use is not the one that has excluded middle as an axiom, which appears in some early papers by Routley and Meyer and in Fine's [10]. But excluded middle can be made valid using the postulates of section 6 below.

I use the replacement property (stated in the following theorem) very often in this paper, sometimes without acknowledgement.

Theorem 13 (Replacement Property). *The following rule is derivable in* **LB***:*

$$\frac{\vdash A \leftrightarrow B}{\vdash C \leftrightarrow C'}$$

where C' results by the replacement in C of zero or more occurrences of A with B.

Proof. Assume that $\vdash A \leftrightarrow B$. The proof is by induction on the complexity of C.

Case 1. C is a propositional variable, p. Subcase (a) A is p. Then, by assumption, $\vdash p \leftrightarrow B$ and this is what needs to be proven. Subcase (b). A is not p. Then A does not occur in C and so C' is just C, i.e., p, and by axiom 1 and adjunction $\vdash p \leftrightarrow p$.

Case 2. C is t. This case is similar to case 1.

Case 3. C is a conjunction, $D \wedge E$. By the inductive hypothesis, $D \leftrightarrow D'$ and $E \leftrightarrow E'$ are both theorems. By axiom 3 and modus ponens, $(D \wedge E) \to D'$ and $(D \wedge E) \to E'$ are both provable. By adjunction and axiom 4 and modus ponens, $(D \wedge E) \to (D' \wedge E')$. The other direction follows in the same manner.

Case 4. C is a disjunction, $D \vee E$. By the inductive hypothesis, $D \leftrightarrow D'$ and $E \leftrightarrow E'$ are both theorems. By the usual fiddling and adjunction, $(D \to (D' \vee E')) \wedge (E \to (D' \vee E'))$ and, by axiom 5 and modus ponens, $(D \vee E) \to (D' \vee E')$. The other direction is proven in the same way.

Case 5. C is a negation, $\neg D$. By the inductive hypothesis, $\vdash D \leftrightarrow D'$ and so $\vdash D \to D'$. By fact 12(2), $\vdash D' \to \neg\neg D'$ and so, by affixing, $\vdash D \to \neg\neg D'$. By contraposition, $\vdash \neg D' \to \neg D$. The other direction follows in the same manner.

Case 6. C is an implication, $D \to E$. By the inductive hypothesis, $\vdash D \leftrightarrow D'$ and $\vdash E \leftrightarrow E'$ and so $\vdash D' \leftrightarrow D$ and $\vdash E \to E'$. We have the following application of the affixing rule:

$$\frac{\vdash D' \to D \qquad \vdash E \to E'}{\vdash (D \to E) \to (D' \to E')}$$

Therefore $\vdash (D \to E) \to (D' \to E')$. The converse direction follows in the same manner.

Case 7. C is a fusion, $D \circ E$. By the inductive hypothesis, $\vdash D \leftrightarrow D'$ and $\vdash E \leftrightarrow E'$, hence, by affixing, $\vdash (E' \to (D' \circ E')) \to (E \to (D' \circ E'))$ and $\vdash (D' \to (E' \to (D' \circ E'))) \to (D \to (E \to (D' \circ E')))$. But, $\vdash D' \to (E' \to (D' \circ E'))$, and so $\vdash D \to (E \to (D' \circ E'))$. By importation, $\vdash (D \circ E) \to (D' \circ E')$. The proof of the converse direction is similar. $\square$

4 Soundness

In this section, I prove that all of the axioms of **LB** are valid and that all of its rules preserve validity in the class of state models.

It is obvious that $A \to A$ is valid in the class of state models. The proof of the following lemma is easily done by appealing to the semantic entailment lemma.

Lemma 14. *The following schemata are valid in the class of state models: (a) $A \to (A \lor B)$; (b) $(A \to C) \to ((A \land B) \to C)$; (c) $((A \to C) \land (B \to C)) \to ((A \lor B) \to C)$.*

Theorem 15. $\neg\neg A \to A$ *is valid in the class of state models.*

Proof. Suppose that $a \vDash \neg\neg A$. By lemma 8, there is some a' that is the least state in $[\![\neg\neg A]\!]$, so $a' \leq a$. By lemma 10, $-a'$ is the least state in $[\![\neg A]\!]$ and $--a' = a'$ is the least state in $[\![A]\!]$. Therefore, by the hereditariness lemma, $a \vDash A$. By the semantic entailment lemma, $\neg\neg A \to A$ is valid on the model. $\square$

Theorem 16. *For any state a and any formulae A, B, $a \vDash A \lor B$ if and only if $a \vDash \neg(\neg A \land \neg B)$.*

Proof. $\Rightarrow$ Suppose that $a \vDash A \lor B$. Let b be the minimum state in $[\![A]\!]$ and c be the minimum state in $[\![B]\!]$. By the truth condition for disjunction, $b \sqcap c \leq a$. By semantic postulates 4 and 5, $a \perp -(b \sqcap c)$. By lemma 9, $-b$ and $-c$ are the minimum states in $[\![\neg A]\!]$ and $[\![\neg B]\!]$ respectively and so $-b \sqcup -c$ is the minimum state in $[\![\neg A \land \neg B]\!]$. But $-(-b \sqcup -c) = b \sqcap c$, hence, $a \perp -b \sqcup -c$. By semantic postulate 9 and the truth condition for negation, $a \vDash \neg(\neg A \land \neg B)$.

$\Leftarrow$ Suppose that $a \vDash \neg(\neg A \land \neg B)$. Let b be the minimum state in $[\![A]\!]$ and c the minimum state in $[\![B]\!]$. By lemma 9, $a \perp -b \sqcup -c$. By the symmetry of $\perp$, $-b \sqcup -c \perp a$ and, by semantic postulate 6, $-(-b \sqcup -c) \leq a$. But $-(-b \sqcup -c) = b \sqcap c$, and so $b \sqcap c \leq a$. By the truth condition for disjunction, $a \vDash A \lor B$. $\square$

Corollary 17. $(A \lor B) \leftrightarrow \neg(\neg A \land \neg B)$ *is valid in the class of state models.*

Proof. Follows from theorem 16 and the semantic entailment lemma. $\square$

Lemma 18. *(a) If $\Sigma \vDash A \to B$ and $\Sigma \vDash C \to D$, then $\Sigma \vDash (B \to C) \to (A \to D)$; (b) if $\Sigma \vDash A \to \neg B$ then $\Sigma \vDash B \to \neg A$; (c) if $\Sigma \vDash A \to (B \to C)$ then $\Sigma \vDash (A \circ B) \to C$; (d) if $\Sigma \vDash (A \circ B) \to C$ then $\Sigma \vDash A \to (B \to C)$; (e) if $\Sigma \vDash A$, then $\Sigma \vDash t \to A$.*

Proof. (a) Assume that $\Sigma \vDash A \to B$ and $\Sigma \vDash C \to D$. Also suppose that $a \vDash B \to C$ and $b \vDash A$. By the truth condition for implication, $\Sigma \circ b \vDash B$, and by semantic postulate 2, $b \vDash B$. So, $a \circ b \vDash C$. But $\Sigma \circ (a \circ b) \vDash D$ and so, $a \circ b \vDash D$. Thus, $a \vDash A \to D$. By the semantic entailment lemma, $\Sigma \vDash (B \to C) \to (A \to D)$.

(b) Suppose that $\Sigma \vDash A \to \neg B$. Suppose also that $a \vDash B$ and $b \vDash A$. I show that $a \perp b$. $\Sigma \circ b \vDash \neg B$, i.e., $b \vDash \neg B$, and so $b \perp a$. By the symmetry of $\perp$, $a \perp b$. Thus, $a \vDash \neg A$. By the semantic entailment lemma, $\Sigma \vDash B \to \neg A$.

(c) Suppose that $\Sigma \vDash A \to (B \to C)$ and that $a \vDash A \circ B$. By the truth condition for fusion, there are b, c such that $b \vDash A$, $c \vDash B$, and $b \circ c \leq a$. By the truth condition for implication, $\Sigma \circ b \vDash B \to C$ and so $b \vDash B \to C$ and $b \circ c \vDash C$. By hereditariness, $a \vDash C$. By the semantic entailment lemma, $\Sigma \vDash (A \circ B) \to C$.

(d) Suppose that $\Sigma \vDash (A \circ B) \to C$ and $a \vDash A$. Suppose also that $b \vDash B$. By the truth condition for fusion, $a \circ b \vDash A \circ B$. By the truth condition for implication, $\Sigma \circ (a \circ b) \vDash C$. By semantic postulate 2, $a \circ b = \Sigma \circ (a \circ b)$, so $a \circ b \vDash C$. Thus, $a \vDash B \to C$ and so, by the semantic entailment lemma, $\Sigma \vDash A \to (B \to C)$.

(e) Assume that $\Sigma \vDash A$. Suppose that $a \vDash t$. By the truth condition for t, $\Sigma \leq a$. By the hereditariness lemma, $a \vDash A$. Thus, by the semantic entailment lemma, $\Sigma \vDash t \to A$. $\qquad\square$

It follows from the fact that $\Sigma \circ \Sigma = \Sigma$ that modus ponens preserves truth in a model and the truth condition for conjunction implies that the truths at Σ are closed under conjunction. These facts together with the preceding lemmas, theorems, and corollaries show that the logic is sound. Thus, I can state the following theorem.

Theorem 19 (Soundness of **LB**). *If A is a theorem of* **LB***, then A is valid in the class of state models.*

5 Adding Distribution

The distribution thesis is

$$(A \wedge (B \vee C)) \to ((A \wedge B) \vee (A \wedge C))$$

One way to make distribution valid is to require that $\langle S, \sqcap, \sqcup \rangle$ satisfy the following distributive law:

$$a \sqcup (b \sqcap c) = (a \sqcup b) \sqcap (a \sqcup c)$$

It is easy to show, using the De Morgan properties of $\sqcap$ and $\sqcup$ and semantic postulate 7, that the other distributive law also holds:

$$a \sqcap (b \sqcup c) = (a \sqcap b) \sqcup (a \sqcap c)$$

I call state models that satisfy the distributive law, "distributive state models".[7]

Theorem 20. *The thesis of distribution is valid on the class of distributive state models.*

Proof. Suppose a is an arbitrary state in a distributive state model and that $a \vDash A \wedge (B \vee C)$ and so $a \vDash A \vee B$. By the truth condition for disjunction, $b \sqcap c \leq a$, where b is the minimal state of $[\![B]\!]$ and c is the minimal state of $[\![C]\!]$. Moreover, $a' \leq a$, where a' is the minimal sate of $[\![A]\!]$. By simple lattice properties, $a' \sqcup (b \sqcap c) \leq a$. By the distributive law, $(a' \sqcup b) \sqcap (a' \sqcup c) \leq a$. By hereditariness, $a' \sqcup b \vDash A$ and $a' \sqcup b \vDash B$ and so $a' \sqcup b \vDash A \wedge B$. Similarly, $a' \sqcup c \vDash A \wedge C$. Thus, $a \vDash (A \wedge B) \vee (A \wedge C)$. Thus, by the semantic entailment lemma, $(A \wedge (B \vee C)) \rightarrow ((A \wedge B) \vee (A \wedge C))$ is valid on the class of distributive state models. $\qquad\square$

The logic that results from the addition of the distribution axiom schema to the axiomatic basis for **LB** is Routley and Meyer's system **B**.

6 Extensions of LB and B

Other substructural logics (and classical logic) are produced by adding axioms to **LB** and **B**. The table below is a list of well-known schemata and the corresponding semantic postulates. (A list that gives the correlated structural proof-theoretic rules for most of these schemata can be found in [20, p. 26] and these conditions follow those set out for relevant groupoids in [17] and for theories in [10].)

The correspondences between the implication theses and their semantic postulates are quite well known (see, e.g., [10, 23]). In order to illustrate the technique of proving soundness, I prove that $(B \rightarrow C) \rightarrow ((A \rightarrow B) \rightarrow (A \rightarrow C))$ is valid in the class of state models in which $\circ$ is associative. Suppose that $a \vDash B \rightarrow C$ and $b \vDash A \rightarrow B$. I show that $a \circ b \vDash A \rightarrow C$. Suppose that $c \vDash A$. Then $b \circ c \vDash B$ and $a \circ (b \circ c) \vDash C$. By the B postulate, $(a \circ b) \circ c \vDash C$, and so, $a \circ b \vDash A \rightarrow C$ as required. Thus, $a \vDash (A \rightarrow B) \rightarrow (A \rightarrow C)$. Thus, by the semantic entailment lemma, $(B \rightarrow C) \rightarrow ((A \rightarrow B) \rightarrow (A \rightarrow C))$ is valid on the model.

---

[7]Another way to make distribution valid is to add the following semantic postulate and alter the truth condition for disjunction.

(Dist$\leq$) If $a = b \sqcap c$ and $a \leq a'$, then there are b', c' such that $b \leq b'$, $c \leq c'$ and $a' = b' \sqcap c'$.

This postulate, I think, is independently plausible. The new truth condition is

$\qquad a \vDash A \vee B$ if and only if there are $b \vDash A$, $c \vDash B$, and $b \sqcap c = a$.

The semantic postulate (Dist$\leq$) is added to ensure that hereditariness holds in models that satisfy distribution.

Thesis	Semantic Postulate	Name
$(A \to (A \to B)) \to (A \to B)$	$(a \circ b) \circ b \le a \circ b$	Contraction (W)
$((A \to B) \wedge A) \to B$	$a \circ a \le a$	Pseudo-MP (WI)
$(B \to C) \to ((A \to B) \to (A \to C))$	$a \circ (b \circ c) \le (a \circ b) \circ c$	Prefixing (B)
$(A \to B) \to ((B \to C) \to (A \to C))$	$a \circ (b \circ c) \le (b \circ a) \circ c$	Suffixing (B$'$)
$A \to ((A \to B) \to B)$	$a \circ b \le b \circ a$	Assertion (CI)
$(A \to (B \to C)) \to (B \to (A \to C))$	$(a \circ b) \circ c \le (a \circ c) \circ b$	Permutation (C)
$((A \to B) \wedge (B \to C)) \to (A \to C)$	$a \circ (a \circ b) \le a \circ b$	Conjunctive Syllogism
$A \to (B \to A)$	$a \le a \circ b$	Positive Paradox (K)
$A \to (B \to B)$	$b \le a \circ b$	K$'$
$A \to (A \to A)$	$a \le a \circ a$	Mingle
$(A \to \neg A) \to \neg A$	$a \circ b \perp b$ implies $a \perp b$	Reductio
$(A \to \neg B) \to (B \to \neg A)$	$a \circ b \perp c$ implies $a \circ c \perp b$	Contraposition
$A \vee \neg A$	$a \sqcap -a \le \Sigma$	Excluded Middle

Among the logics that we can construct from the schemata on this list are

- **MALL** = **LB** + B + B$'$ + C + Contraposition;

- **LR** = **MALL** + W + Reductio;

- **DJ** = **B** + Conjunctive Syllogism + Contraposition;

- **DK** = **DJ** + Excluded Middle;

- **RW** = **MALL** + Distribution;

- **R** = **LR** + Distribution;

- **RM** = **R** + Mingle;

- **PC** = **R** + K.

The logic **E** of relevant entailment can also be constructed, but it is somewhat more complicated than the others and needs explanation. In order to state the postulate of specialised assertion, I define a *denecessitation operator*:

$$\Box^{-} a \ =_{df} \ a \circ \Sigma$$

I call this a denecessitation operator because in **E** and weaker logics, one can interpret $t \to A$ as a form of necessitation of A. If $a \vDash t \to A$, then $\Box^{-} a \vDash A$. In order to characterise **E**, I need to add a condition corresponding to the relevant version of the modal T-axiom, i.e., $(t \to A) \to A$:

$$\Box^{-} a \le a$$

One apparently unique feature of **E** is that it contains schema of specialised assertion, viz.,

$$(A \to B) \to (((A \to B) \to C) \to C).$$

Specialised assertion can be made valid by adding a special commutativity postulate:

$$\Box^- a \circ \Box^- b \; = \; \Box^- b \circ \Box^- a$$

The commutativity postulate can be derived from the T-axiom, **B**, and **B**′. Here is the derivation of one direction (the derivation of the other is the same):[8]

1.	$((a \circ \Sigma) \circ (b \circ \Sigma)) \circ \Sigma$	$\leq$	$(a \circ \Sigma) \circ (b \circ \Sigma)$	T-axiom
2.	$(b \circ \Sigma) \circ ((a \circ \Sigma) \circ \Sigma)$	$\leq$	$((a \circ \Sigma) \circ (b \circ \Sigma)) \circ \Sigma$	1, B′
3.	$a \circ (\Sigma \circ \Sigma)$	$\leq$	$(a \circ \Sigma) \circ \Sigma$	B
4.	$a \circ \Sigma$	$=$	$a \circ (\Sigma \circ \Sigma)$	
5.	$(b \circ \Sigma) \circ (a \circ \Sigma)$	$\leq$	$(a \circ \Sigma) \circ (b \circ \Sigma)$	$2, 3, 4$

The incompatibility treatment of negation is not usually combined with an application operator. Because these proofs are not as widely known, I give proofs of the soundness of the reductio principle and contraposition.

Fact 21. *If a state model satisfies the reductio postulate, then it makes valid* $(A \to \neg A) \to \neg A$.

Proof. Assume that a model satisfies, for all $a, b \in S$, if $a \circ b \perp b$, then $a \perp b$. Suppose that $a \vDash A \to \neg A$ and that $b \vDash A$. By the truth condition for implication, $a \circ b \vDash \neg A$ and so $a \circ b \perp b$. By the reductio postulate, $a \perp b$. Thus, $a \vDash \neg A$. By the semantic entailment lemma, $(A \to \neg A) \to \neg A$ is valid in the class of state models that satisfy reductio. $\Box$

Fact 22. *If a state model satisfies the contraposition postulate, then it makes valid* $(A \to \neg B) \to (B \to \neg A)$.

Proof. Assume that the model satisfies, for all $a, b, c \in S$, if $a \circ b \perp c$ then $a \circ c \perp b$. Suppose that $a \vDash A \to \neg B$ and $b \vDash B$. I show that $a \circ b \vDash \neg A$. Suppose that $c \vDash A$. Then $a \circ c \vDash \neg B$ and $a \circ c \perp b$. By the contraposition condition, $a \circ b \perp c$ and so $a \circ b \vDash \neg A$ as required. By the truth condition for implication, $a \vDash B \to \neg A$ and by the semantic entailment lemma, $(A \to \neg B) \to (B \to \neg A)$ is valid in the model. $\Box$

---

[8]I am grateful to Katalin Bimbó for this derivation.

The proof that the law of excluded middle is made valid by the condition that for all states a, $a \sqcap -a \leq \Sigma$ is extremely easy and is omitted.

Some combinations of these theses added to **LB** yield logics that have the distribution thesis as a theorem. In particular **LB+WI+K+K$'$** contains distribution. A proof-theoretic derivation can be found in [20, p. 36]. The following is a model theoretic argument that even with the non-distributive truth condition for disjunction, a model that satisfies those postulates makes distribution valid.

Fact 23. *If a state model satisfies the postulates for* WI, K, *and* K$'$, *it makes distribution valid.*

Proof. Suppose that $a \vDash A \wedge (B \vee C)$. Let b be the minimal state in $[\![B]\!]$ and c be the minimal state in $[\![C]\!]$. By the K postulate and hereditariness, $a \circ b \vDash A$ and by the K$'$ postulate and hereditariness, $a \circ b \vDash B$. Thus, $a \circ b \vDash A \wedge B$ and so, by lemma 14, $a \circ b \vDash (A \wedge B) \vee (A \wedge C)$. By a similar argument, $a \circ c \vDash (A \wedge B) \vee (A \wedge C)$. Let e be the minimal state in $[\![(A \wedge B) \vee (A \wedge C)]\!]$. Thus, $e \leq a \circ b$ and $e \leq a \circ c$. By semantic postulate 10, $e \leq a \circ (b \sqcap c)$ and so $a \circ (b \sqcap c) \vDash (A \wedge B) \vee (A \wedge C)$. By hypothesis, $a \vDash B \vee C$ and so $b \sqcap c \leq a$. By semantic postulate 3 and hereditariness, $a \circ a \vDash (A \wedge B) \vee (A \wedge C)$ and by the WI postulate and hereditariness $a \vDash (A \wedge B) \vee (A \wedge C)$. $\square$

6.1   Adding Backwards Implication

In treatments of some weak substructural logics, like Full Lambek (**FL**), there is usually an additional implication connective, $\leftarrow$ (the "backwards arrow"). This connective is usually represented by a backslash ($\backslash$) but in logics that contain the backslash, implication is represented by a forward slash. I follow Greg Restall [20] in using a left arrow.[9]

Backwards implication has the following truth condition:

$$a \vDash A \leftarrow B \text{ if and only if for all } b \vDash B, \ b \circ a \vDash A.$$

In order to add the backwards arrow, I also add the following semantic postulates:

for all states a, $a = a \circ \Sigma$;

for all states a and b, there is a least state c such that $a \leq b \circ c$.

Adding the first postulate allows a semantic entailment fact like the one given below for the backwards arrow and adding the second postulate enables me to show that, for all formulae A and B, $[\![B \leftarrow A]\!]$ has a least member. Adding the first condition

---

[9]Robert Goldblatt suggested that I do not call this connective "noitacilpmi".

makes Σ both a left and a right identity for fusion. In the context of weak relevant logics, this is an unusual stipulation. I add it to prove the semantic entailment lemma for backwards implication. How to prove that in the present framework without this condition, I am uncertain.[10]

Adding $\leftarrow$ to the language requires adding the following rules to the axiomatic basis of **LB**.

$$\frac{\vdash (A \circ B) \to C}{\vdash B \to (C \leftarrow A)} \qquad \frac{\vdash B \to (C \leftarrow A)}{\vdash (A \circ B) \to C}$$

$$\frac{\vdash A \to B}{\vdash B \leftarrow A} \qquad \frac{\vdash A \leftarrow B}{\vdash B \to A}$$

I call the base logic with backwards implication, $\mathbf{FL}^{\neg}$, the "Full Lambek logic with negation".

Additional backwards implication theses can be made valid by adding conditions from the list given in section 6. For example, if the conjunctive syllogism condition is added to the semantics, it makes valid the following version of contraction:

$$((B \leftarrow A) \leftarrow A) \to (B \leftarrow A)$$

I leave the derivation of this thesis to the reader. If we add CI, then we make $\leftarrow$ equivalent to $\to$. That is, we can then prove $(B \leftarrow A) \leftrightarrow (A \to B)$. Adding backwards implication to some logics, such as **E**, results in a non-conservative extension. In **E**, adding the postulate that $a = a \circ \Sigma$ results in the identification of $\Box^{\neg}a$ and a for all states a, and it makes valid $A \to (t \to A)$. Identifying $\Box^{\neg}a$ with a in a model of **E** yields a model for the logic **R**.

Lemma 24 (Semantic Entailment for Backwards Implication). $\Sigma \vDash B \leftarrow A$ *if and only if for all* $a \in S$, $a \vDash A$ *implies* $a \vDash B$.

Proof. $\Rightarrow$ Suppose that $\Sigma \vDash B \leftarrow A$ and $a \vDash A$. Then $a \circ \Sigma \vDash B$. But $a \circ \Sigma = a$ so $a \vDash B$.

$\Leftarrow$ Suppose that for all states a, $a \vDash A$ implies $a \vDash B$. Then $a \vDash A$ implies that $a \circ \Sigma \vDash B$, and so $\Sigma \vDash B \leftarrow A$. $\qquad\square$

Fact 25 follows directly from the semantic entailment lemma and the semantic entailment lemma for backwards implication.

Fact 25. $\Sigma \vDash A \to B$ *if and only if* $\Sigma \vDash B \leftarrow A$.

---

[10]Bimbó only has t as a left identity in her semantics for $\mathbf{T}^{\circ t}_{\leftarrow}$ in [2]. Her method might transfer to the present semantics.

Lemma 26. *For all formulae A, $[\![A]\!]$ contains a unique least state.*

Proof. Follows from lemma 8 with the addition of the following proof for the case of backwards implication.

By the inductive hypothesis, $[\![B]\!]$ and $[\![C]\!]$ contain unique minimal states. Let b be the minimal state of $[\![B]\!]$ and c the minimal state of $[\![C]\!]$. By the semantic postulates for backwards implication, there is a minimal state a such that $c \leq b \circ a$. This a is the minimal state in $[\![C \leftarrow B]\!]$. $\quad\square$

Lemma 27. *If $\Sigma \vDash (A \circ B) \rightarrow C$ then $\Sigma \vDash B \rightarrow (C \leftarrow A)$.*

Proof. Suppose that $\Sigma \vDash (A \circ B) \rightarrow C$ and that $b \vDash B$ and $a \vDash A$. By exportation for forwards implication, $\Sigma \vDash A \rightarrow (B \rightarrow C)$, so $\Sigma \circ a \vDash B \rightarrow C$ and $a \vDash B \rightarrow C$. Thus, $a \circ b \vDash C$. By the truth condition for backwards implication, $b \vDash C \leftarrow A$ and so $\Sigma \circ b \vDash C \leftarrow A$. By the truth condition for implication, $\Sigma \vDash A \rightarrow (C \leftarrow B)$. $\quad\square$

Lemma 28. *If $\Sigma \vDash B \rightarrow (C \leftarrow A)$ then $\Sigma \vDash (A \circ B) \rightarrow C$.*

Proof. Suppose that $\Sigma \vDash B \rightarrow (C \leftarrow A)$ and that $a \vDash A \circ B$. Let b be the minimal state in $[\![A]\!]$ and c be the minimal state in $[\![B]\!]$. Then $b \circ c \leq a$ and $\Sigma \circ c \vDash C \leftarrow A$. But $\Sigma \circ c = c$ so $c \vDash C \leftarrow A$. By the truth condition for backwards implication, $b \circ c \vDash C$. By hereditariness, $a \vDash C$ so $\Sigma \circ a \vDash C$. By the semantic entailment lemma, $\Sigma \vDash (A \circ B) \rightarrow C$. $\quad\square$

The preceding lemmas imply the following soundness theorem.

Theorem 29. *All of the theorems of $\mathbf{FL}^{\neg}$ are valid in the class of state models with backwards implication.*

6.2   Negation and Ackermann Constants

I include the Ackermann constant, t in the language largely to aid in the construction of the canonical model in sections 7.1 and 7.2. But this constant can be made to do a lot more work. Following standard practice, I define f (the "falsum") as $\neg t$. The truth condition for the falsum is

$$a \vDash f \text{ if and only if } -\Sigma \leq a.$$

It is common in the literature on intuitionist and substructural logics to add f as a primitive and use it to define negation as $\neg A =_{df} A \rightarrow f$. I don't do this, but I do consider the effects on the logics and semantics of adding the corresponding biconditional schema,

$$\neg A \leftrightarrow (A \rightarrow f),$$

and a semantic postulate that corresponds to it,

$$(\text{F}) \quad a \perp b \text{ if and only if } -\Sigma \leq a \circ b.$$

Adding (F) also makes valid certain formulae that do not contain f, such as $\neg(\neg p \circ p)$. The logics **LB**, **B**, and **DJ** do not have this formula as a theorem.[11]

Adding (F) to the conditions on frames for some extensions of **LB** can also allow us to derive other semantic postulates, such as reductio and contraposition.

Fact 30. *Every state model that satisfies (F) and* W *also satisfies reductio.*

Proof. Suppose that $a \circ b \perp b$. By (F), $-\Sigma \leq (a \circ b) \circ b$. Thus, by W, $-\Sigma \leq a \circ b$, and by (F), $a \perp b$. $\square$

Fact 31. *Every state model that satisfies (F) and* B′ *also satisfies contraposition.*

Proof. Suppose that $a \circ b \perp c$. By the symmetry of $\perp$, $c \perp a \circ b$. By (F), $-\Sigma \leq c \circ (a \circ b)$. By B′, $-\Sigma \leq (a \circ c) \circ b$, and so by (F), $a \circ c \perp b$. $\square$

Theorem 32. LB^t *and its extensions are sound over the corresponding class of models.*

I can also add the Church constants T and F to the model, but with some modification. Following [18], I can add a state ω to the frame and add the postulate that for all states a, $a \leq \omega$. I also have to add a state $\alpha = -\omega$ such that for all a, $\alpha \leq a$. I then add the conditions, that $a \vDash T$, for all states a, and $a \vDash F$ if and only if $a = \omega$.

7 Completeness

Let L be **LB** or any of the logics that extend **LB** by means of the postulates given in sections 5 and 6. I use '$\vdash_L$' to mean 'is provable in L'. The (weak) completeness theorem given in this section shows that only the theorems of L are valid in the appropriate class of state models.

---

[11]I am grateful to John Slaney and his wonderful program, MaGIC (Matrix Generator for Implication Connectives), for proving that this formula is not in these logics. I am also grateful to Shawn Standefer for pointing out MaGIC proves that this formula is not a theorem of these logics.

7.1 Canonical Frame

The construction of the canonical frame is, for the most part, rather conventional. What is different about it is that it does not admit arbitrary theories of the logics as states. In order to satisfy semantic postulate 6 — for each state a in the canonical frame $a^\perp$ has a least member — only principal theories are admitted.[12] That is to say, for each a state a in the canonical model there is some formula A such that, for all formulae B, $B \in a$ if and only if $\vdash_L A \to B$. I use the notation '$[A]_L$' to denote the set of formulae entailed by A according to the logic L. The set of states for L, S_L, is the set of $[A]_L$ for all formulae A. For each logic L, Σ_L is the set of theorems of L.

The operators $\circ_L$ and $\sqcap_L$ are defined as

$$[A]_L \circ_L [B]_L =_{df} [A \circ B]_L \quad \text{and} \quad [A]_L \sqcap_L [B]_L =_{df} [A \vee B]_L.$$

The relation $\perp_L$ is defined as

$[A]_L \perp_L [B]_L$ if and only if there is some formula $C \in [B]_L$ such that $\neg C \in [A]_L$.

The ordering relation is defined as subset, that is, $[A]_L \leq_L [B]_L$ if and only if $[A]_L \subseteq [B]_L$.

I first prove that the canonical frame has the properties of a semilattice. I do this by showing that $\sqcap_L$ is set-theoretic intersection. Any set of sets that is closed under intersection is a meet semilattice.

Lemma 33. $[A]_L \sqcap_L [B]_L = [A]_L \cap [B]_L$.

Proof. Suppose first that C is in both $[A]_L$ and $[B]_L$. Then both $A \to C$ and $B \to C$ are theorems of L. Therefore, $(A \vee B) \to C$ is a theorem of L and $C \in [A \vee B]_L$. Therefore, $[A]_L \cap [B]_L \subseteq [A \vee B]_L$.

For the converse, suppose that $C \in [A \vee B]_L$. Thus, $(A \vee B) \to C$ is a theorem of L. So, by fact 12(3), $A \to C$ and $B \to C$ are also theorems. Therefore, C is in both $[A]_L$ and $[B]_L$ and is also in $[A]_L \cap [B]_L$. $\square$

It follows from lemma 33 and the definition of $\leq$ for the canonical frame that $[A]_L \leq [B]_L$ if and only if $[A]_L = [A]_L \sqcap_L [B]_L$.

Facts 34, 35, and 35 below are used often in the proofs that follow.

Fact 34. $[B]_L \leq [A]_L$ *if and only if* $\vdash_L A \to B$.

---

[12]The use of principal theories rather than arbitrary theories to prove completeness is also found in [5]. For more details about the use of principal theories to represent lattices, see [4].

Proof. $\Rightarrow$ Assume that $[B]_L \leq [A]_L$. Then, $[B]_L \subseteq [A]_L$ and so $B \in [A]_L$. By the definition of $[A]_L$, $\vdash A \to B$.

$\Leftarrow$ Assume that $\vdash_L A \to B$ and that $C \in [B]_L$. Then $\vdash_L B \to C$ and by affixing, $\vdash_L A \to C$. Thus, $C \in [A]_L$. Thus, $[B]_L \subseteq [A]_L$ and so $[B]_L \leq [A]_L$. $\square$

Fact 35. *(a) If $C \to D \in [A]_L$ and $C \in [B]_L$, then $D \in [A]_L \circ_L [B]_L$ and (b) for any formulae A, B and C, $C \in [A]_L \circ_L [B]_L$ if and only if there is some formula $D \in [B]_L$ such that $D \to C \in [A]_L$.*

Proof. (a) Suppose that $C \to D \in [A]_L$ and $C \in [B]_L$. Then $\vdash_L B \to C$. By affixing $\vdash_L (C \to D) \to (B \to D)$. Again by affixing, $\vdash_L (A \to (C \to D)) \to (A \to (B \to D))$. By hypothesis, $\vdash_L A \to (C \to D)$, so $\vdash_L A \to (B \to D)$ and by importation, $\vdash_L (A \circ B) \to D$. Therefore, $D \in [A]_L \circ_L [B]_L$.

(b) $\Rightarrow$ Suppose that $C \in [A]_L \circ_L [B]_L$. Since $[A]_L \circ_L [B]_L = [A \circ B]_L$, $\vdash_L (A \circ B) \to C$. By the exportation rule, $\vdash_L A \to (B \to C)$. Thus, $B \to C \in [A]_L$. Since $B \in [B]_L$, there is some formula $D \in [B]_L$ such that $D \to C \in [A]_L$.

$\Leftarrow$ Follows from (a). $\square$

The following lemma shows that the canonical frame obeys semantic postulate 2.

Lemma 36. $\Sigma \circ_L [A]_L = [A]_L$ *for all formulae A.*

Proof. Σ is the set of theorems of L, and so $B \to B \in \Sigma$ for all formulae B. By fact 35, If $B \in [A]_L$, then $B \in \Sigma \circ_L [A]_L$. So, $[A]_L \subseteq \Sigma \circ_L [A]_L$.

Now, suppose that $B \in \Sigma \circ_L [A]_L$. By the construction of the canonical frame, $\vdash_L (t \circ A) \to B$. By the exportation rule, $\vdash_L t \to (A \to B)$, and by the fact that t is an axiom and the theorems of L are closed under modus ponens, $\vdash_L A \to B$ and so $B \in [A]_L$. Generalising on B, $\Sigma \circ_L [A]_L \subseteq [A]_L$. $\square$

The next lemma shows that semantic postulate 3 holds in the canonical frame.

Lemma 37. *If $[A]_L \leq [A']_L$ and $[B] \leq [B']$, then $[A]_L \circ_L [B]_L \leq [A']_L \circ_L [B']_L$.*

Proof. Suppose that $[A]_L \leq [A']_L$ and $[B] \leq [B']$. Then, by fact 34, $\vdash_L A' \to A$ and $\vdash_L B' \to B$. By affixing, $\vdash_L (A \to (B \to (A \circ B))) \to (A' \to (B \to (A \circ B)))$ and since $\vdash_L A \to (B \to (A \circ B))$, $\vdash_L A' \to (B \to (A \circ B))$. Again by affixing, $\vdash_L (B \to (A \circ B)) \to (B' \to (A \circ B))$ and once again by affixing, $\vdash_L A' \to (B' \to (A \circ B))$. Thus, by importation, $\vdash_L (A' \circ B') \to (A \circ B)$. Thus, by fact 34 and the construction of the canonical frame, $[A]_L \circ_L [B]_L \leq [A']_L \circ_L [B']_L$. $\square$

From lemma 38 below and the contraposition rule, it follows that the incompatibility relation is symmetrical (and hence the canonical frame satisfies semantic postulate 4).

Lemma 38. $[A]_L \perp [B]_L$ *if and only if* $\vdash_L A \to \neg B$.

Proof. $\Rightarrow$ Assume $[A]_L \perp [B]_L$. By the definition of $\perp$ for the canonical model, there is some formula C such that $\vdash_L A \to \neg C$ and $\vdash_L B \to C$. By contraposition, double negation and a bit of fiddling, $\vdash_L \neg C \to \neg B$. By affixing, $\vdash_L A \to \neg B$.
$\Leftarrow$ Obvious. $\square$

The next lemma shows that the canonical frame satisfies semantic postulate 5.

Lemma 39. *If* $[A]_L \perp [B]_L$ *and* $[B]_L \leq [B']_L$, *then* $[A]_L \perp [B']_L$.

Proof. Suppose that $[A]_L \perp [B]_L$ and $[B]_L \leq [B']_L$. By lemma 38, $\vdash_L A \to \neg B$ and, by fact 34, $\vdash_L B' \to B$. By contraposition and double negation, $\vdash_L \neg B \to \neg B'$ and by affixing, $\vdash_L A \to \neg B'$. Therefore, by lemma 38, $[A]_L \perp [B']_L$. $\square$

Semantic postulate 6 follows directly from the lemma 40.

Lemma 40. $[\neg A]_L$ *is the unique least member of* $[A]_L^{\perp}$.

Proof. $[A]_L = [\neg\neg A]_L$, therefore by the definition of $\perp$ in the canonical model, $[A]_L \perp [\neg A]_L$. Suppose that $[A]_L \perp [B]_L$. By lemma 38, $\vdash_L A \to \neg B$. By contraposition, $\vdash_L B \to \neg A$. By fact 34, $[\neg A]_L \leq [B]_L$. Thus, $[\neg A]_L$ is the unique least member of $[A]_L^{\perp}$. $\square$

Semantic postulate 7 follows from the fact that A and $\neg\neg A$ are equivalent in **LB**. Semantic postulate 8 is shown to hold in the next lemma.

Lemma 41. *For all formulae B and C, there is some formula A such that $[A]_L$ is the least state for which* $[C]_L \leq_L [A]_L \circ_L [B]_L$.

Proof. Let B and C be arbitrary formulae. Let A be $B \to C$. $\vdash_L ((B \to C) \circ B) \to C$, and so by fact 34, $[C]_L \leq_L [A]_L \circ_L [B]_L$. Now suppose that for some $[D]_L$, $[C]_L \leq_L [D]_L \circ_L [B]_L$. By fact 34, $\vdash_L (D \circ B) \to C$, and so $\vdash_L D \to (B \to C)$. Thus, by fact 34, $[B \to C]_L \leq_L [D]_L$, i.e., $[A]_L \leq_L [D]_L$. $\square$

Lemmas 42 and 43 show that semantic postulates 9 and 10 hold in the canonical frame.

Lemma 42. *If* $[A]_L \subseteq [B]_L \circ_L [C]_L$ *and* $[A]_L \subseteq [B']_L \circ_L [C]_L$, *then* $[A]_L \subseteq ([B]_L \cap [B']_L) \circ_L [C]_L$.

Proof. Assume that $[A]_L \subseteq [B]_L \circ_L [C]_L$ and $[A]_L \subseteq [B']_L \circ_L [C]_L$. Then $C \to A$ is in both $[B]_L$ and $[B']_L$, and so $B \to (C \to A)$ and $B' \to (C \to A)$ are theorems of L. Thus, $(B \vee B') \to (C \to A)$ is a theorem, hence $[A]_L \subseteq [B \vee B']_L \circ_L [C]_L$. By the construction of the canonical frame, $[A]_L \subseteq ([B]_L \sqcap_L [B']_L) \circ_L [C]_L$. $\square$

Lemma 43. *If $[A]_L \subseteq [B]_L \circ_L [C]_L$ and $[A]_L \subseteq [B]_L \circ_L [C']_L$, then $[A]_L \subseteq [B]_L \circ_L ([C]_L \cap [C']_L)$.*

Proof. Suppose that $[A]_L \subseteq [B]_L \circ_L [C]_L$ and $[A]_L \subseteq [B]_L \circ_L [C']_L$. Then both $C \to A$ and $C' \to A$ are in $[B]_L$. Thus, $(C \vee C') \to A$ is in $[B]_L$ and so $[A]_L \subseteq [B]_L \circ_L ([C]_L \cap [C']_L)$. $\square$

Lemma 44. $([A]_L \sqcap_L [A']_L) \circ_L ([B]_L \sqcap_L [B']_L) \subseteq ([A]_L \circ_L [B]_L) \sqcap_L ([A']_L \circ_L [B']_L)$.

Proof. Suppose that $C \in ([A]_L \sqcap_L [A']_L) \circ_L ([B]_L \sqcap_L [B']_L)$. By fact 35, there is a formula $D \in ([B]_L \sqcap_L [B']_L)$ such that $D \to C \in ([A]_L \sqcap_L [A']_L)$. Then, by lemma 33, $B \to D$, $B' \to D$, $A \to (D \to C)$ and $A' \to (D \to C)$ are all theorems of L. Hence, $(A \circ B) \to C$ and $(A' \circ B') \to C$ are all theorems of L and so $C \in [A \circ B]_L \sqcap_L [A' \circ B']_L$. $\square$

Finally, the lemma below shows that the canonical frame satisfies semantic postulate 12.

Lemma 45. *For all formulae A and B, $([A]_L \sqcap_L [B]_L)^{\perp} = [A]_L^{\perp} \cap [B]_L^{\perp}$.*

Proof. Suppose that $[C]_L \in ([A]_L \sqcap_L [B]_L)^{\perp}$. By the definition of the intersection operator for the canonical frame, $[A]_L \sqcap_L [B]_L = [A \vee B]_L$, so by lemma 38, $\vdash_L (A \vee B) \to \neg C$, so $\vdash_L A \to \neg C$ and $\vdash_L B \to \neg C$. Thus, $[C]_L \in [A]_L^{\perp}$ and $[C]_L \in [B]_L^{\perp}$.
The proof of the converse is similar. $\square$

The following theorem follows from the preceding lemmas.

Theorem 46. *The canonical frame for L is a state frame.*

7.2   The Canonical Model

To construct a canonical model, I add a canonical variable assignment, v_L, such that, for any propositional variable p, $v_L(p)$ is the set of principal theories that contain p. Clearly, $v_L(p)$ is closed upwards under $\subseteq$, is closed under set-theoretic intersection, and has a unique least member $[p]_L$. v_L determines a satisfaction relation $\vDash_L$ in the manner set out in section 2.2.

Theorem 47 (Truth). *For all formulae A and B, $B \in [A]_L$ if and only if $[A]_L \vDash_L B$.*

Proof. By induction on the complexity of B.

Case 1. B is a propositional variable, p. Assume that $p \in [A]_L$. Then $[A]_L \in v_L(p)$ and so $[A]_L \vDash_L p$. Now assume that $[A]_L \vDash_L p$. By the truth condition for propositional variables, $[A]_L \in v_L(p)$ and so $p \in [A]_L$.

Case 2. B is t. $t \in [A]_L$ if and only if $[t]_L \subseteq [A]_L$. Since Σ_L is just $[t]_L$, $t \in [A]_L$ if and only if $\Sigma_L \leq [A]_L$. By the truth condition for t, $[A]_L \vDash_L t$ if and only if $t \in [A]_L$.

Case 3. B is a conjunction, $C \wedge D$. Straightforward.

Case 4. B is a disjunction, $C \vee D$. Assume that $C \vee D$ is in $[A]_L$. Thus, $[C \vee D]_L \subseteq [A]_L$. By definition, $[C \vee D]_L = [C]_L \sqcap_L [D]_L$ and so $[C]_L \sqcap_L [D]_L \leq [A]_L$. Therefore, $[A]_L \vDash C \vee D$.

Now assume that $[A]_L \vDash_L C \vee D$. By the truth condition for disjunction and the inductive hypothesis, $[C]_L \sqcap_L [D]_L \leq [A]_L$. By definition, $[C]_L \sqcap_L [D]_L = [C \vee D]_L$ and so $C \vee D \in [A]_L$.

Case 5. B is a negation, $\neg C$. Assume that $\neg C \in [A]_L$. $\vdash_L A \to \neg C$. By lemma 38, $[A]_L \perp [C]_L$. Thus, if $[C]_L \leq [D]_L$, then $[A]_L \perp [D]_L$. By the inductive hypothesis, if $[D]_L \vDash_L C$, then $[A]_L \perp [D]_L$, i.e., $[D]_L \vDash_L \neg C$.

Assume that $[A]_L \vDash_L \neg C$. Consider $[C]_L$. By the inductive hypothesis, $[C]_L \vDash C$ and by the truth condition for negation, $[A]_L \perp [C]_L$. By the definition of $\perp$ for the canonical model, there is some $D \in [C]_L$ such that $\neg D \in [A]_L$. Thus, $\vdash A \to \neg D$ and $\vdash_L C \to D$. By contraposition, $\vdash_L \neg D \to \neg C$. By affixing, $\vdash_L A \to \neg C$ and so $\neg C \in [A]_L$.

Case 6. B is an implication, $C \to D$. Assume that $C \to D \in [A]_L$. Also assume that $[E]_L \vDash C$. By the inductive hypothesis, $C \in [E]_L$. By fact 35, $D \in [A]_L \circ_L [E]_L$ and the inductive hypothesis, $[A]_L \circ_L [E]_L \vDash D$. Generalising on E, $[A]_L \vDash_L C \to D$.

Now assume that $[A]_L \vDash_L C \to D$. By the inductive hypothesis, $[C]_L \vDash_L C$. By the truth condition for implication, $[A]_L \circ_L [C]_L \vDash_L D$. Thus, by the inductive hypothesis, $D \in [A]_L \circ_L [C]_L$. By the construction of the canonical frame, $\vdash_L (A \circ C) \to D$, and by exportation, $\vdash_L A \to (C \to D)$. Therefore, $C \to D \in [A]_L$.

Case 7. B is a fusion, $C \circ D$. Assume that $(C \circ D) \in [A]_L$. Then $\vdash_L A \to (C \circ D)$. Thus, $[C \circ D]_L \leq [A]_L$ and by the definition of $\circ_L$, $[C]_L \circ_L [D]_L \leq_L [A]_L$. By the inductive hypothesis, $[C]_L \vDash_L C$ and $[D]_L \vDash_L D$ and so, $[A]_L \vDash C \circ D$.

Now assume that $[A]_L \vDash_L C \circ D$. Then there are E, F such that $[E]_L \vDash_L C$, $[F]_L \vDash_L D$, and $[E]_L \circ_L [F]_L \leq_L [A]_L$. By the inductive hypothesis, $\vdash_L E \to C$ and $\vdash_L F \to D$. Thus, $\vdash_L (E \circ F) \to (C \circ D)$. By definition of fusion in the canonical frame, $[E]_L \circ_L [F]_L = [E \circ F]_L$ and so $\vdash_L A \to (E \circ F)$. By affixing, $\vdash_L A \to (C \circ D)$, so $C \circ D \in [A]_L$. $\qquad\square$

The weak completeness theorem for **LB** follows from the fact that $\Sigma_{LB} \vDash_{LB} A$ if and only if $A \in \Sigma_{LB}$ if and only if $\vdash_{LB} A$.

Theorem 48 (Completeness of **LB**). *For all formulae A, A is valid in the class of state models only if A is a theorem of* **LB**.

7.3 Completeness for Extensions of LB

In order to prove the completeness of **B** and its extensions, I use the following lemma.

Lemma 49. *For all formulae A and B, $[A]_L \sqcup [B]_L = [A \wedge B]_L$.*

Proof. By lemma 40 and the definition of $\sqcup$, $[A]_L \sqcup [B]_L = -(-[A]_L \sqcap_L -[B]_L) = -([\neg A]_L \sqcap_L [\neg B]_L) = [\neg(\neg A \vee \neg B)]_L$. But $\neg(\neg A \vee \neg B)$ is equivalent to $A \wedge B$ and so $[A]_L \sqcup [B]_L = [A \wedge B]_L$. $\qquad\qquad\square$

Here are the proofs that the various logics are complete over their corresponding class of models. I skip the proofs for logics that contain assertion, K, and K′, because they are rather simple.

Theorem 50. *If L is an extension of **B**, then $\langle S_L, \sqcap_L, \sqcup_L \rangle$ satisfies the distributive law.*

Proof. Assume that L is an extension of **B**, and so contains the thesis of distribution. By the construction of the canonical frame and lemma 49, $[A]_L \sqcup ([B]_L \sqcap_L [C]_L) = [A \wedge (B \vee C)]_L$. But $\vdash_L (A \wedge (B \vee C)) \leftrightarrow ((A \wedge B) \vee (A \wedge C))$, hence $[A \wedge (B \vee C)]_L = [(A \wedge B) \vee (A \wedge C)]_L$. By the definition of the canonical frame, therefore, $[A]_L \sqcup ([B]_L \sqcap_L [C]_L) = ([A]_L \sqcup [B]_L) \sqcap_L ([A]_L \sqcup [C]_L)$. $\qquad\square$

The preceding theorem together with the completeness theorem for **LB** tells us that **B** is complete over the class of distributive state models.

In order to prove that the other extensions of **LB** are complete, I note that for any extension of **LB**, for all formulae A and B, $[A]_L \circ_L [B]_L$ is the set of formulae C such that $\vdash_L A \to (B \to C)$. I use this fact often in the proofs that follow.

Theorem 51. *If L contains the prefixing schema as a theorem, then the canonical model satisfies the condition that, for arbitrary A, B and C, $[A]_L \circ_L ([B]_L \circ_L [C]_L) \leq ([A]_L \circ_L [B]_L) \circ_L [C]_L$.*

Proof. Assume L contains the prefixing schema. Suppose that $D \in [A]_L \circ_L ([B]_L \circ_L [C]_L)$. Then there is some formula E such that $\vdash_L A \to (E \to D)$ and $\vdash_L B \to (C \to E)$. By prefixing, $\vdash_L (E \to D) \to ((C \to E) \to (C \to D))$. So, $(C \to E) \to (C \to D) \in [A]_L$. And $C \to E \in [B]_L$. So, $C \to D \in [A]_L \circ_L [B]_L$ and $D \in ([A]_L \circ_L [B]_L) \circ_L [C]_L$. Generalising on D, $[A]_L \circ_L ([B]_L \circ_L [C]_L) \leq ([A]_L \circ_L [B]_L) \circ_L [C]_L$. $\qquad\square$

Theorem 52. *If L contains the suffixing schema as a theorem, then the canonical model satisfies the condition, for arbitrary A, B, C, that $[A]_L \circ_L ([B]_L \circ_L [C]_L) \leq ([B]_L \circ_L [A]_L) \circ_L [C]_L$.*

Proof. Assume that L contains the suffixing schema. Suppose that $D \in [A]_L \circ_L ([B]_L \circ_L [C]_L)$. Then there is some formula E such that $\vdash_L A \to (E \to D)$ and $\vdash_L B \to (C \to E)$. By suffixing, $\vdash_L (C \to E) \to ((E \to D) \to (C \to D))$. By affixing, $(E \to D) \to (C \to D) \in [B]_L$. Thus, $C \to D \in [B]_L \circ_L [A]_L$ and $D \in ([B]_L \circ_L [A]_L) \circ_L [C]_L$. Generalising on D, $[A]_L \circ_L ([B]_L \circ_L [C]_L) \leq ([B]_L \circ_L [A]_L) \circ_L [C]_L$. $\square$

Theorem 53. *If L contains the permutation schema as a theorem, then the canonical model satisfies the condition, for arbitrary A, B, C that $([A]_L \circ_L [B]_L) \circ_L [C]_L \leq ([A]_L \circ_L [C]_L) \circ_L [B]_L$.*

Proof. Assume that L contains the permutation schema. Suppose that $D \in ([A]_L \circ_L [B]_L) \circ_L [C]_L$. Then there is some formula E, $\vdash_L A \to (B \to (E \to D))$ and $\vdash C \to E$. By permutation, $\vdash_L A \to (E \to (B \to D))$. Then $E \to (B \to D) \in [A]_L$ and $E \in [C]_L$ so $B \to D \in [A]_L \circ_L [C]_L$. Therefore, $D \in ([A]_L \circ_L [C]_L) \circ_L [B]_L$. Generalising on D, $([A]_L \circ_L [B]_L) \circ_L [C]_L \leq ([A]_L \circ_L [C]_L) \circ_L [B]_L$. $\square$

The proof regarding CI is very similar to the proof concerning C.

Theorem 54. *If L contains conjunctive syllogism, then for all A, B, $[A]_L \circ_L ([A]_L \circ_L [B]_L) \leq [A]_L \circ_L [B]_L$.*

Proof. Assume that L contains conjunctive syllogism. Assume that $C \in [A]_L \circ_L ([A]_L \circ_L [B]_L)$. Then there is some formula D such that $\vdash_L A \to (D \to C)$ and $\vdash_L A \to (B \to D)$. Then both $D \to C$ and $B \to D$ are in $[A]_L$. By conjunctive syllogism, $B \to C \in [A]_L$. So, $C \in [A]_L \circ_L [B]_L$. Generalising on C, $[A]_L \circ_L ([A]_L \circ_L [B]_L) \leq [A]_L \circ_L [B]_L$. $\square$

Theorem 55. *If L contains the contraction schema, then the canonical model satisfies the condition that $([A]_L \circ_L [B]_L) \circ_L [B]_L \leq [A]_L \circ_L [B]_L$.*

Proof. Assume that L contains the contraction schema. Suppose that $C \in ([A]_L \circ_L [B]_L) \circ_L [B]_L$. Thus, $\vdash_L (A \circ B) \to (B \to C)$. Then, by exportation, $\vdash_L A \to (B \to (B \to C))$. By the contraction schema, $\vdash_L (B \to (B \to C)) \to (B \to C)$. By affixing, $\vdash A \to (B \to C)$ and so $C \in [A]_L \circ_L [B]_L$. $\square$

Theorem 56. *If L contains the mingle schema as a theorem, then the canonical model satisfies the condition that $[A]_L \leq [A]_L \circ_L [A]_L$.*

Proof. Assume that L contains the mingle schema and suppose that $B \in [A]_L$. Then $\vdash_L A \to B$. By mingle, $\vdash_L B \to (B \to B)$, and by affixing, $\vdash_L A \to (B \to B)$. Thus, $B \in [A]_L \circ_L [B]_L$. But $\vdash_L A \to B$ and so $[B]_L \leq [A]_L$, hence by lemma 37, $B \in [A]_L \circ_L [A]_L$. $\square$

Theorem 57. *If L contains the reductio schema, then the canonical model satisfies the condition that $[A]_L \circ_L [B]_L \perp [B]_L$ implies $[A]_L \perp [B]_L$.*

Proof. Assume that L contains the reductio schema and suppose that $[A]_L \circ_L [B]_L \perp [B]_L$. Thus, $\vdash_L (A \circ B) \to \neg B$. By exportation, $\vdash_L A \to (B \to \neg B)$. By reductio, $\vdash_L (B \to \neg B) \to \neg B$. By affixing, $\vdash A \to \neg B$ and therefore, $[A]_L \perp [B]_L$. $\square$

Theorem 58. *If L contains the contraposition schema, then the canonical model satisfies the condition that $[A]_L \circ_L [B]_L \perp [C]$ implies $[A]_L \circ_L [C]_L \perp [B]_L$.*

Proof. Assume that L contains the contraposition schema. Suppose that $[A]_L \circ_L [B]_L \perp [C]_L$. Thus, $\vdash_L (A \circ B) \to \neg C$ and so $\vdash_L A \to (B \to \neg C)$. By contraposition, $\vdash_L (B \to \neg C) \to (C \to \neg B)$ and so $\vdash_L (A \circ C) \to \neg B$. Thus, $[A]_L \circ_L [C]_L \perp [B]_L$. $\square$

Theorem 59. *If L contains the pseudo-modus ponens schema, then the canonical model satisfies $[A]_L \circ_L [A]_L \leq_L [A]_L$.*

Proof. Suppose that L contains the pseudo-modus ponens schema. Let $B \in [A]_L \circ_L [A]_L$. Then, by lemma 35, there is some formula $C \in [A]_L$ such that $C \to B \in [A]_L$. Thus, $(C \to B) \wedge C \in [A]_L$. By pseudo-modus ponens, $B \in [A]_L$. Thus, $[A]_L \circ_L [A]_L \leq_L [A]_L$, for all formulae A. $\square$

The proof for the postulate corresponding to excluded middle is easy.

7.4 Completeness for **FL$^\neg$** and its Extensions

For the purposes of this section, I assume that backwards implication is in the language and that L is **FL$^\neg$** or one of its extensions.

Lemma 60. *If L is an extension of **FL$^\neg$**, then (i) $[A]_L \circ_L \Sigma_L = [A]_L$ and (ii) for all $[A]_L, [B]_L \in S_L$, there is a least state $[C]_L$ such that $[A]_L \leq [B]_L \circ_L [C]_L$.*

Proof. Assume that L is an extension of **FL$^\neg$**.

(i) Let A be an arbitrary formula. $\vdash_L (A \leftarrow A) \to (A \leftarrow A)$. By importation for the backwards arrow, $\vdash_L (A \circ (A \leftarrow A)) \to A$. By exportation for implication, $\vdash_L A \to ((A \leftarrow A) \to A)$. So, $(A \leftarrow A) \to A \in [A]_L$. But $A \leftarrow A \in \Sigma_L$, so $A \in [A]_L \circ_L \Sigma_L$. Therefore, $[A]_L \subseteq [A]_L \circ_L \Sigma_L$.

Now suppose that $B \in [A]_L \circ_L \Sigma_L$. By the definition of the canonical frame, $\vdash_L (A \circ t) \to B$. By exportation for backwards implication, $\vdash_L t \to (B \leftarrow A)$. By modus ponens, $\vdash_L B \leftarrow A$ and so $\vdash_L A \to B$. Therefore, $B \in [A]_L$, and so $[A]_L \circ_L \Sigma_L \subseteq [A]_L$.

(ii) First I show that $[A]_L \subseteq [B]_L \circ_L [A \leftarrow B]_L$. By the definition of application for the canonical model, $[B]_L \circ_L [A \leftarrow B]_L = [B \circ (A \leftarrow B)]_L$. $\vdash_{\mathbf{FL}^\neg} (B \circ (A \leftarrow B)) \to A$ and, since $[B \circ (A \leftarrow B)]_L$ is an L theory, $A \in [B \circ (A \leftarrow B)]_L$, i.e. $A \in [B]_L \circ_L [A \leftarrow B)]_L$. Therefore, $[A]_L \subseteq [B]_L \circ_L [A \leftarrow B]_L$.

To show that $[A \leftarrow B]_L$ is the least such theory, I suppose that there is some formula C for which $[A]_L \subseteq [B]_L \circ_L [C]_L$. Then $\vdash_L (B \circ C) \to A$ and so $\vdash_L C \to (A \leftarrow B)$. Hence, $[C]_L \subseteq [A \leftarrow B]_L$. $\qquad\square$

Lemma 61. $D \leftarrow C \in [A]_L$ *if and only if, for all formulae* B, *if* $C \in [B]_L$, *then* $D \in [B]_L \circ_L [A]_L$.

Proof. $\Rightarrow$ Suppose that $D \leftarrow C \in [A]_L$ and for an arbitrary formula B, $C \in [B]_L$. As shown in the proof of lemma 60, $\vdash_L C \to ((D \leftarrow C) \to D)$. So, $(D \leftarrow C) \to D \in [B]_L$. Thus, $D \in [B]_L \circ_L [A]_L$.

$\Leftarrow$ Suppose that for all formulae B, if $C \in [B]_L$, then $D \in [B]_L \circ_L [A]_L$. Instantiate B to C; therefore, $D \in [C]_L \circ_L [A]_L$. Thus, $\vdash_L (C \circ A) \to D$ and so $\vdash_L A \to (D \leftarrow C)$. Therefore, $D \leftarrow C \in [A]_L$. $\qquad\square$

The previous lemma shows that the truth lemma goes through even when the language is extended to include backwards implication and the logic is an extension of $\mathbf{FL}^\neg$.

8 Conclusion

I have shown that this semantical theory can characterise a wide range of logical systems, but clearly further work can be done. I have only been concerned in this paper with weak completeness. I have not explored the nature of the semantic consequence relation. Where Γ is a set of formulae, let us define $\Gamma \vDash A$ as saying that any state in any model that satisfies every member of Γ also satisfies A. To convert the current results to strong completeness theorems, we need to show that $\vDash$ is finitary, that is, if $\Gamma \vDash A$, then there is some finite subset Γ' of Γ such that $\Gamma' \vDash A$. We cannot extract that theorem from the current completeness proof as is usually done, because the canonical model is based on principal theories rather than arbitrary theories.

Other than showing strong completeness, the current semantics stands in need of additions to treat modalities (including the exponentials of linear logic) and the quantifiers. Adding an operator to the semantics to treat necessity is rather straightforward, but it is not clear to me yet how to give a philosophically reasonable treatment of possibility.

Perhaps more interesting is the issue of adding quantification. One way to add quantifiers is to develop the current frame theory into an *admissible semantics* in the sense of [13]. In admissible semantics, a set of propositions and propositional functions is part of the frame. In such a semantics we can place a condition on frames that for all propositional functions φ there is a proposition that is the least upper bound of the set of $\varphi(i)$ where i is a member of the domain. This in turn will enable us to show that each universally quantified statement expresses a proposition that has a minimal state. This will enable us to accept the standard clause for the universal quantifier, that is, $a \vDash_f \forall x A$ if and only if $a \vDash_{f[i/x]} A$ for all i in the domain.

References

[1] Gerard Allwein and J. Michael Dunn. Kripke models for linear logic. *Journal of Symbolic Logic*, 58:514–545, 1993.

[2] Katalin Bimbó. Relevance logics. In D. Jacquette, editor, *Philosophy of Logic: The Handbook of the Philosophy of Science, vol. 5*, pages 723–789. Elsevier, Amsterdam, 2007.

[3] Katalin Bimbó. Dual gaggle semantics for entailment. *Notre Dame Journal of Formal Logic*, 50(1):23–41, 2009.

[4] Katalin Bimbó and J. Michael Dunn. *Generalized Galois Logics: Relational Semantics of Nonclassical Logical Calculi*, volume 188 of *CSLI Lecture Notes*. CSLI Publications, Stanford, CA, 2008.

[5] Katalin Bimbó and J. Michael Dunn. St. Alasdair on lattices everywhere. In I. Düntsch and E. Mares, editors, *Alasdair Urquhart on Nonclassical and Algebraic Logic and Complexity of Proofs*, pages 323–346. Springer Verlag, Basel, 2021.

[6] Garrett Birkhoff. *Lattice Theory*. American Mathematical Society, Providence, RI, 1940.

[7] Ross Brady. *Universal Logic*, volume 109 of *CSLI Lecture Notes*. CSLI Publications, Stanford, CA, 2006.

[8] Kosta Došen. Sequent systems and groupoid models, II. *Studia Logica*, 48:41–65, 1989.

[9] J. Michael Dunn. Star and perp. *Philosophical Perspectives*, 7:331–357, 1993.

[10] Kit Fine. Models for entailment. *Journal of Philosophical Logic*, 3:347–372, 1974.

[11] Nikolaos Galatos, Peter Jipsen, Tomasz Kowalski, and Hiroakira Ono. *Residuated Lattices: An Algebraic Glimpse at Substructural Logics*. Elsevier, Amsterdam, 2007.

[12] Robert Goldblatt. Semantic analysis of orthologic. *Journal of Philosophical Logic*, 3:19–35, 1974.

[13] Robert Goldblatt. *Quantifiers, Propositions and Identity: Admissible Semantics for Quantified Modal and Substructural Logics*. Cambridge University Press, Cambridge, 2011.

[14] I. Lloyd Humberstone. Operational semantics for positive R. *Notre Dame Journal of Formal Logic*, 29:61–80, 1987. doi: 10.1305/ndjfl/1093637771.

[15] Mark Jago. Truthmaker semantics for relevant logic. *Journal of Philosophical Logic*, 49:691–702, 2020.

[16] Edwin Mares. *The Logic of Entailment and its History*. Cambridge University Press, Cambridge, 2024.

[17] Robert K. Meyer and Richard Routley. Algebraic analysis of entailment I. *Logique et Analyse*, 15(59/60):407–428, 1972.

[18] Hiroakira Ono. Semantics for substructural logics. In K. Došen and P. Schröder-Heister, editors, *Substructural Logics*, pages 259–291. Oxford University Press, Oxford, 1993.

[19] Stephen Read. Truthmakers and the disjunction thesis. *Mind*, 109:67–80, 2000.

[20] Greg Restall. *Introduction to Subsructural Logics*. Routlege, London, 2000.

[21] Richard Routley and Robert K. Meyer. Semantics for entailment. In H. Leblanc, editor, *Truth, Syntax, and Modality*. North Holland, Amsterdam, 1973.

[22] Richard Routley and Val Routley. The semantics of first-degree entailment. *Noûs*, 6:335–395, 1972.

[23] Richard Routley, Robert K. Meyer, Ross Brady, and Val Plumwood. *Relevant Logics and their Rivals*. Ridgeview, Atascadero, CA, 1983.

[24] Alasdair Urquhart. Semantics for relevance logics. *Journal of Symbolic Logic*, 37:159–169, 1972.

[25] Alasdair Urquhart. A topological representation theory for lattices. *Algebra Universalis*, 8:45–58, 1978.

Received 27 November 2025

RECAPTURE, AMBIGUITY AND CONFLATION

ELLIE RIPLEY

Monash University and the University of Melbourne
`<ripley@negation.rocks>`

Abstract

Many theorists who reject that classical logic gives a good theory of validity are still at pains to point to some other thing classical logic is correct about. Such theorists are engaged in a project sometimes called *classical recapture*. This paper focuses on an approach to classical recapture, which is based on the *ambiguity thesis*: the connectives are ambiguous. Ambiguity has been cashed out via translations, however, I will show—drawing on independent motivations—that classical logic is better thought of as *conflated*.

2020 *Mathematics Subject Classification.* Primary: 03B47, Secondary: 03B60.

1 Introduction

A number of phenomena, among them semantic paradoxes like the liar and curry, have led many theorists to conclude that classical logic cannot be the final word on which arguments are valid. Such theorists offer other nonclassical logics to fill this role: paracomplete [11], paraconsistent [2, 23], substructural [12, 27, 28, 35], some combination of multiple of these [7], or something else. But even with a logic identified as giving a correct story about validity, there is still an explanatory question that these theorists face: if classical logic is not right, why does it work so well so much of the time?

Here we can divide our nonclassical theorists into two camps. One camp rejects the need to offer any such explanation—classical logic, they maintain, does *not* work so well at all! Here, for example, is [32, p. xi]:

> The contemporary state of complacency with respect to the manifold deficiencies of classical logic and classical theories reflects ... the fact that classical logic is not greatly subject to the testing process ... If more attempt were made to apply the classical theory in reasoning, its inadequacy, and especially that of its treatment of such central notions as deducibility, valid argument, and sufficiency, would be much more widely noticed. ... [Classical logic] is simply inadequate for many basic reasoning tasks ...

On such a view, there is no need to explain the successes of classical logic, because there is little if any success to explain.

However, many advocates of nonclassical logics are not ready to go this far, and want to acknowledge and endorse successful uses of classical logic, while still maintaining that it is in some sense incorrect about real validity. Such advocates, it seems, owe some kind of story about classical logic's success. After all, an advocate of classical logic as a theory of validity has an immediate explanation here: classical logic is successful, where it is, because it gets the facts about validity right. If classical logic does *not* get the facts about validity right, it must be successful, where it is, for some other reason.

In principle, this explanatory debt could be discharged in any number of ways. But there is a particular family of ways to discharge this debt that have come to be known as 'classical recapture'. Here's how I'll use this term. An advocate of a certain nonclassical logic L as a theory of validity is engaged in classical recapture when they point to some logical fact F connecting L to classical logic, and then use the claim that L captures real validity, plus the fact F, to argue that classical logic in fact has some other, success-explaining property.

An example might help. Suppose someone thinks that the real story about validity is given by the paracomplete logic K3.[1] For any sentence A, let $?A$ be the sentence $A \vee \neg A$, and given a set Γ of sentences, let $?\Gamma$ be the set containing $?A$ for each $A \in \Gamma$. Then we have the following fact: An argument $[\Gamma \succ A]$ is classically valid iff $[?\Gamma, \Gamma \succ A]$ is valid in K3. Now, suppose we are looking at some argument $[\Gamma \succ A]$ that is classically valid but not valid in K3, and where reasoning according to that argument seems crucial to classical logic's success in some domain. Then an advocate of K3 can explain this success by appealing to the fact, saying that all the sentences in $?\Gamma$ are in fact true, or perhaps even guaranteed to be true by the nature of the domain in question, or something like that. On this view, it is the real validity of $[?\Gamma, \Gamma \succ A]$ doing the work, but because we can take $?\Gamma$ for granted in some (nonlogical!) sense, appealing to the invalid $[\Gamma \succ A]$ instead does no harm. Moreover, since this connection between K3 and classical logic is systematic, such an approach has some hope to explain the alleged successes of classical logic more generally. (For an example of this kind of strategy, see in particular [3].)

This paper focuses on a different kind of classical recapture, based on an alleged *ambiguity* in the connectives of classical logic. Ideas like this have been around among relevant logicians for as long as there have been relevant logicians; see for example [1] and [25]. Here, I will focus on a statement of this idea found in [16]. Section 2 presents this version of the idea, and section 3 argues that it cannot be

---

[1]For details on K3, see for example [5, 6, 24].

made to work. There is certainly some intuitive appeal here, but I will argue that ambiguity is not a helpful way to develop the idea, and the logical facts cited in [16] do not do the right work. Then, in section 4, I develop a variation on this idea, based on conflation rather than ambiguity, and citing slightly different logical facts. This conflation-based approach is novel, and I believe is the most promising approach yet developed for this kind of approach to classical recapture; but as I will show, it only works given certain assumptions about what real validity is like. Section 5 concludes.

2 Recapture I: Ambiguity

According to the ambiguity thesis, some subclassical logic gives the right story about validity, and yet classical logic is *not really wrong*, not even about validity. Instead, the language of classical logic is *ambiguous*. However, when correctly disambiguated, the pronouncements on validity made by classical logic really are *correct*. To see this, we just need to figure out which pronouncements they are!

The alleged ambiguity in question is in the binary *connectives* of classical logic: conjunction, disjunction, and implication.[2] According to the ambiguity thesis, these are ambiguous between *additive* and *multiplicative* readings, of the sort distinguished in some substructural logics.[3] Before exploring the ambiguity thesis in any detail, then, I'll pause to lay out the additive/multiplicative distinction that is crucial to it.

2.1 Linear and Affine Logics

Two related substructural logics will be useful here; I'll call them simply *linear logic* (or MALL) and *affine logic* (or MAAL).[4] These logics use the same propositional language, which I will call the *linear language* (or $\mathcal{L}_L$). From a countably infinite stock of atomic sentences $p, q, r, \ldots$, the language $\mathcal{L}_L$ builds up complex sentences with a range of connectives. These are zeroary $\top, t, \bot, f$; unary $\neg$; and binary $\sqcap, \otimes, \sqcup, \,\invamp, \sqsupset, \multimap$. Of these, $\top, \bot, \sqcap, \sqcup,$ and $\sqsupset$ are *additive*; $t, f, \otimes, \invamp,$ and $\multimap$ are

---

[2] The paper [21] extends the allegations to cover quantifiers as well, but I leave the quantifiers to the side in this paper.

[3] The additive/multiplicative distinction and its relatives travel under a few names: 'extensional'/'intensional' and 'lattice-theoretic'/'group-theoretic' are other pairs to look out for. I'll stick to 'additive' and 'multiplicative', though.

[4] The logics I'm talking about are more standardly called 'multiplicative-additive linear logic' and 'multiplicative-additive affine logic', which is why I abbreviate them 'MALL' and 'MAAL', each with an 'MA'. But I'll still use 'linear logic' and 'affine logic' to talk about these logics, just to save words—in this paper, I am always talking about the 'multiplicative-additive' version of these. For more on these differences (it's a matter of which connectives are included), see for example [13].

multiplicative; and ¬ alone is neither. The connectives that participate in the additive/multiplicative distinction come in pairs. ⊤ and t are truth constants; ⊥, f falsity constants; ⊓, ⊗ conjunctions, ⊔, ⅋ disjunctions, and ⊐, ⊸ implications.[5]

In all the logics I will look at in this paper, negation can be used to interdefine conjunction, disjunction, and implication, in any of the ways familiar from classical logic. In $\mathcal{L}_{\mathrm{L}}$, this preserves the additive/multiplicative distinction; for example, $A \,⅋\, B$ can be understood as $¬(¬A \otimes ¬B)$, and $A \sqsupset B$ can be understood as $¬(A \sqcap ¬B)$. Similarly, negation can be used to interdefine truth and falsity constants, again preserving the additive/multiplicative distinction; so ⊥ can be understood as ¬⊤, and f can be understood as ¬t. As such, I will focus on the connectives ⊤, t, ¬, ⊓, ⊗; the rest will take care of themselves. (Everything I have to say here plays out the same whether the other connectives are understood as primitive connectives analogous to these or as defined from these via negation; I'm just doing this to save space.)

A *linear sequent* is $[\Gamma \succ \Delta]$, where Γ and Δ are finite multisets drawn from $\mathcal{L}_{\mathrm{L}}$. Fig. 1 presents two sequent calculi **MAAL** and **MALL** for a fragment of this language. These differ only in the presence or absence of the rule D (for dilution). The proof system **MAAL** determines (classical) affine logic without exponentials: a sequent is valid in MAAL iff it can be derived in **MAAL**. The proof system **MALL** determines (classical) linear logic without exponentials: a sequent is valid in MALL iff it can be derived in **MALL**. While I cannot give a full introduction to linear or affine logics here, this proof system bears some comment.

First, **MAAL** and **MALL** include the Id axiom only for *atomic* formulas. This makes no difference to the logic they determine; $[A \succ A]$ is derivable for every A, in either system. But it is still a choice that matters: it is handy in some of the proofs that follow to know that in every derivation of a sequent $[\Gamma \succ \Delta]$, each occurrence of each connective must have been introduced by one of the connective rules.

Second, **MAAL** and **MALL** alike feature no rules of *contraction*, and their derivable sequents are not closed under contraction either. Rules of contraction have it that if a premise or conclusion is needed twice, it suffices to have it once:

$$\text{WL:} \quad \frac{[A, A, \Gamma \succ \Delta]}{[A, \Gamma \succ \Delta]} \qquad\qquad \text{WR:} \quad \frac{[\Gamma \succ \Delta, A, A]}{[\Gamma \succ \Delta, A]}$$

---

[5]There is no single standard notation here, but this is not too far off some usual suspects. It has the virtue of making the additive/multiplicative distinction visible: the additive connectives are squarish, while the multiplicatives are roundish. It has some vices as well (for example, what is sometimes called 'polarity' (see [38]) is masked, and ⊥ is used here as an additive constant, while the same symbol is used in [13] as a multiplicative constant). Unfortunately, past notation choices have left us in a bit of a pickle: all the most natural symbols have been used, and many of them have been used in different sources to mean different ones of these connectives.

Id: $$\overline{[p \succ p]}$$

D: $$\boxed{\frac{[\Gamma \succ \Delta]}{[\Gamma', \Gamma \succ \Delta, \Delta']}}$$

$\neg$L: $$\frac{[\Gamma \succ \Delta, A]}{[\neg A, \Gamma \succ \Delta]}$$

$\neg$R: $$\frac{[A, \Gamma \succ \Delta]}{[\Gamma \succ \Delta, \neg A]}$$

$\sqcap$L: $$\frac{[A/B, \Gamma \succ \Delta]}{[A \sqcap B, \Gamma \succ \Delta]}$$

$\sqcap$R: $$\frac{[\Gamma \succ \Delta, A] \quad [\Gamma \succ \Delta, B]}{[\Gamma \succ \Delta, A \sqcap B]}$$

$\otimes$L: $$\frac{[A, B, \Gamma \succ \Delta]}{[A \otimes B, \Gamma \succ \Delta]}$$

$\otimes$R: $$\frac{[\Gamma \succ \Delta, A] \quad [\Gamma' \succ \Delta', B]}{[\Gamma, \Gamma' \succ \Delta, \Delta', A \otimes B]}$$

$\top$R: $$\overline{[\Gamma \succ \Delta, \top]}$$

tL: $$\frac{[\Gamma \succ \Delta]}{[t, \Gamma \succ \Delta]}$$

tR: $$\overline{[\ \succ t]}$$

MAAL: all rules   **MALL**: without D

Figure 1: **MAAL** and **MALL**

Looked at the other way around, these rules allow for *duplication* of premises and conclusions in searching for a derivation of a particular sequent. (Dilution, when present, allows for premises and conclusions to be *discarded* when read in the same way.) Since these rules are not admissible in **MALL** or **MAAL**, when we think about valid arguments in MALL or MAAL it is important to keep in mind how many times each premise or conclusion appears, as this can affect validity.

2.2 Additive and Multiplicative

It is in this keeping track of number that the additive/multiplicative distinction becomes visible. To get the idea, let's start with conjunction. Think of an occurrence of an additive conjunction $A \sqcap B$ like this: it's a thing that, if you look at it one way it's an occurrence of A, and if you look at it another way it's an occurrence of B.[6] To

---

[6]The famous cover photo of [15] with the intricately-carved wooden blocks is a good analogy to keep in mind, if you're familiar with it.

draw on an occurrence of an additive conjunction successfully as a premise, then, you should be in a place to draw successfully on an occurrence of one of its conjuncts. Then you can use the occurrence of the additive conjunction as an occurrence of that conjunct. This is just what $\sqcap$L gives us. (Actually, this oversimplifies things a bit, since $\sqcap$L isn't an invertible rule. For example, while $p \sqcap q \vdash_{\text{MALL}} p \sqcap q$, still $p \nvdash_{\text{MALL}} p \sqcap q$ and $q \nvdash_{\text{MALL}} p \sqcap q$. That is, sometimes an $A \sqcap B$ will do where neither an A alone nor a B alone will: it can be that what's needed is precisely something that can play either role.)

And to reach an additive conjunction as a conclusion, one must be in a place to reach *either* conjunct as a conclusion. Importantly, there is no need to be able to reach *both* conjuncts—this is why the schematic Γ, Δ in $\sqcap$R appear fully in both premise-sequents. If we conclude $A \sqcap B$ via this rule, and we want to look at it as an A, we can draw on the premise-sequent $[\Gamma \succ \Delta, A]$; if we want to look at it as a B, we can draw on the premise-sequent $[\Gamma \succ \Delta, B]$. Either way, we only use Γ and Δ once each.[7]

On the other hand, think of an occurrence of a multiplicative conjunction $A \otimes B$ as one occurrence of *each* of A and B. It's just a pair; no funny business. To draw on a multiplicative conjunction successfully as a premise, one must be in a place to draw successfully on *the pair* of its conjuncts; this is what $\otimes$L gives us. And to reach a multiplicative conjunction as a conclusion, one must be in a place to reach *both* conjuncts as conclusions. Since there is a need to reach both conjuncts, one must draw on everything needed to reach the first *together with* everything needed to reach the second. This is why the premise-sequents in $\otimes$R feature different contexts, combined in the conclusion-sequent. It is only by reaching an A (using Γ, Δ) and *also* reaching a B (using Γ', Δ') that we can reach an $A \otimes B$.

A similar distinction is afoot in the truth constants. t can be concluded from *nothing* at all, as seen in tR, and correspondingly counts for nothing when drawn on as a premise, as seen in tL. $\top$, on the other hand, can be concluded from *anything* at all, and correspondingly has no left rule.

In MALL, the additive and multiplicative conjunctions are independent of each other. But in MAAL, with dilution, we get some connection. Dilution is a principle that allows us to carry around *extra* premises and conclusions, ones we do not use.

---

[7]As this discussion hopefully makes clear, there is no contraction at all—no duplication or double-use—either explicit or implicit in the $\sqcap$R rule. This is so despite the fact that the rule 'takes in' two Γ's and 'gives out' one, and similarly for Δ, which can create a misleading appearance. For example, [35, p. 510] writes of 'problematic' forms of 'premise and conclusion combination' in this rule; but in fact no combination is occurring at all. The premise-sequents of $\sqcap$R simply note two things: that Γ (as premises) and Δ (as conclusions) would suffice for A as a conclusion, and also that they would suffice for B as a conclusion. Under these circumstances, they do not need to be 'combined' at all, even with themselves, to suffice for $A \sqcap B$ as a conclusion.

Seen from the other end, it allows us to *discard* these extra premises and conclusions as we look at a derivation from bottom to top. If I can discard things, though, $A \otimes B$ suffices for $A \sqcap B$. The reason is simple: with discarding as an option, a pair of an A and a B turns out to be something that if looked at one way gives an A and if looked at another way gives a B. If I want it to give an A I simply use the A from the pair, *discarding* the B; and if I want it to give a B I simply use the B from the pair, *discarding* the A. Reasoning in this way, via D, we can show that $\otimes$ obeys the rule corresponding to $\sqcap$L in MAAL, and that $\sqcap$ obeys the rule corresponding to $\otimes$R in MAAL. In MAAL, then, although not in MALL, we have $A \otimes B \vdash A \sqcap B$.

The situation is slightly different for the truth constants. Since everything implies $\top$ by $\top$R, we have $t \vdash \top$ even in MALL. Adding dilution gives us the other direction, allowing us to show that $\top$ obeys the rule corresponding to tL in MAAL (which is just an instance of D), and that t obeys the rule corresponding to $\top$R in MAAL. In MAAL, then, although not in MALL, the distinction between t and $\top$ disappears.

For applications of these and related logics to the paradoxes, see [4, 14, 22, 27, 29, 33, 35], as well as [16], which comes in for particular discussion in what follows.

2.3 The Ambiguity Thesis

With the distinction between additive and multiplicative in view, we can return to the ambiguity thesis. This thesis comes in two general forms: as a thesis about conjunction, disjunction, etc. *in natural language*, and as a thesis about conjunction, disjunction, etc. *in the language of classical logic*. (See for example [20, §2.1].) For my purposes here, I will set aside the version that focuses on natural language, and deal exclusively with the version of the thesis that focuses on the language of classical logic.[8]

The *classical language* $\mathcal{L}_{\mathrm{C}}$ I'll work with here has the very same stock $p, q, r, \ldots$ of atomic formulas as $\mathcal{L}_{\mathrm{L}}$, but builds compound formulas with different connectives: zeroary $\dagger, \ddagger$, unary $\neg$, and binary $\wedge, \vee, \rightarrow$. Here $\dagger$ is a truth constant, $\ddagger$ a falsity constant, $\neg$ negation, $\wedge$ conjunction, $\vee$ disjunction, and $\rightarrow$ implication; there is no additive/multiplicative distinction to worry about. Just as in the case of $\mathcal{L}_{\mathrm{L}}$, we can use negation to define $\vee$ and $\rightarrow$ from $\wedge$, and $\ddagger$ from $\dagger$, so I will focus on the connectives $\dagger, \neg, \wedge$.

A *classical sequent* is $[\Gamma \succ \Delta]$, with Γ and Δ finite multisets of $\mathcal{L}_{\mathrm{C}}$ formulas. (Note that although classical logic is closed under contraction (and its converse), I still use multisets for classical sequents. So for example $[p \wedge q, p \wedge q \succ r]$ and $[p \wedge q \succ r]$ are distinct sequents.)

---

[8] For remarks about natural language along these lines, see [21], [20, §2.1], or [37, pp. 306ff].

Although the consequence relation of classical logic is familiar, I will need to fix a few proof systems on this language in what follows, and this is as good a place as any to do it. So Fig. 2 gives four proof systems in $\mathcal{L}_C$. Here **NOD⁻** serves as a base system; it contains only the unboxed rules in Fig. 2, and these rules are common to all four systems to be discussed. In **NOD⁻**, there is no dilution, and the connectives $\wedge$ and $\dagger$ are governed by multiplicative-like rules on the left and additive-like rules on the right. (Compare, for example, $\wedge^{\otimes}$L in Fig. 2 to $\otimes$L in Fig. 1.) Starting from this base system, I will consider two selections of rules that can be added: first, we can add dilution or not; and second, we can add the 'missing' rules for $\wedge$ and $\dagger$—that is, additive-like rules on the left and multiplicative-like rules on the right—or not. (These 'missing' rules—$\wedge^{\sqcap}$L, $\wedge^{\otimes}$R, $\dagger^t$R—are given in a single box in Fig. 2.) Those two independent choices then produce four proof systems:

	as is	+D
NOD⁻		
as is	**NOD⁻**	**G3C**⋆
$+\wedge^{\sqcap}$L, $\wedge^{\otimes}$R, $\dagger^t$R	**NOD**	**G3C**⁺

Despite the fact that none of these four systems includes any rule of contraction, two of them—**G3C**⁺ and **G3C**⋆—determine classical logic.[9] The others—**NOD** and **NOD⁻**—do not; I'll come back to these later.

The central idea for recapture by ambiguity is this: no classically valid sequent is *mistaken*. If properly understood, it really is valid. But this proper understanding can require disambiguation. In particular, the classical connective $\wedge$ should be seen as ambiguous between the linear connectives $\sqcap$ and $\otimes$. (The same for $\vee$ between $\sqcup$ and $⅋$, and $\rightarrow$ between $\sqsupset$ and $\multimap$.)

As [16, p. 456] puts the point:

> Differently from what most relevant logicians claim, we do not maintain that there are *invalid* inferential principles in classical propositional logic; if properly disambiguated, i.e. given the right interpretation of the logical constants and of the propositional variables contained therein, all laws of classical logic can be salvaged in MALL.

> Failing to disambiguate, however, can yield paralogisms.

I note here, just for the sake of setting it aside, that [16] diagnoses an additional ambiguity, in the notion of validity, between what they call 'internal consequence'

---

[9]To see this, note that **G3C**⋆ is closely related to the contraction-free proof system **G3cp** for classical logic studied in [18, pp. 49ff]. All the rules of **G3C**⋆ are sound for classical logic, and all the rules of **G3cp** are derivable in **G3C**⋆, and that suffices for **G3C**⋆ to determine classical logic. The additional rules in **G3C**⁺ over and above those in **G3C**⋆ are already admissible in **G3cp** (since also sound for classical logic), and indeed already derivable in **G3C**⋆.

$$\text{Id:} \quad \frac{}{[p \succ p]} \qquad\qquad \text{D:} \quad \boxed{\frac{[\Gamma \succ \Delta]}{[\Gamma', \Gamma \succ \Delta, \Delta']}}$$

$$\neg\text{L:} \quad \frac{[\Gamma \succ \Delta, A]}{[\neg A, \Gamma \succ \Delta]} \qquad\qquad \neg\text{R:} \quad \frac{[A, \Gamma \succ \Delta]}{[\Gamma \succ \Delta, \neg A]}$$

$$\wedge^{\sqcap}\text{L:} \quad \boxed{\frac{[A/B, \Gamma \succ \Delta]}{[A \wedge B, \Gamma \succ \Delta]}} \qquad\qquad \wedge^{\sqcap}\text{R:} \quad \frac{[\Gamma \succ \Delta, A] \quad [\Gamma \succ \Delta, B]}{[\Gamma \succ \Delta, A \wedge B]}$$

$$\wedge^{\otimes}\text{L:} \quad \frac{[A, B, \Gamma \succ \Delta]}{[A \wedge B, \Gamma \succ \Delta]} \qquad\qquad \wedge^{\otimes}\text{R:} \quad \boxed{\frac{[\Gamma \succ \Delta, A] \quad [\Gamma' \succ \Delta', B]}{[\Gamma, \Gamma' \succ \Delta, \Delta', A \wedge B]}}$$

$$\dagger^{\top}\text{R:} \quad \frac{}{[\Gamma \succ \Delta, \dagger]}$$

$$\dagger^{\mathrm{t}}\text{L:} \quad \frac{[\Gamma \succ \Delta]}{[\dagger, \Gamma \succ \Delta]} \qquad\qquad \dagger^{\mathrm{t}}\text{R:} \quad \boxed{\frac{}{[\ \succ \dagger]}}$$

Four systems: $\mathbf{G3C^+}$ = full system; $\quad \mathbf{G3C^\star} = \mathbf{G3C^+}$ − single-box rules; $\mathbf{NOD} = \mathbf{G3C^+}$ − D; $\quad \mathbf{NOD^-}$ = unboxed rules

Figure 2: Systems for the classical language

and 'external consequence' with respect to a sequent system. I do not discuss that idea in this paper; the ambiguity in connectives alleged is entirely with respect to 'internal' validity, and that is the only notion of validity I work with here.

2.4 Translations: Grishin and Ono

But how can we be sure that classical logic really does not go wrong, that for any classical sequent there really is a disambiguation that renders it really valid? To make this claim precise, we need to choose a notion of 'real validity' for $\mathcal{L}_{\mathrm{L}}$. Depending on whether we choose MAAL or MALL, different ways of proceeding are appropriate.[10]

---

[10]Other choices of 'real validity' for $\mathcal{L}_{\mathrm{L}}$ are not in view in this paper.

The paper [16] goes with MALL, and appeals to what is there called the 'Grishin–Ono translations'; see [19] and [34, p. 48ff]. For my purposes here, it is important to distinguish Grishin's original translations from Ono's modifications of them; I'll accordingly refer to them as the 'Grishin translations' and the 'Ono translations' respectively.

Each of these approaches works with two translations—negative and positive—from $\mathcal{L}_\mathrm{C}$ to $\mathcal{L}_\mathrm{L}$. The Grishin translations $^\pm$ and the Ono translations $^{o\pm}$ are given in Fig. 3.[11] They differ from each other only on atomic formulas: the recursive clauses are the same. The Grishin translations are the identity on atomic formulas, while the Ono translations are not; they involve additively combining multiplicative constants.

Grishin ($^\pm$) and Ono ($^{o\pm}$) translations ($\mathcal{L}_\mathrm{C}$ to $\mathcal{L}_\mathrm{L}$)

C	C^-	C^+	C^{o-}	C^{o+}
p	p	p	$\mathsf{t} \sqcap p$	$\mathsf{f} \sqcup p$
$\dagger$	t	$\top$	t	$\top$
$\ddagger$	$\bot$	f	$\bot$	f
$\neg A$	$\neg A^+$	$\neg A^-$	$\neg A^{o+}$	$\neg A^{o-}$
$A \wedge B$	$A^- \otimes B^-$	$A^+ \sqcap B^+$	$A^{o-} \otimes B^{o-}$	$A^{o+} \sqcap B^{o+}$
$A \vee B$	$A^- \sqcup B^-$	$A^+ \mathbin{\rotatebox[origin=c]{180}{\&}} B^+$	$A^{o-} \sqcup B^{o-}$	$A^{o+} \mathbin{\rotatebox[origin=c]{180}{\&}} B^{o+}$
$A \to B$	$A^+ \sqsupset B^-$	$A^- \multimap B^+$	$A^{o+} \sqsupset B^{o-}$	$A^{o-} \multimap B^{o+}$

Figure 3: The Grishin and Ono translations

These translations are important because of the following facts:

Fact 1 (Grishin). $\Gamma \vdash_\mathrm{CL} \Delta$ *iff* $\Gamma^- \vdash_\mathrm{MAAL} \Delta^+$.

Fact 2 (Ono). $\Gamma \vdash_\mathrm{CL} \Delta$ *iff* $\Gamma^{o-} \vdash_\mathrm{MALL} \Delta^{o+}$.

While these facts are known, it is useful for what follows to rehearse the proof of Fact 1. (See again [34, pp. 48ff] for a closely related proof of Fact 2.) This proof draws on the proof system **G3C***, presented above. It also draws on a translation β I will call the *blurring translation* from $\mathcal{L}_\mathrm{L}$ to $\mathcal{L}_\mathrm{C}$; since the blurring translation turns out to be important later, I pause to present it separately, in Fig. 4. The blurring

---

[11] Since disjunctions, conditionals, and falsity constants are all here officially understood as defined, some lines in Fig. 3 are redundant; I include them nonetheless to save interested readers some time.

L	p	$\top$	t	$\neg B$	$B \sqcap C$	$B \otimes C$
$\beta(L)$	p	$\dagger$	$\dagger$	$\neg\beta(B)$	$\beta(B) \wedge \beta(C)$	$\beta(B) \wedge \beta(C)$

Figure 4: The blurring translation β

translation simply collapses additive and multiplicative connectives together: either conjunction in $\mathcal{L}_\mathrm{L}$ is translated as the one conjunction in $\mathcal{L}_\mathrm{C}$, and the same for all other kinds of connectives.

With this translation in hand, we can proceed to the proof of Fact 1. The proof proceeds by using $\pm$ to translate $\mathbf{G3C}^\star$ derivations into $\mathbf{MAAL}$ derivations, and conversely using β to translate $\mathbf{MAAL}$ derivations into $\mathbf{G3C}^\star$ derivations.

Proof. LTR: if $[\Gamma \succ \Delta]$ is classically valid, it can be derived in $\mathbf{G3C}^\star$; so take a $\mathbf{G3C}^\star$ derivation of it. For each sequent $[\Sigma \succ \Theta]$ occurring in this derivation, replace it with $[\Sigma^- \succ \Theta^+]$. The result is a derivation in $\mathbf{MAAL}$ of $[\Gamma^- \succ \Delta^+]$, as can be determined by inspecting the rules of both systems. For example, when a conjunction is inserted on the left in $\mathbf{G3C}^\star$, it must use the rule $\wedge^\otimes\mathrm{L}$. The negative translation interprets a conjunction on the left as $\otimes$, which obeys the corresponding $\otimes\mathrm{L}$ in $\mathbf{MAAL}$.

RTL: if $[\Gamma^- \succ \Delta^+]$ is valid in MAAL, it can be derived in $\mathbf{MAAL}$; so take a $\mathbf{MAAL}$ derivation of it. For each sequent $[\Sigma \succ \Theta]$ occurring in this derivation, replace it with $[\beta[\Sigma] \succ \beta[\Theta]]$. The result is a derivation in $\mathbf{G3C}^\star$ of $[\beta[\Gamma^-] \succ \beta[\Delta^+]]$, as can again be determined by inspecting the rules of both systems. But for any $A \in \mathcal{L}_\mathrm{C}$, it holds that $\beta(A^\pm) = A$; so this is a $\mathbf{G3C}^\star$ derivation of $[\Gamma \succ \Delta]$. $\qquad\square$

Proof of Fact 2 is similar, with some extra fiddling to take care of the absence of the dilution rule in $\mathbf{MALL}$. It is this extra fiddling that requires fiddling with atomic formulas in the translation: while we cannot dilute in general in $\mathbf{MALL}$, we *can* 'dilute' with t on the left (thanks to $\mathsf{t}\mathrm{L}$) and f on the right, and so with $\mathsf{t} \sqcap p$ on the left (thanks to $\sqcap\mathrm{L}$) and $\mathsf{f} \sqcup p$ on the right. Moreover, if we can dilute with the components of a compound sentence, we can dilute with the compound sentence itself. So we *do* have dilution in $\mathbf{MALL}$ for all those formulas in the image of the Ono translations. That takes care of the LTR direction. For the RTL direction, note that while in general $\beta \circ o^\pm (A) \neq A$, still $\beta \circ o^\pm (A) \dashv\vdash_\mathrm{CL} A$.

Note in particular—this will come back later—that the translations $\pm$ or $o^\pm$ are at the philosophical heart of the ambiguity thesis, since these disambiguation translations are the key formal representation of ambiguity we are offered. The blurring function β, on the other hand, is a mere technical auxiliary, invoked simply to demonstrate that $\pm$ and $o^\pm$ behave as advertised.

3 Problems for Ambiguity Recapture

3.1 Atoms

One problem for this approach to recapture involves the difference between the Grishin translations and the Ono translations. This difference itself forms a dilemma for an ambiguity approach based on MALL, like the approach of [16]. A hypothetical ambiguity approach based on MAAL instead of MALL would be constrained by this issue, but would not run into such serious trouble. Let me explain.

According to the ambiguity thesis, it is the *connectives* of $\mathcal{L}_C$ that are ambiguous; we have no reason to suspect any ambiguity in the *atomic formulas*. Even if it is true that the classical connective $\wedge$ is ambiguous between the linear connectives $\sqcap$ and $\otimes$, this is no reason to think that the atomic formula p is ambiguous between the linear formulas $\mathsf{t} \sqcap p$ and $\mathsf{f} \sqcup p$.

Indeed, the idea that atomic formulas are ambiguous in this way threatens to lead to a regress. We might think that the ps in $\mathsf{t} \sqcap p$ and $\mathsf{f} \sqcup p$ must themselves be ambiguous in the same way; so p turns out to be four-ways ambiguous, between $\mathsf{t} \sqcap (\mathsf{t} \sqcap p)$, $\mathsf{t} \sqcap (\mathsf{f} \sqcup p)$, $\mathsf{f} \sqcup (\mathsf{t} \sqcap p)$, and $\mathsf{f} \sqcup (\mathsf{f} \sqcup p)$. But then all these ps too are presumably ambiguous in the same way, so we have an eight-way ambiguity, and then a sixteen-way ambiguity, and so on.[12] Now, there's nothing absurd about the idea of something's being ambiguous between infinitely many readings. But in this case, we have no way of unambiguously stating *any* of these readings; all we can do is use the ambiguous p again, wrapped up in more and more complex formulas. Moreover, the claim has to be that this is the case for *every* atomic sentence. And none of this is motivated at all by the idea that the classical connectives are ambiguous.

If we are to fill in the ambiguity thesis in a motivated way, then, we should use the Grishin translations rather than the Ono translations. The Grishin translations 'disambiguate' the classical connectives, but leave atomic formulas completely alone, as they should. For an advocate of MAAL, this is all well and good (until §3.2); Fact 1 provides the desired recapture.

But this just pushes an advocate of MALL onto an equally pointy horn: Fact 2 crucially requires the Ono translations, and will not work if we substitute the Grishin translations instead. The paper [16], in advocating ambiguity-based recapture for MALL, acknowledges this issue but tries to sidestep it; I think the point deserves a closer look. Here is the complete discussion offered there:

> [I]f properly disambiguated; i.e. given the right interpretation of the logical
> connectives and of the propositional variables contained therein, all laws of

---

[12] As the editor pointed out, if we count ambiguities only up to logical equivalence, these numbers are lower, since for example $\mathsf{t} \sqcap (\mathsf{t} \sqcap p)$ is equivalent in MALL (and so also in MAAL) to $\mathsf{t} \sqcap p$; and $\mathsf{f} \sqcup (\mathsf{f} \sqcup p)$ is equivalent to $\mathsf{f} \sqcup p$; and so on. However, as the editor also noted, the point is unaffected by this, since the ambiguities would still head off to infinity, just at a slower pace.

classical logic can be salvaged in MALL [p. 456].

There is an important footnote to this sentence:

> The qualification 'and of the propositional variables' cannot be dispensed with. [Statement of Fact 2 here.] However, if we confine ourselves to the classical tautologies that play a role in the known versions of the paradoxes, you do not need to replace propositional variables in order to get to a theorem of MALL.

This footnote alludes to the distinction between the Grishin and Ono translations, and acknowledges that the Grishin translations applied to MALL do not succeed in recapturing classical logic.

The footnote also, in its last sentence, suggests a different tactic: apply the Grishin translations to *linear* logic, and don't worry about recapturing full classical logic. Instead, the suggestion seems to be that it is enough if we recapture 'the classical tautologies that play a role in the known versions of the paradoxes'. This is a suggestion worth considering. When we follow through on it, here is what we get:

Fact 3. $\Gamma \vdash_{\mathrm{NOD}^-} \Delta$ *iff* $\Gamma^- \vdash_{\mathrm{MALL}} \Delta^+$.

Proof. Just as in the proof of Fact 1, but now using the proof systems **MALL** and **NOD$^-$** instead of **MAAL** and **G3C***. The only difference is the absence here of the rule D in each system, but D in each system is only used in the above proof when there is already a use of D in the other system, so removing D from both systems at once creates no problems. $\square$

So if we take Mares & Paoli's development of ambiguity recapture seriously, it seems that what gets recaptured is not classical logic at all, but instead NOD$^-$. NOD$^-$, it's worth noticing, is not terribly classically-flavoured at all. Here are some examples. First, NOD$^-$ is not reflexive; while $p \vdash_{\mathrm{NOD}^-} p$, we also have $p \wedge q \nvdash_{\mathrm{NOD}^-} p \wedge q$, since $p \otimes q \nvdash_{\mathrm{MALL}} p \sqcap q$. This same example shows that NOD$^-$ is not closed under uniform substitution. As a second example, conjunctions do not always entail their conjuncts in NOD$^-$: $p \wedge q \nvdash_{\mathrm{NOD}^-} p$, since $p \otimes q \nvdash_{\mathrm{MALL}} p$. And as a third example, NOD$^-$ lacks distribution of conjunction over disjunction: $p \wedge (q \vee r) \nvdash_{\mathrm{NOD}^-} (p \wedge q) \vee (p \wedge r)$, since $p \otimes (q \sqcup r) \nvdash_{\mathrm{MALL}} (p \sqcap q) \,\mathscr{B}\, (p \sqcap r)$. So this is not classical recapture at all, but instead NOD$^-$ recapture. Whence the success, then, of classical logic? We are left with no explanation.

But the last quoted sentence from [16] suggests a different goal: not to explain the success of *all* of classical logic, but rather just the success of those principles that make trouble when paradoxes are around. This is a more limited goal, but could still be of some importance. However, even such a limited goal can't be achieved in this way.

The paper [16] does not say which classical tautologies we should be thinking of as the ones that play a role in the paradoxes, but it is not hard to fill in some usual suspects. One place we might look is at classically valid explosion arguments, like $[A, \neg A \succ\ ]$ or $[A \wedge \neg A \succ\ ]$. These are indeed valid in NOD$^-$; the first because it itself is valid in MALL, and the second because $[A \otimes \neg A \succ\ ]$ is. Or we might look at classically valid excluded middle arguments, like $[\ \succ A, \neg A]$ or $[\ \succ A \vee \neg A]$. These are likewise valid in NOD$^-$, for the same reasons, mutatis mutandis.[13]

But although those kinds of classical validity do play an important role in getting trouble to arise from the paradoxes, there are two more that also leap immediately to mind: $\rightarrow$ contraction, in the form $[\ \succ (A \rightarrow (A \rightarrow B)) \rightarrow (A \rightarrow B)]$, and pseudo modus ponens, in the form $[\ \succ (A \wedge (A \rightarrow B) \rightarrow B]$.[14]

With these $\rightarrow$-involving classical tautologies, the situation is different than it was with explosion or excluded middle. Applying the Grishin translations to $\rightarrow$ contraction gives $[\ \succ (A^+ \sqsupset (A^+ \sqsupset B^-)) \multimap (A^- \multimap B^+)]$; starting from the instance $[\ \succ (p \rightarrow (p \rightarrow q)) \rightarrow (p \rightarrow q)]$ thus gives $[\ \succ (p \sqsupset (p \sqsupset q)) \multimap (p \multimap q)]$. This, though, is *not* valid in MALL. And applying the Grishin translations to pseudo modus ponens gives $[\ \succ (A^+ \otimes (A^+ \sqsupset B^-)) \multimap B^+]$; starting from the instance $\succ (p \wedge (p \rightarrow q)) \rightarrow q]$ thus gives $[\ \succ (p \otimes (p \sqsupset q)) \multimap q]$. This isn't valid in MALL either. So [16]'s claim seems to be false: these are paradigms of 'classical tautologies that play a role in the known versions of the paradoxes', but disambiguating them with the Grishin translations does not yield linear tautologies.

Combining the Grishin translations with MALL, then, is no way at all to recapture classical logic; it recaptures only the weak (and somewhat bizarre) logic NOD$^-$. In addition, contrary to what [16] claims, NOD$^-$ is too weak to get any grip on at least some standard principles that lead to truth-theoretic paradox. To my mind, this all suggests that the strategy of recapture by ambiguity is a better fit for advocates of MAAL than it is for advocates of MALL: these advocates can stick to the Grishin translations and still arrive at full classical recapture.

3.2   How Ambiguity Doesn't Work

But there is a bigger problem with this whole way of thinking, one not sensitive to these details. Whether we are looking at MAAL or MALL, at $^\pm$ or $^{o\pm}$, *this isn't*

---

[13]The editor pointed out that derivations of these explosion arguments might cause no trouble in the absence of right weakening, and that derivations of these excluded middle arguments might cause no trouble in the absence of left weakening. Exactly so; see also [10].

So we might prefer to consider argument forms like $[A, \neg A \succ B]$ or $[B \succ A, \neg A]$ instead. These, though, are not valid in NOD$^-$, so I have stuck with the unweakened forms in the main text, so that we can see some examples where the claim of [16] holds.

[14]See [26, 36] for discussion and details.

how ambiguity works! The Grishin and Ono translations both require care around negative and positive occurrences of formulas: occurrences in negative position in a sequent are translated with $^-$ or $^{o-}$, and occurrences in positive position with $^+$ or $^{o+}$. Ambiguous terms, though, do not interact with the polarity of their positions in this way.

'Bat' is ambiguous between a type of mammal and a type of sporting equipment. So if I say to you 'I sure do like bats', I've made an ambiguous claim; absent context, you may well not know how to interpret my utterance. But it's not as though: when we draw on 'I sure do like bats' as a *premise* it's about mammals, but when we *conclude* it it's about sporting equipment. And in claims like 'Either I like bats or I don't like bats', the negation doesn't force us to interpret the second occurrence *differently* from the first. Ambiguous words are simply not sensitive to whether they occur positively or negatively; their ambiguity is orthogonal to this.

But this means that whatever phenomenon the Grishin and Ono translations target, it isn't ambiguity. These translations do need to take care of positive and negative occurrences, and they translate connectives differently in positive and in negative positions. Such a discipline does not fit with the ambiguity story; it's introduced in order to take advantage of Fact 2. But that's just to say that appealing to Fact 2 is not an appropriate way to fill in the ambiguity story.

Drawing these problems together, then, I conclude that the ambiguity story, at least as it is developed in [16], is in trouble. It requires either an unmotivated and implausible translation of atomic formulas or else does not work to recapture crucial classical principles, and anyway depends on careful attention to positive and negative occurrences unlike any familiar case of ambiguity. The first problem can be avoided by using MAAL rather than MALL as our story about 'real validity', but the second problem hits MAAL and MALL alike. This is not, then, how to go about recapture.

4   Recapture II: Conflation

In the rest of the paper, I will suggest and develop an alternate path to classical recapture. This path is immediately inspired by the ambiguity approach criticized above, and is immensely similar in spirit. My goal is to spell out something in the area in a more plausible and workable way. The key will be this: the translation β, in the above a mere technical auxiliary useful for proving Fact 1, will here bear the central philosophical weight. This is motivated by understanding the classical connectives as *conflations* of their linear counterparts, and drawing on the theory of conflation developed and defended in [30, 31]. Let's see how it works.

4.1 Conflation

To *conflate* is to treat distinct things as one. For example, in many ordinary conversations it is harmless to treat weight and mass as though they were the same, say by converting between pounds (a measure of weight) and kilograms (a measure of mass), and so this is what we often do. Weight and mass are not the same, but since in the situations we usually find ourselves in gravity doesn't vary all that much, we do no harm, and save ourselves some trouble, by conflating in this way. (Conflation, of course, is not always harmless. Treating mass and weight as the same thing would be a disaster for calculating spacecraft trajectories, for example.)

Conflation is a different phenomenon from ambiguity. 'Case', for example, is ambiguous. When we read and understand a sentence like 'Of course his case attracted attention although it was wrinkled and worn', part of our understanding involves retrieving the appropriate disambiguation of 'case'.[15] This is not what happens in cases of conflation. Take, for example, the case of Fred and the ant farm, taken from [8, 9]:

> Fred has an ant colony in his kitchen in which there are two big ants, which we will call 'Ant A' and 'Ant B'. Big ants make every effort to avoid conflict and so our two ants arrange to split their time between running around on the surface performing feats of strength, and napping down in the bowels of the ant colony. Fred never catches on to the fact that there are in fact two big ants and decides to name 'the' big ant 'Charley' [9, p. 692].

Now, when Fred says 'Charley never seems to sleep', this is not in any way an *ambiguous* utterance. It is not that, to understand what he's said, we retrieve the meaning that Ant A never seems to sleep, or the meaning that Ant B never seems to sleep. Rather, it seems that to understand what Fred's said here, we need to know that Fred means to be talking about a single ant that he calls 'Charley'.[16]

So ambiguity and conflation are different phenomena. Here I will develop and explore the idea that $\mathcal{L}_C$ offers a *conflated* view of $\mathcal{L}_L$, conflated exactly along additive/multiplicative lines. On this view, it's not that $\wedge$ is an ambiguous connective, sometimes meaning $\sqcap$ and sometimes $\otimes$. Rather, both meanings are always present; $\wedge$ *runs together* these distinct meanings. Users of $\wedge$ are ignoring the difference between $\sqcap$ and $\otimes$, treating them as the same. They conflate the linear connectives.

---

[15]This sentence is taken from [17].

[16]In this case, Fred's conflation results from a false belief—and no doubt will result *in* further false beliefs. But this is an inessential feature of the story, not necessary for conflation. The weight/mass example shows this: we can (and often do) conflate weight and mass simply because the difference between them does not matter for purposes at hand, and not because of any kind of error.

4.2   The Blurring Apparatus

The formal treatment of conflation I develop in [31] is limited in at least two ways. First, it restricts itself to conflation of *propositions*, not objects, properties, or operations on propositions like $\sqcap$ and $\otimes$. Second, it works with sequents involving *sets* of premises and conclusions, not any structures with more texture. For present purposes, it's clearly crucial to go beyond the second limitation: we're working with multisets here, and need to draw more distinctions than can be captured with just sets. This section will lay out a multiset-based version of [31]'s approach; for more details on everything here other than the specifically multiset-related stuff, see that paper.

I won't here, though, go beyond the first limitation: I will still consider only conflation of full propositions. This is enough to capture the phenomena of interest here, thanks to the translation β, which shows how the proposed conflation in connectives lifts to conflation of propositions. Given $A, B \in \mathcal{L}_\mathrm{L}$, I will say $\mathcal{L}_\mathrm{C}$ conflates them iff $\beta(A) = \beta(B)$.

Blurring begins with four ingredients: two languages, a relation between them, and a consequence relation on the first language. From these ingredients, we brew up a consequence relation for the second language. The first language, and the consequence relation on it, should be understood as the unconflated language, with validity on this language; the second language should be understood to contain the conflation in question; and the relation tells us whether and how propositions from the first language are conflated in the second. The consequence relation that blurring outputs then is to be understood as validity for the conflated language. This consequence relation counts an argument as valid in the conflated language iff the argument is related to an argument valid in the unconflated language in the right way, with the needed relation between arguments to be determined by the given relation between languages.

In this case, then, our two languages are $\mathcal{L}_\mathrm{L}$ and $\mathcal{L}_\mathrm{C}$, the relation between them is given by β (in this case, it happens to be a function, but this is not crucial to the blurring process), and we can consider either MAAL or MALL as the consequence relation from which we begin. Blurring will yield a consequence relation for $\mathcal{L}_\mathrm{C}$, according to which a $\mathcal{L}_\mathrm{C}$ argument is valid iff it is β-related to an $\mathcal{L}_\mathrm{L}$ argument valid in the starting consequence relation (either MAAL or MALL).

To make this work, we need to lift β from formulas to multisets of formulas, taking a multiset Γ of formulas from $\mathcal{L}_\mathrm{L}$ to a multiset $\beta[\Gamma]$ of formulas from $\mathcal{L}_\mathrm{C}$. Here's how: when Γ is the empty multiset, then $\beta[\Gamma]$ is also the empty multiset; and when Γ is Σ, A, then $\beta[\Gamma] = \beta[\Sigma], \beta(A)$.[17] So, for example, $\beta[\neg p, q \sqcap (p \otimes r), q \otimes (p \otimes$

---

[17]Thanks to an anonymous referee for simplifying what I had here in an earlier version; as the

$r), q \sqcap (p \sqcap r)] = [\neg p, q \wedge (p \wedge r), q \wedge (p \wedge r), q \wedge (p \wedge r)]$. With this lifting in hand, the blurring is straightforward:

Definition 4. For $\mathcal{L}_C$ multisets Γ, Δ and a notion $\vdash_X$ of validity for $\mathcal{L}_L$: $\Gamma \Vdash_X \Delta$ iff there are $\mathcal{L}_L$ multisets Γ', Δ' such that 1) $\Gamma' \vdash_X \Delta'$; 2) $\beta[\Gamma'] = \Gamma$; and 3) $\beta[\Delta'] = \Delta$.

That is, $\Vdash_{\mathrm{MAAL}}$ is the consequence relation for $\mathcal{L}_C$ we arrive at by blurring $\vdash_{\mathrm{MAAL}}$ according to β as above, and likewise $\Vdash_{\mathrm{MALL}}$ is the consequence relation for $\mathcal{L}_L$ we arrive at by blurring $\vdash_{\mathrm{MALL}}$ according to β as above.

4.3 The Results

So if MALL is the correct story of validity for $\mathcal{L}_L$, and if the connectives of $\mathcal{L}_C$ really do conflate the connectives of $\mathcal{L}_L$ in the imagined way, then on this view $\Vdash_{\mathrm{MALL}}$ really is the correct story for validity in $\mathcal{L}_C$. It does not ever go wrong; it simply gives its correct verdicts in a blurry way. Similarly, if MAAL is the correct story of validity for $\mathcal{L}_L$, then it is $\Vdash_{\mathrm{MAAL}}$ that is correct for $\mathcal{L}_C$. This gives us a way to motivate consequence relations for $\mathcal{L}_C$ based on a story very like the original ambiguity picture. But what consequence relations have we thereby motivated?

Theorem 5. $\Vdash_{\mathrm{MALL}}$ *is* NOD. $\Vdash_{\mathrm{MAAL}}$ *is* CL.

Proof. Here I give the proof for $\Vdash_{\mathrm{MAAL}}$ and CL; the proof for $\Vdash_{\mathrm{MALL}}$ and NOD is just the same, mutatis mutandis.

First, that $\Gamma \Vdash_{\mathrm{MAAL}} \Delta$ implies $\Gamma \vdash_{\mathrm{CL}} \Delta$. Suppose $\Gamma \Vdash_{\mathrm{MAAL}} \Delta$. Then there is a derivation of $[\Gamma \succ \Delta]$ in **MAAL**. Take this derivation, and apply β to every formula in it. The result is a derivation of $[\beta[\Gamma] \succ \beta[\Delta]]$ in **G3C**$^+$, as can be determined by inspection of the rules of these systems.[18]

Second, that $\Gamma \vdash_{\mathrm{CL}} \Delta$ implies $\Gamma \Vdash_{\mathrm{MAAL}} \Delta$. Suppose $\Gamma \vdash_{\mathrm{CL}} \Delta$. Then there is a derivation $\mathcal{D}$ of $[\Gamma \succ \Delta]$ in **G3C**$^+$. We want to show that there is some derivation $\mathcal{E}$ in **MAAL** of some $[\Gamma' \succ \Delta']$ such that $\beta[\Gamma'] = \Gamma$ and $\beta[\Delta'] = \Delta$; this suffices to show that $\Gamma \Vdash_{\mathrm{MAAL}} \Delta$.[19]

---

referee notes, everything in this paper is finite, so working on one formula at a time like this is enough.

[18]Note that the extra rules that **G3C**$^+$ has beyond **G3C*** are put to use here: the **MAAL** derivation we begin from is arbitrary, and so can apply rules like $\sqcap$L and $\otimes$R that are never needed in the (similar) RTL direction of the proof of Fact 1. (These extra rules of **G3C*** are derivable in **G3C***, so the difference is one of ease, not anything more.)

[19]All we need is that the conclusion sequent of $\mathcal{E}$ is related to that of $\mathcal{D}$ in this way. In fact, the derivation $\mathcal{E}$ we arrive at will be much more closely related than this to $\mathcal{D}$; but this extra closeness is not appealed to in the proof.

We can show this by induction on the derivation $\mathcal{D}$. Note at the outset (since this comes in handy a few times below) that β is surjective, and so given some $A \in \mathcal{L}_{\mathrm{C}}$ we can always help ourselves to an $A' \in \mathcal{L}_{\mathrm{L}}$ with $\beta(A') = A$, and similarly for entire multisets instead of single formulas.

- In the base case, $\mathcal{D}$ is either a single application of Id, of $\dagger^{\top}\mathrm{R}$, or of $\dagger^{\mathrm{t}}\mathrm{R}$. If it is Id or $\dagger^{\mathrm{t}}\mathrm{R}$, then $\mathcal{E}$ can consist of Id or $\mathrm{t}R$ straightforwardly. If it is $\dagger^{\top}\mathrm{R}$, then its conclusion sequent is some $[\Sigma \succ \Theta, \dagger]$; we need to choose Σ', Θ' from $\mathcal{L}_{\mathrm{L}}$ such that $\beta[\Sigma'] = \Sigma$ and $\beta[\Theta'] = \Theta$, and then E can consist of $\top\mathrm{R}$ with conclusion sequent $[\Sigma' \succ \Theta', \top]$.

- For the rule $\neg\mathrm{L}$, $\mathcal{D}$ is of the form:

$$\mathcal{D}'$$
$$\vdots$$
$$\neg\mathrm{L}: \quad \frac{[\Sigma \succ \Theta, A]}{[\neg A, \Sigma \succ \Theta]}$$

By the inductive hypothesis, there is some derivation $\mathcal{E}'$ in **MAAL** with conclusion sequent $[\Sigma' \succ \Theta', A']$ such that $\beta[\Sigma'] = \Sigma$, $\beta[\Theta'] = \Theta$, and $\beta(A') = A$. We can then extend $\mathcal{E}'$ with an application of $\neg\mathrm{L}$ to reach an **MAAL** derivation of $[\neg A', \Sigma' \succ \Theta']$; and since $\beta(\neg A') = \neg A$, this is all we need.

- For the rule $\wedge^{\sqcap}\mathrm{L}$, $\mathcal{D}$ is of the form:

$$\mathcal{D}'$$
$$\vdots$$
$$\wedge^{\sqcap}\mathrm{L}: \quad \frac{[A/B, \Sigma \succ \Theta]}{[A \wedge B, \Sigma \succ \Theta]}$$

Wlog, let the conclusion sequent of $\mathcal{D}'$ be $[A, \Sigma \succ \Theta]$. By the inductive hypothesis, there is some derivation $\mathcal{E}'$ in **MAAL** with conclusion sequent $[A', \Sigma' \succ \Theta']$ such that $\beta[\Sigma'] = \Sigma$, $\beta[\Theta'] = \Theta$, and $\beta(A') = A$. We can then extend $\mathcal{E}'$ with an application of $\sqcap\mathrm{L}$ to reach an **MAAL** derivation of $[A' \sqcap B', \Sigma' \succ \Theta']$, choosing some B' such that $\beta(B') = B$. Since $\beta(A' \sqcap B') = A \wedge B$, this is all we need.

These cases are enough to give the general flavour of the proof.[20] $\qquad\square$

---

[20]Importantly, this proof strategy would fail if we add a rule like contraction to **G3C^{+}**, *even if* we also add contraction to the underlying system **MAAL**. This is because a derivation of a sequent like $[A, A, \Sigma \succ \Theta]$ might have arrived at the two As in two different ways, and so yield an underlying **MAAL** derivation of some $[A', A'', \Sigma' \succ \Theta']$ where A' is a distinct formula from A''; in such a case, even if contraction were available it couldn't be applied.

5 Conclusion

I've explored the ambiguity-based approach to classical recapture advocated in [16], and argued that it cannot be made to work as they present it, for two main reasons. First, to recapture classical logic from MALL as they aim to do, their approach uses the Ono translations; but these translations mess with all atomic sentences in unmotivated ways. An advocate of MAAL rather than MALL could avoid this, however, by using the Grishin translations instead. Second, and more importantly, this entire approach to recapture, based as it is on positive and negative translation pairs, does not fit with the phenomenon of ambiguity, which is simply insensitive to polarity. So the theoretical picture developed in [16] does not fit the logical facts invoked there.

I've tried to resolve this issue by slightly changing the theoretical picture and pointing to slightly different logical facts. On the approach to recapture I explore here, the important phenomenon is not ambiguity but conflation: running two things together. Rather than saying that classical conjunction is ambiguous between additive and multiplicative conjunction, meaning additive conjunction in some occurrences and multiplicative in others, I say that classical conjunction *conflates* additive and multiplicative conjunction. To use classical conjunction is precisely to *not distinguish* additive from multiplicative.

I've drawn on the treatment of conflation in [30, 31] to develop this idea, extending this treatment for the first time to multisets. On this approach, if the connectives of $\mathcal{L}_\mathrm{C}$ conflate those of $\mathcal{L}_\mathrm{L}$, then validity in $\mathcal{L}_\mathrm{C}$ depends on validity for $\mathcal{L}_\mathrm{L}$. As I've shown, if validity for $\mathcal{L}_\mathrm{L}$ is given by MAAL, then the resulting validity for $\mathcal{L}_\mathrm{C}$ is exactly classical. This gives a potential explanation for the success of classical logic: the classical language runs together additive and multiplicative, but the difference between additive and multiplicative does not always matter. Classical logic gives a blurry picture of affine validity, and sometimes a blurry picture is enough.

If, however, validity for $\mathcal{L}_\mathrm{L}$ is given by MALL instead of MAAL, then we do not have full classical recapture. Instead, we have the same kind of 'recapture' story, but around the logic determined by **NOD** instead, which is weaker than classical. This is an interesting logic, and I'm curious to explore it more in the future, but for now I'll leave it there.

References

[1] Alan Ross Anderson and Nuel D. Belnap. *Entailment: The Logic of Relevance and Necessity*, volume 1. Princeton University Press, Princeton, NJ, 1975.

[2] Jc Beall. *Spandrels of Truth*. Oxford University Press, Oxford, 2009.

[3] Jc Beall. LP+, K3+, FDE+, and their 'classical collapse'. *Review of Symbolic Logic*, 6(4):742–754, 2013.

[4] Jc Beall and Julien Murzi. Two flavors of Curry's paradox. *Journal of Philosophy*, 110(3):143–165, 2013.

[5] Jc Beall and Bas C. van Fraassen. *Possibilities and Paradox: An Introduction to Modal and Many-valued Logic*. Oxford University Press, Oxford, 2003.

[6] Jc Beall, Michael Glanzberg, and Ellie Ripley. *Formal Theories of Truth*. Oxford University Press, Oxford, 2018.

[7] Ross T. Brady. *Universal Logic*. Number 109 in CSLI Lecture Notes. CSLI Publications, Stanford, CA, 2006.

[8] Joseph L. Camp, Jr. *Confusion: A Study in the Theory of Knowledge*. Harvard University Press, Cambridge, MA, 2002.

[9] Joseph L. Camp, Jr. Précis of *Confusion*. *Philosophy and Phenomenological Research*, 74(3):692–699, 2007.

[10] Bruno Da Re. Structural weakening and paradoxes. *Notre Dame Journal of Formal Logic*, 62(2):369–398, 2021.

[11] Hartry Field. *Saving Truth from Paradox*. Oxford University Press, Oxford, 2008.

[12] Rohan French. Structural reflexivity and the paradoxes of self-reference. *Ergo*, 3(5):113–131, 2016.

[13] Jean-Yves Girard. Linear logic: Its syntax and semantics. In J.-Y. Girard, Y. Lafont, and L. Regnier, editors, *Advances in Linear Logic*, pages 1–42. Cambridge University Press, 1995.

[14] Vyacheslav Nikolaevich Grishin. Predicate and set-theoretic calculi based on logic without contractions [in Russian]. *Izvestiya Akademii Nauk SSSR, Ser. Mat.*, 45(1):47–68, 1981. doi: 10.1070/IM1982v018n01ABEH001382.

[15] Douglas Hofstadter. *Gödel, Escher, Bach: An Eternal Golden Braid*. Basic Books, 1979.

[16] Edwin Mares and Francesco Paoli. Logical consequence and the paradoxes. *Journal of Philosophical Logic*, 43(2–3):439–469, 2014.

[17] Robert A. Mason and Marcel Adam Just. Lexical ambiguity in sentence comprehension. *Brain Research*, 1146:115–127, 2007.

[18] Sara Negri and Jan von Plato. *Structural Proof Theory*. Cambridge University Press, Cambridge, 2001.

[19] Hiroakira Ono. Structural rules and a logical hierarchy. In P. P. Petkov, editor, *Mathematical Logic*, pages 95–104. Plenum Press, New York, 1990.

[20] Francesco Paoli. *Substructural Logics: A Primer*. Kluwer Academic Publishing, Dordrecht, 2002.

[21] Francesco Paoli. The ambiguity of quantifiers. *Philosophical Studies*, 124(3): 313–330, 2005.

[22] Uwe Petersen. Logic without contraction as based on inclusion and unrestricted abstraction. *Studia Logica*, 64(3):365–403, 2000.

[23] Graham Priest. *In Contradiction*. Oxford University Press, Oxford, 2006.

[24] Graham Priest. *An Introduction to Non-classical Logic: From If to Is*. Cambridge University Press, Cambridge, 2nd edition, 2008.

[25] Stephen Read. *Relevant Logic: A Philosophical Examination of Inference*. Basil Blackwell, Oxford, 1988.

[26] Greg Restall. How to be *really* contraction free. *Studia Logica*, 52(3):381–391, 1993.

[27] Greg Restall. *On logics without contraction*. PhD thesis, The University of Queensland, 1994.

[28] Ellie Ripley. Paradoxes and failures of cut. *Australasian Journal of Philosophy*, 91(1):139–164, 2013.

[29] Ellie Ripley. Comparing substructural theories of truth. *Ergo*, 2(13):299–328, 2015.

[30] Ellie Ripley. Vagueness is a kind of conflation. *Logic and Logical Philosophy*, 26(1):115–135, 2017.

[31] Ellie Ripley. Blurring: An approach to conflation. *Notre Dame Journal of Formal Logic*, 59(2):171–188, 2018.

[32] Richard Routley, Robert K. Meyer, Val Plumwood, and Ross T. Brady. *Relevant Logics and their Rivals 1*. Ridgeview, Atascadero, CA, 1982.

[33] Lionel Shapiro. Deflating logical consequence. *Philosophical Quarterly*, 61(243): 320–342, 2011.

[34] Anne Sjerp Troelstra. *Lectures on Linear Logic*. CSLI Publications, Stanford, CA, 1992.

[35] Elia Zardini. Truth without contra(di)ction. *Review of Symbolic Logic*, 4(4): 498–535, 2011.

[36] Elia Zardini. Naive modus ponens. *Journal of Philosophical Logic*, 42(4):575–593, 2013.

[37] Elia Zardini. Restriction by noncontraction. *Notre Dame Journal of Formal Logic*, 57(2):287–327, 2016.

[38] Noam Zeilberger. On the unity of duality. *Annals of Pure and Applied Logic*, 153(1):66–96, April 2008.

Received 13 March 2025

Algebras for Relevant Reasoners

Igor Sedlár

Institute of Computer Science, Czech Academy of Sciences
`<sedlar@cs.cas.cz>`

Abstract

The informational interpretations of relevant logic suggest that it provides a natural framework for the development of epistemic logic. This paper proposes a simple extension of relevant modal logic to formalise reasoning about relevant reasoners situated in classical worlds. This approach avoids many of the technical challenges of previous proposals and allows for straightforward algebraic generalisation. The main technical result is a representation theorem for a class of relevant algebras, which provides a solid foundation for further exploration of relevant epistemic logic.

2020 *Mathematics Subject Classification.* Primary: 03B47, Secondary: 03B42, 03B45.

1 Introduction

The informational interpretations of relevant logic [12, 13, 16, 19] suggest that it provides a natural framework within which epistemic logic can be developed. Compared to epistemic logic based on classical logic, relevant logic offers a more fine-grained framework tracking relevance relations between individual pieces of information. Versions of epistemic logic based on relevant modal logic have been proposed, for example, in [3, 2, 15, 27, 28].

An interesting avenue of research, with the potential to integrate relevant logic into mainstream classical epistemic logic, is to consider versions of relevant epistemic logic in which agents are modelled as reasoning relevantly while situated in classical possible worlds. This line of research goes back at least to [11], with more recent

---

A preliminary version of this article was presented at the *2nd Third Workshop*, at the University of Alberta, Canada, on May 13th, 2024. I am grateful to Katalin Bimbó for organising the workshop and inviting me, and to the workshop audience for their many insightful comments. This work was supported by the long-term strategic development funding of the Institute of Computer Science (RVO:67985807) and the Czech Science Foundation project 22-01137S.

contributions being those in [7, 22, 23, 24, 25, 26]. Although intuitively appealing, these versions of relevant epistemic logic usually present a number of technical challenges.

In this paper, we propose a simple extension of relevant modal logic that is able to formalise reasoning about relevant reasoners situated in classical worlds, while avoiding many of the subtleties of previous approaches. The advantage of our approach is also that it allows for straightforward algebraic generalisation. Indeed, the main technical result presented in this paper is a representation result for a class of relevant algebras.

The paper is structured as follows. In Section 2, we recall the backbone of our contribution, namely, frames and algebras for relevant modal logics. In Section 3, we introduce relevant epistemic frames, a variant of frames for relevant modal logics with a subset of normal situations comprising classical possible worlds. This semantics is in the same spirit as [24], but it avoids a number of complications encountered there. In Section 4, we introduce a class of algebras that generalises our relational semantics and we prove our main result — namely, that every algebra is isomorphic to a complex algebra of a relational frame.

2 Modal Relevant Frames and Algebras

In this section, we recall modal relevant frames and their algebraic counterparts, modal relevant algebras. We recall the notions of a complex algebra and a prime filter frame, which allow us to move between the relational world of frames and the operational world of algebras. Finally, we recall the representation result for modal relevant algebras, namely, that every modal relevant algebra is isomorphic to a complex modal relevant algebra. The results presented in this section are well known and are presented here for the benefit of the non-expert reader. We will build on them in the following sections.

2.1 Frames

We begin by recalling modal relevant frames [8, 9] and their informal interpretation.

Definition 1. *A* modal relevant frame *is a relational structure*

$$\mathfrak{R} = \langle S, \leqslant, N, R, {}^*, Q \rangle, \quad where$$

- $\langle S, \leqslant \rangle$ *is a pre-ordered set ($\leqslant$ is reflexive and transitive);*

- $N \subseteq S$ *such that $s \in N$ and $s \leqslant t$ only if $t \in N$ (N is an up-set);*

- R is a ternary relation on S such that, if $Rstu$, then $s' \leqslant s$, $t' \leqslant t$ and $u \leqslant u'$ only if $Rs't'u'$ (R is antitone in the first two positions and monotone in the third position);

- $s \leqslant t$ iff $\exists v\,(v \in N\ \&\ Rvst)$;

- $*$ is a unary operation on S such that $s \leqslant t$ only if $t^* \leqslant s^*$ ($*$ is antitone) and

- $s^{**} = s$ ($*$ is involutive);

- Q is a binary relation on S such that, if Qst, then $s' \leqslant s$ and $t \leqslant t'$ only if $Qs't'$ (Q is antitone in its first position and monotone in its second position).

Structures $\langle S, \leqslant, N, R \rangle$ that satisfy the first four conditions are the positive frames of [20], which provide relational semantics for the basic positive relevant logic $\mathbf{B^+}$. Modal relevant frames without the assumption that $s^{**} = s$ are frames for modal relevant logics studied by Fuhrmann [8, 9]; the class of all such frames determines the basic conjunctively regular relevant modal logic $\mathbf{BM.C}$. Modal relevant frames as defined here determine the modal relevant logic $\mathbf{B.C}$. We assume that $*$ is involutive for the sake of simplicity; see Remark 1 in Subsection 2.2.

Let us discuss the intuitive ideas behind the definition of modal relevant frames.

Situations. Elements of S represent *situations* (or theories or bodies of information). These contain, or *support*, information about the world. This information is not required to be truthful, not even consistent, complete or closed under the laws of classical modal logic. Situations are not possible worlds. Situations can be naturally *ordered* by the amount of information they support. In particular, $s \leqslant t$ means that t supports at least as much information as s. In turn, all possible *pieces of information* that can be thought of as being supported by a situation in a given frame can be modelled as up-sets of situations in that frame. More about situations and their relation to the semantics of relevant logics can be found in [12, 13, 19].

The ternary relation. The ternary relation R on situations represents *information combination*. In particular, $Rstu$ can be read as 'u supports all information that results from combining information supported by s with information supported by t'. Naturally, s supports 'x implies y', where x and y are pieces of information, iff $Rstu$ and t supports x only if u supports y, for all situations t and u. More about interpretations of the ternary relation can be found in [1, 13, 16, 19].

Normal situations. Situations are in general not closed under classical modal logic, but they are closed under the *informational constraints* of the frame. Informational constraints are represented by pairs of pieces of information $\langle x, y \rangle$ such that every situation that supports x supports y. The set N of *normal situations* has the characteristic feature that a pair $\langle x, y \rangle$ is an informational constraint iff 'x implies y' is supported by all situations in N.[1]

The Routley star. Given the nature of situations, supporting x is independent from supporting 'not x'. Situations may be incomplete and even inconsistent. Nevertheless, the definition of modal relevant frames assumes that the set of situations is 'negation-complete' in the sense that for each s there is a unique s^* such that s^* supports exactly those x such that s does not support 'not x'. The condition $s = s^{**}$ corresponds to the assumption that x is the same information as 'not not x'. More on the Routley star operation $*$ can be found in [6, 17, 21].

The modal accessibility relation. We assume that situations can contain information about a fixed agent, in particular about what the *agent's information about the world* is. The binary relation Q then expresses the second-order information about the agent's information in an indirect way, namely, Qst means that all information available to the agent according to s is supported by t. The set $Q[s] = \{t : Qst\}$ is an up-set thanks to the assumption that Q is monotone in its second position; therefore, it expresses a piece of information, namely, the information the agent has according to s (the agent's *information state* according to s).

2.2 Algebras

We proceed by recalling the notion of a modal relevant algebra. While frames are intuitively understood as structures comprising situations that support pieces of information defined in terms of situations, algebras are structures consisting of pieces of information taken as primitives.

---

[1]This explains the frame condition linking N, R and $\leqslant$. Firstly, assume that $s \leqslant t$ and consider the following pieces of information: $x = \{s' : s \leqslant s'\}$ and $y = \{t' : t' \not\leqslant t\}$. Obviously, $\langle x, y \rangle$ is not an informational constraint since s supports x but not y. Therefore, there is a $v \in N$ such that v does not support 'x implies y'. In other words, there are u and u' such that $Rvuu'$, u supports x but u' does not support y. But this means that $s \leqslant u$ and $u' \leqslant t$. Consequently, $Rvst$. Secondly, assume that $v \in N$ and $Rvst$. If s supports x, then t supports x as well since presumably 'x implies x' expresses a reasonable informational constraint.

Definition 2. *A modal relevant algebra is an algebra*

$$X = \langle X, \vee, \wedge, \cdot, \rightarrow, \neg, 1, \square \rangle \quad such \ that$$

- $\langle X, \vee, \wedge \rangle$ *is a distributive lattice (with $x \leq y$ defined as $x \vee y = y$);*

- $\langle X, \cdot, 1 \rangle$ *is a groupoid ($\cdot$ is an arbitrary binary operation) with left identity, that is, $1 \cdot x = x$;*

- *multiplication $\cdot$ distributes over $\vee$ from both sides;*

- $x \cdot y \leq z$ *iff $x \leq y \rightarrow z$ (residuation);*

- $\neg(x \vee y) = \neg x \wedge \neg y$ *and* $\neg\neg x = x$;

- $\square(x \wedge y) = \square x \wedge \square y$.

Note that $\bigwedge_{i=1}^{n} x_i \leq y$ implies $\bigwedge_{i=1}^{n} \square x_i \leq \square y$. $\{\neg, \square\}$-free reducts of modal relevant algebras are known as positive Ackermann groupoids, and are studied in [14]. These algebras provide an algebraic semantics for $\mathbf{B}^+$.[2] To the best of our knowledge, this particular definition of a modal relevant algebra is new, but it is a straightforward variation on well-established definitions.

Intuitively, modal relevant algebras can be thought of as algebras of *pieces of information* (or propositions), with operations generating 'x and y' ($x \wedge y$), 'x or y' ($x \vee y$), 'x combined with y' ($x \cdot y$), 'x implies y' ($x \rightarrow y$) and 'the agent has information that x' ($\square x$) from pieces of information x and y. The partial order $\leq$ can be seen as a representation of the *entailment* relation between pieces of information. The fact that $\bigwedge_{i=1}^{n} x_i \leq y$ implies $\bigwedge_{i=1}^{n} \square x_i \leq \square y$ means that the information available to the agent is closed under entailment in the given algebra. The element 1 can be thought of as the piece of information that entails informational constraints. In particular, by left identity and residuation, $x \leq y$ iff $1 \cdot x \leq y$ iff $1 \leq x \rightarrow y$. Given this interpretation, and the previously discussed interpretation of modal relevant frames, it is not surprising that the prime example of modal relevant algebra arises from up-sets in modal relevant frames.

Definition 3. *The* full complex algebra *of a modal relevant frame $\mathfrak{R} = \langle S, \leqslant, N, R, {}^*, Q \rangle$ is the algebra*

$$\mathfrak{R}^+ = \langle \mathcal{U}(S), \cup, \cap, \cdot, \rightarrow, \neg, N, \square \rangle, \quad where$$

---

[2]'Relevant algebras' studied by Urquhart [29] do not contain $\square$, they do not require $\neg\neg x = x$, but they do require the dual De Morgan law $\neg(x \wedge y) = \neg x \vee \neg y$ which is derivable in the presence of $\neg\neg x = x$. Importantly Urquhart's relevant algebras are based on *bounded* distributive lattices and it is assumed that $x \cdot \bot = \bot = \bot \cdot x$, $\neg\top = \bot$ and $\neg\bot = \top$.

- $\mathcal{U}(S)$ *is the collection of all up-sets on* $\langle S, \leqslant \rangle$;

- $x \cdot y = \{t \colon \exists s, t \, (Rstu \,\&\, s \in x \,\&\, t \in y)\}$;

- $x \to y = \{s \colon \forall t, u \, (Rstu \,\&\, t \in x \Rightarrow u \in y)\}$;

- $\neg x = \{s \colon s^* \notin x\}$;

- $\Box x = \{s \colon \forall t \, (Qst \Rightarrow t \in x)\}$.

Note that full complex algebras are well defined thanks to the assumption that R in modal relevant frames is antitone in the first position (which takes care of $x \to y$) and monotone in the third position (which takes care of $x \cdot y$), the assumption that $*$ is antitone (which takes care of $\neg x$), the assumption that Q is antitone in the first position (which takes care of $\Box x$), and the assumption that N is an up-set.

The full complex algebra of a frame is used in the definition of a *model* based on the frame. A model is, in fact, a homomorphism from a suitable algebra of formulas to the full complex algebra of the frame.

Lemma 1. $\mathfrak{R}^+$ *is a modal relevant algebra for every modal relevant frame* $\mathfrak{R}$.

Proof. Essentially [14] and [8]. $\qquad\qquad\square$

Definition 4. *A complex modal relevant algebra is a modal relevant algebra that is a subalgebra of the full complex algebra of a modal relevant frame.*

Recall that a *prime filter* on a distributive lattice $\langle D, \vee, \wedge \rangle$ is a set $P \subseteq D$ such that $P \notin \{\emptyset, D\}$ (P is non-empty and 'proper'), $x \wedge y \in P$ iff $x \in P$ and $y \in P$, and $x \vee y \in P$ iff $x \in P$ or $y \in P$. Note that prime filters are up-sets on $\langle D, \leq \rangle$.

Definition 5. *The* prime filter frame *of a modal relevant algebra* $\boldsymbol{X} = \langle X, \vee, \wedge, \cdot, \to, \neg, 1, \Box \rangle$ *is the relational structure*

$$\boldsymbol{X}_+ = \langle \mathcal{F}_P(\boldsymbol{X}), \subseteq, N, R, {}^*, Q \rangle, \quad \text{where}$$

- $\mathcal{F}_P(\boldsymbol{X})$ *is the set of all prime filters on* $\boldsymbol{X}$;

- $s \in N$ *iff* $1 \in s$;

- *Rstu iff, for all* x *and* y*, if* $x \to y \in s$ *and* $x \in t$*, then* $y \in u$;

- $s^* = \{x \colon \neg x \notin s\}$;

- *Qst iff, for all* x*, if* $\Box x \in s$*, then* $x \in t$.

Remark 1. Note that we need to show that $*$ is well defined, that is, we have to show that s^* is a prime filter on $\boldsymbol{X}$. This follows from the De Morgan laws that hold in $\boldsymbol{X}$ and $x = \neg\neg x$. In particular, the double negation law makes sure that s^* is non-empty and proper for all s. If we did not assume the law of double negation, we would have two options. First, we could work with a pair of *split negations* $\neg$ and $\sim$ such that $\sim\neg x = x$ and $x = \neg\sim x$; see [18, pp. 59–72] for example. This would force us to consider modal relevant frames with a *pair* of Routley stars. Second, we could extend the set of states in the prime filter frame to include the empty set and the carrier set of the underlying modal relevant algebra.[3] This option would cause problems in the proof of the representation result of Section 4; see Remark 2.

An example of prime filter frames familiar to logicians are *canonical frames*, that is, prime filter frames of Lindenbaum–Tarski algebras of (some) logics extending the Distributive Lattice Logic.

Lemma 2. $\boldsymbol{X_+}$ *is a modal relevant frame for each modal relevant algebra* $\boldsymbol{X}$.

Proof. Essentially [14] and [8]. $\qquad\square$

2.3 Representation

Now we are ready to recall an interesting fact, namely, that every 'abstract' modal relevant algebra is isomorphic to a complex modal relevant algebra. This is a representation result similar to Stone's representation theorem for Boolean algebras and Jónsson and Tarski's representation result for Boolean algebras with operators. (See [4, 5].)

A central tool for representation of expansions of distributive lattices is the well-known Prime Filter Lemma.

Lemma 3 (Prime Filter Lemma). *Let* $\langle D, \vee, \wedge, \{o_i\}_{i \in J}\rangle$ *be an algebra with a distributive lattice reduct. Assume that* $F, I \subseteq D$ *such that* F *is a filter,* I *is an ideal and* $F \cap I = \emptyset$. *Then there is a prime filter* P *such that* $F \subseteq P$ *and* $P \cap I = \emptyset$. *In particular, if* $x \notin F$, *then there is a prime filter* $P \supseteq F$ *such that* $x \notin P$.

Proof. See [10, p. 84], where the lemma is stated in a dual form (prime ideals instead of prime filters). $\qquad\square$

Theorem 1. *Every modal relevant algebra is isomorphic to a complex modal relevant algebra.*

---

[3]Call s a 'weak prime filter' on a distributive lattice D iff s is a prime filter on D or $s \in \{\emptyset, D\}$. Prime filters are sometimes defined in such a way that they can be non-proper (e.g., in [5]), or even empty.

Proof. Let $X = \langle X, \vee, \wedge, \cdot, \rightarrow, \neg, \Box, 1 \rangle$ be a modal relevant algebra. By Lemmas 1 and 2, $(X_+)^+$ is a complex modal relevant algebra, and so is each of its subalgebras. Let $C(X)$ be the subalgebra of $(X_+)^+$ consisting of sets $e(x) = \{s \in \mathcal{U}(\mathcal{F}_P(X)) : x \in s\}$, with $e\left(\star(x_1, \ldots, x_n)\right) =_{df} \star^{C(X)}\left(e(x_1), \ldots, e(x_n)\right)$. It can be shown that $e : x \mapsto e(x)$ is an isomorphism between X and $C(X)$. The argument showing that e is injective uses the Prime Filter Lemma. For more details see [4, 5, 9, 14]. $\qquad\square$

3 Relevant Epistemic Frames

In this section, we introduce relevant epistemic frames (Subsection 3.1). We discuss how they can be used to represent reasoning about relevant reasoners situated in classical possible worlds and how they relate to the framework of [24] (Subsection 3.2).

3.1 Worlds

Definition 6. *A relevant epistemic frame is a relational structure*

$$\mathfrak{F} = \langle S, \leqslant, N, K, R, {}^*, Q \rangle \quad \text{such that}$$

- $\langle S, \leqslant, N, R, {}^*, Q \rangle$ *is a modal relevant frame;*

- $K \subseteq S$ *is a non-empty up-set such that, for all $s \in K$ and all $t, u \in S$:*

(1) Rs^*s^*s

(2) $Rtus^* \implies (t \leqslant s \,\&\, u \leqslant s)$

(3) $Rstu \implies (t \leqslant s \,\&\, s^* \leqslant u)$

We define $W =_{df} \{s : s \in K \,\&\, s^* \in K\}$.

Before we motivate the definition, let us note some of its immediate consequences.

Lemma 4. *The following hold in each relevant epistemic frame:*

1. $K \subseteq N$;

2. if $s \in K$, then $s^* \leqslant s$;

3. if $s^* \in K$, then $s \leqslant s^*$.

Proof. 1. For all s there is a $t \in N$ such that Rts^*s^*. If $s \in K$ then, by (2), $t \leqslant s$, which means that $s \in N$. 2. As before, we have Rts^*s^* and so, by (2), $s^* \leqslant s$. 3. If $s^* \in K$ then, by the previous point, $s^{**} \leqslant s^*$, which entails $s \leqslant s^*$ by the fact that $*$ is involutive. $\qquad\square$

The conditions (1)–(3) are motivated by the following lemma.

Lemma 5. *For all $w \in W$ in a relevant epistemic frame and all up-sets x, y on that frame:*

   1. $w \in \neg x$ iff $w \notin x$;

   2. $w \in x \cdot y$ iff $w \in x$ and $w \in y$;

   3. $w \in x \to y$ iff $w \notin x$ or $w \in y$.

Proof. 1. If $w \in \neg x$, then $w^* \notin x$. Since we have $w \leqslant w^*$ by Lemma 4, we obtain $w \notin x$. If $w \notin x$, then $w^* \notin x$ since we have $w^* \leqslant w$ by Lemma 4. Hence, we obtain $w \in \neg x$.

   2. If $w \in x \cdot y$, then there are s and t such that $Rstw$, $s \in x$ and $t \in y$. Since $w^* \in K$ by the definition of W, we can use Lemma 4(3.) to infer that $Rstw^*$. By (2), $s \leqslant w$ and $t \leqslant w$, which means that $w \in x$ and $w \in y$. Conversely, if $w \in x$ and $w \in y$, then $w^* \in x$ and $w^* \in y$ by Lemma 4(3.) and $w^* \in K$. Hence, by (1), $w \in x \cdot y$.

   3. If $w \notin x \to y$, then there are s and t such that $Rwst$, $s \in x$ and $t \notin y$. By (3), $s \leqslant w$ and $w^* \leqslant t$. By Lemma 4, $w \leqslant t$. It follows that $w \in x$ and $w \notin y$. Conversely, if $w \in x$ and $w \notin y$, then $w^* \in x$ by Lemma 4. Hence, $w \notin x \to y$ by (1). $\qquad\square$

It follows from Lemma 5 that situations $w \in W$ behave like classical possible worlds. The following example shows that relevant epistemic frames appear in familiar places.

Example 1. An example of a relevant epistemic frame is the canonical frame for **B.C** where K is the set of prime **B.C**-theories that contain all substitution instances of classical tautologies.[4] K is clearly an up-set. Note that $s^* \in K$ iff s does not contain any negation of a classical tautology. Hence, W is the set of prime **B.C**-theories which are *classical* and are *not contra-classical*. Clearly, W is not an up-set.

   Assume that $s \in K$. To show that (1) is satisfied, we need to show that $\varphi \to \psi \in s^*$ and $\varphi \in s^*$ implies $\psi \in s$. The assumption that $\varphi \to \psi \in s^*$ and $\varphi \in s^*$ entails

---

[4]Assuming that $(\varphi \cdot \psi) \leftrightarrow (\varphi \wedge \psi)$ is taken to be a classical tautology.

that $\neg(\varphi \to \psi) \notin s$ and $\neg\varphi \notin s$. If $\psi \notin s$, then $\neg\varphi \vee \psi \notin s$. Since $\neg(\varphi \to \psi) \vee (\neg\varphi \vee \psi)$ is a classical tautology and $\chi \vee \neg\chi \in s$ for all χ (LEM), we obtain $\neg(\varphi \to \psi) \in s$. This is a contradiction, so $\psi \in s$. To show that (2) is satisfied, we have to show that if $Rtus^*$, then $t \cup u \subseteq s$. Assume that $\varphi \in t$. We know that $\psi \in u$ for some ψ. It follows that $\varphi \cdot \psi \in s^*$ and so $(\neg\varphi \cdot \psi) \notin s$. Note that $\neg(\varphi \cdot \psi) \vee (\varphi \wedge \psi)$ is a classical tautology. By LEM, $\varphi \wedge \psi \in s$. Hence, $\varphi \in t$ implies $\varphi \in s$. The fact that $\varphi \in u$ implies $\varphi \in s$ can be shown analogously. Finally, to show that (3) is satisfied, assume that $Rstu$. We have to show that $t \subseteq s$ and $s^* \subseteq u$. Assume that $\varphi \in t$. We know that $\psi \notin u$ for some ψ. Hence, $\varphi \to \psi \notin s$. Since $\varphi \vee (\varphi \to \psi)$ is a classical tautology, we obtain $\varphi \in s$ by LEM. It follows that, in general, $t \subseteq s$. Now assume that $\varphi \in s^*$. Hence, $\neg\varphi \notin s$. We know that $\psi \in t$ for some ψ. Since $\neg\varphi \vee (\psi \to \varphi)$ is a classical tautology, we obtain $\psi \to \varphi \in s$ by LEM. It follows from the assumption that $Rstu$ that $\varphi \in u$. Hence, in general, $s^* \subseteq u$.

3.2 Discussion

In our discussion of the epistemic interpretation of relevant epistemic frames, we will make use of their complex algebras.

Definition 7. *The* full complex algebra *of a relevant epistemic frame* $\mathfrak{F} = \langle S, \leqslant, N, K, R, {}^*, Q \rangle$ *is the algebra*

$$\mathfrak{F}^+ = \langle \mathcal{U}(S), \cup, \cap, \cdot, \to, \neg, \Box, N, K \rangle,$$

where the operations $\cdot, \to, \neg$ *and* $\Box$ *on up-sets are defined as for full complex algebras of modal relevant frames.*

Let Pr be a countable set of propositional variables. The set Fm of *epistemic formulas* is defined using the grammar

$$\varphi := p \mid 0 \mid 1 \mid \neg\varphi \mid \varphi \wedge \varphi \mid \varphi \vee \varphi \mid \varphi \cdot \varphi \mid \varphi \to \varphi \mid \Box\varphi$$

where $p \in Pr$. The set of formulas Fm forms an algebra of the same type as complex algebras of frames. An *interpretation* of Fm in $\mathfrak{F}$ is a homomorphism $I \colon Fm \to \mathfrak{F}^+$. A formula φ is *valid* in $\mathfrak{F}$ iff $N \subseteq I(\varphi)$ for all $I \colon Fm \to \mathfrak{F}^+$. Validity in a class of frames is defined as expected. The *logic* of a class of frames H, denoted as $L(\mathsf{H})$, is the set of formulas that are valid in H.

Given a model $\mathfrak{M} = \langle \mathfrak{F}, I \rangle$, we say that a situation s supports φ iff $s \in I(\varphi)$. We will also write $s \vDash \varphi$ to indicate this. Lemma 5 implies that situations $w \in W$ are classical possible worlds. So we can read $w \vDash \varphi$ as saying that φ *holds in the world* w. Hence, $w \vDash \Box\varphi$ means that in the word w the agent has information that supports

φ. Even though w is a classical possible world, the information the agent has in w, represented by their epistemic state $Q[w] = \{s : Qws\}$, does not have to respect classical logic. Looking at Example 1 again, one easily finds worlds supporting $\Box\varphi$ without supporting $\Box(\psi \to \varphi)$, for instance.

We say that φ is *classically valid* in a class of frames H iff $W \subseteq I(\varphi)$ for all interpretations $I \colon Fm \to \mathfrak{F}^+$ in all $\mathfrak{F} \in \mathsf{H}$. Note that $W \subseteq I(\varphi)$ iff $K \cap (S \setminus \neg K) \subseteq I(\varphi)$ iff $K \subseteq I(\varphi) \cup \neg K$. The *classical logic* of H, denoted as $CL(\mathsf{H})$, is the set of formulas that are classically valid in H. Clearly, $L(\mathsf{H}) \subseteq CL(\mathsf{H})$. Moreover,

$$(4) \qquad \varphi \to \psi \in L(\mathsf{H}) \implies \Box\varphi \to \Box\psi \in CL(\mathsf{H})$$

for all H. If CPL is the set of substitution instances of classical propositional tautologies in the language of Fm, then we clearly also have

$$(5) \qquad CPL \subseteq CL(\mathsf{H})$$

for all H. As we noted earlier, $\varphi \to \psi \in CL(\mathsf{H})$ does not imply $\Box\varphi \to \Box\psi \in CL(\mathsf{H})$. Consequently, $CL(\mathsf{H})$ is an epistemic logic extending CPL modelling *relevant reasoners*, that is, agents whose epistemic attitudes are closed under the (relevant) logic $L(\mathsf{H})$, but not under the classical logic $CL(\mathsf{H})$ itself. From the viewpoint of this logic, relevant logic is restricted to the scope of the modal operator $\Box$ as shown in Figure 1 on the left.

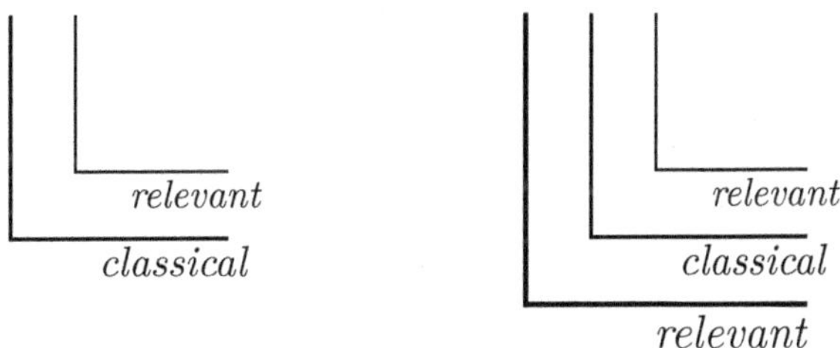

Figure 1: The architecture of the frameworks in [24] (left) and in this paper (right).

In order to axiomatise $CL(\mathsf{H})$, one would need to have some means of representing $L(\mathsf{H})$ inside $CL(\mathsf{H})$. In a similar setting, the paper [24] introduces an auxiliary modal operator $\Box_L$ with the characteristic property (transposed to our setting) that

$$(6) \qquad \Box_L\varphi \in CL(\mathsf{H}) \iff \varphi \in L(\mathsf{H})$$

As we noted in [24], this approach needs to deal with a number of technicalities.

The approach behind the framework presented in this paper is different. Recall that $I(0) = K$ for all interpretations. Note that

$$(7) \qquad \varphi \in CL(\mathsf{H}) \iff 0 \to (\varphi \vee \neg 0) \in L(\mathsf{H})$$

Consequently, the classical logic of H is hidden in the relevant logic of H, as shown in Figure 3.2 on the right.

To highlight the contrast with [24], we can define $\Box_{CL}\varphi := 0 \to (\varphi \vee \neg 0)$ and note that (7) amounts to

$$\varphi \in CL(\mathsf{H}) \iff \Box_{CL}\varphi \in L(\mathsf{H}) \, .$$

Note that $\Box_{CL}$ is a 'proper' modal operator since it distributes over conjunctions: $s \vDash \Box_{CL}(\varphi \wedge \psi)$ iff $s \vDash \Box_{CL}\varphi$ and $s \vDash \Box_{CL}\psi$. Thanks to the constant 0, the logic of all frames is axiomatised quite easily, as shown by the results of the next section.

4 Relevant Epistemic Algebras and Their Representation

In this section we introduce the algebraic counterpart of relevant epistemic frames, namely relevant epistemic algebras, and we prove the corresponding representation result, which states that every relevant epistemic algebra is isomorphic to a complex relevant epistemic algebra arising from a relevant epistemic frame.

4.1 Relevant Epistemic Algebras

Definition 8. *A relevant epistemic algebra is an algebra*

$$\boldsymbol{E} = \langle X, \vee, \wedge, \cdot, \to, \neg, \Box, 1, 0 \rangle$$

such that $\langle X, \vee, \wedge, \cdot, \to, \neg, \Box, 1 \rangle$ *is a modal relevant algebra and, for all* $x, y \in X$:

$$(8) \quad 0 \leq \neg(x \wedge y) \vee (x \cdot y)$$
$$(9) \quad 0 \leq \neg(x \cdot y) \vee (x \wedge y)$$
$$(10) \quad 0 \leq (x \to y) \vee (x \wedge \neg y)$$

In what follows we will write $x \supset y$ instead of $\neg x \vee y$ and $x \sqsubseteq y$ instead of $0 \wedge x \leq y \vee \neg 0$. Note that $\sqsubseteq$ can be seen to be a preorder on X.

Lemma 6. *The following hold in each relevant epistemic algebra:*

1. $0 \le 1$;

2. $x \le y$ implies $x \sqsubseteq y$ and $\Box x \sqsubseteq \Box y$;

3. it is not the case that $x \sqsubseteq y$ implies $\Box x \sqsubseteq \Box y$;

4. $x \sqsubseteq y$ implies $\neg y \sqsubseteq \neg x$;

5. $x \sqsubseteq y$ implies $x \wedge z \sqsubseteq y \wedge z$ and $x \vee z \sqsubseteq y \vee z$;

6. $0 \le x \vee \neg x$ and $x \wedge \neg x \le \neg 0$;

7. $x \sqsubseteq \neg y \vee z$ iff $x \wedge y \sqsubseteq z$;

8. $0 \sqsubseteq \neg x \vee y$ iff $x \sqsubseteq y$;

9. $x \sqsubseteq y$ implies $x \cdot z \sqsubseteq y \cdot z$ and $z \cdot x \sqsubseteq z \cdot y$;

10. $\neg 0 \sqsubseteq x$ and $x \sqsubseteq 0$;

11. $(x \sqsubseteq z \,\&\, y \sqsubseteq z)$ iff $(x \vee y) \sqsubseteq z$;

12. $0 \le \neg(x \to y) \vee (\neg x \vee y)$;

13. $\neg x \vee y \sqsubseteq x \to y$ and $x \to y \sqsubseteq \neg x \vee y$;

14. $x \sqsubseteq y$ implies $y \to z \sqsubseteq x \to z$ and $z \to x \sqsubseteq z \to y$;

15. $x \sqsubseteq y$ iff $0 \sqsubseteq x \to y$.

Proof. Item 1. holds by (9) and the properties of modal relevant algebras, as $0 \le \neg(1 \cdot \neg 1) \vee (1 \wedge \neg 1) = 1 \vee (1 \wedge \neg 1) = 1$. Item 2. is immediate. Item 3. can be established using Example 1. Items 4. and 5. follow straightforwardly from the definition of a modal relevant algebra; (item 4. uses the double negation law). Item 6. follows from (9) and the properties of modal relevant algebras as $0 \le \neg(x \cdot 1) \vee (x \wedge 1) \le \neg x \vee x$; moreover, $0 \le x \vee \neg x$ implies $\neg(x \vee \neg x) \le \neg 0$ and so $x \wedge \neg x \le \neg 0$. Item 7. is established as follows: if $0 \wedge (x \wedge y) \le z \vee \neg 0$, then $(0 \wedge (x \wedge y)) \vee \neg y \le (\neg y \vee z) \vee \neg 0$, and so $((0 \wedge x) \vee \neg y) \wedge (y \vee \neg y) \le (\neg y \vee z) \vee \neg 0$. By item 6., $((0 \wedge x) \vee \neg y) \wedge 0 \le (\neg y \vee z) \vee \neg 0$ and so $0 \wedge x \le (\neg y \vee z) \vee \neg 0$. Conversely, if $0 \wedge x \le \neg y \vee z \vee \neg 0$, then $0 \wedge x \wedge y \le (\neg y \vee (z \vee \neg 0)) \wedge y \le (y \wedge \neg y) \vee (y \wedge (z \vee \neg 0))$. By item 6., $0 \wedge (x \wedge y) \le \neg 0 \vee (y \wedge (z \vee \neg 0)) \le \neg 0 \vee z \vee \neg 0 \le z \vee \neg 0$. Item 8. holds thanks to item 7. and the fact that $0 \wedge x \sqsubseteq y$ iff $x \sqsubseteq y$. Item 9. is established as follows. By (9), $0 \sqsubseteq \neg(x \cdot z) \vee (x \wedge z)$. By item 8., $x \cdot z \sqsubseteq x \wedge z$. By item 5., $x \wedge z \sqsubseteq y \wedge z$. By (8), $0 \sqsubseteq \neg(y \wedge z) \vee (y \cdot z)$. Hence, by item 8., $y \wedge z \sqsubseteq y \cdot z$. It follows that $x \cdot z \sqsubseteq y \cdot z$. The fact that $y \cdot x \sqsubseteq z \cdot y$ is established analogously. Items 10. and 11. are obvious. Item 12. is established as follows. By (8), $0 \le ((x \to y) \cdot x) \vee \neg((x \to y) \wedge x) \le y \vee \neg(x \to y) \vee \neg x$. The first claim of item 13. is established as follows. By (10),

$0 \leq (x \rightarrow y) \vee \neg\neg x$ and so, by item 7., $0 \wedge \neg x \leq x \rightarrow y$. Hence, $\neg x \sqsubseteq x \rightarrow y$. By (10), $0 \leq (x \rightarrow y) \vee \neg y$ and so, by item 7., $0 \wedge y \leq (x \rightarrow y)$. Hence, $y \sqsubseteq (x \rightarrow y)$. By item 11., $\neg x \vee y \sqsubseteq x \rightarrow y$. The second claim of item 13. is established as follows. By item 12., $0 \leq \neg(x \rightarrow y) \vee (\neg x \vee y)$, which implies that $0 \sqsubseteq \neg(x \rightarrow y) \vee (\neg x \vee y)$. Hence, by item 8., $x \rightarrow y \sqsubseteq \neg x \vee y$. Item 14. is a straightforward consequence of the previous item and items 4.–5. Item 15. follows from items 8. and 13. $\qquad\square$

We define $x \equiv y$ as $x \sqsubseteq y$ and $y \sqsubseteq x$ (equivalently, $\neg x \sqsubseteq \neg y$). Let $\mathbf{E}^{-\square}$ be the $\square$-free reduct of $\mathbf{E}$.

Proposition 1. *The relation $\equiv$ is a congruence on $\mathbf{E}^{-\square}$. The quotient $\mathbf{E}^{-\square}/\!\equiv$ is a Boolean algebra.*

Proof. The fact that $\equiv$ is a congruence follows from Lemma 6. A quotient of a distributive lattice is a distributive lattice. Moreover, $\mathbf{E}^{-\square}/\equiv$ is bounded and complemented by Lemma 6. $\qquad\square$

4.2 Representation of Relevant Epistemic Algebras

Definition 9. *The prime filter frame of $\mathbf{E} = \langle X, \vee, \wedge, \cdot, \rightarrow, \neg, \square, 1, 0 \rangle$ is the relational structure*

$$\mathbf{E}_+ = \langle \mathcal{F}_P(\mathbf{E}), \subseteq, N, K, R, {}^*, Q \rangle,$$

where $N, R, {}^*$ and Q are defined as in the definition of a prime filter frame of a modal relevant algebra and $s \in K$ iff $0 \in s$.

Lemma 7. *$\mathfrak{F}^+$ is a relevant epistemic algebra, for all $\mathfrak{F}$.*

Proof. It is sufficient to show that conditions (8)–(10) hold in $\mathfrak{F}^+$. (8) Assume that $s \in K$ and $s \notin \neg(x \cap y)$. This means that $s^* \in x$ and $s^* \in y$. By (1), Rs^*s^*s. Hence, $s \in x \cdot y$. This means that, in general, $s \in K$ implies $s \in \neg(x \cap y) \cup (x \cdot y)$ and so (8) is satisfied. (9) Assume that $s \in K$ and $s \notin \neg(x \cdot y)$. Hence, $s^* \in x \cdot y$. This means that there are t, u such that $Rtus^*$ where $t \in x$ and $u \in y$. By (2), $t \leqslant s$ and $u \leqslant s$. Hence, $s \in x \cap y$. It follows that, in general, $s \in K$ implies $s \in \neg(x \cdot y) \cup (x \cap y)$ and (9) is satisfied. (10) Assume that $s \in K$ and $s \notin x \rightarrow y$. Hence, there are t, u such that $Rstu$ and $t \in x$ and $u \notin y$. By (3), $t \leqslant s$ and $s^* \leqslant u$, meaning that $s \in x$ and $s^* \notin y$, i.e., $s \in \neg y$. It follows that, in general, $s \in K$ implies $s \in (x \rightarrow y) \cup (x \cap \neg y)$ and (10) is satisfied. $\qquad\square$

Lemma 8. *$\mathbf{E}_+$ is a relevant epistemic frame, for all $\mathbf{E}$.*

Proof. It is sufficient to show that (1)–(3) hold for each $s \in \mathcal{F}_P(\mathbf{E})$ such that $0 \in s$. (1) Assume that $x \to y \in s^*$ and $x \in s^*$. Then $\neg(x \to y) \notin s$ and $\neg x \notin s$, meaning that $\neg((x \to y) \wedge x) \notin s$. By (8), $(x \to y) \cdot x \in s$, and so $y \in s$. (2) Assume that $Rtus^*$ and $x \in t$. We have to show that $x \in s$. We know that $y \in u$ for some y. By the properties of modal relevant algebras, $x \leq y \to (x \cdot y)$. Hence, $y \to (x \cdot y) \in t$ and so $x \cdot y \in s^*$. This means $\neg(x \cdot y) \notin s$. By (9), $x \wedge y \in s$. It follows that, in general, $t \subseteq s$. The fact that $u \subseteq s$ is established in a similar way. (3) Assume that $Rstu$. First, we show that $t \subseteq s$. Assume that $x \notin s$. We know that $y \notin u$ for some y. By (10), $(x \to y) \vee x \in s$ and so $x \to y \in s$. This means that, if $x \in t$, then $y \in u$, contradicting the assumption. Hence, $x \notin t$. Second, we show that $s^* \subseteq u$. Assume that $y \in s^*$. This means that $\neg y \notin s$. We know that $x \in t$ for some x. By (10), $(x \to y) \vee \neg y \in s$ and so $x \to y \in s$. Hence, $y \in u$. $\qquad\square$

Remark 2. To continue our discussion from Remark 1, note that the proof of Lemma 8 depends on the assumption that the set of states in $\mathbf{E}_+$ contains neither the empty set nor the carrier set of $\mathbf{E}$. If we allowed these sets to be among the states, then condition (3) would have to be reformulated, since it would no longer hold in its present form. In [24], the problem was solved by invoking a frame condition that explicitly refers to the least and greatest states of the frame. These states must satisfy a number of conditions, which complicates the framework. Our current approach is therefore much simpler.

Definition 10. *A complex relevant epistemic algebra is a relevant epistemic algebra that is isomorphic to a subalgebra of the full complex algebra of a relevant epistemic frame.*

By Lemmas 7 and 8, $(\mathbf{E}_+)^+$ is a complex relevant epistemic algebra and so are its subalgebras.

Theorem 2. *Every relevant epistemic algebra is isomorphic to a complex relevant epistemic algebra.*

Proof. Analogous to the proof of Theorem 1. $\qquad\square$

Let $B.C0$ be the expansion of the axiomatisation $B.C$ of the logic of all modal relevant frames with three new axioms, all variants of (8)–(10):

$$0 \to (\neg(\varphi \wedge \psi) \vee (\varphi \cdot \psi))$$
$$0 \to (\neg(\varphi \wedge \psi) \vee \varphi \cdot \psi)$$
$$0 \to ((\varphi \to \psi) \vee (\varphi \wedge \neg\psi))$$

Theorem 2 entails a completeness result, since φ is not provable in $B.C0$ iff the equation $\varphi \leftrightarrow (\varphi \vee 1)$ is not valid in the Lindenbaum–Tarski algebra of $B.C0$, iff $\varphi \leftrightarrow (\varphi \vee 1)$ is not valid in the complex algebra of some relevant epistemic frame iff φ is not valid in some relevant epistemic frame.

We note that completeness results for logics of classes of frames that are characterised by frame conditions for which there is a canonical formula in the 0-free fragment of Fm (e.g., those discussed in [9]) are easily obtained in a similar manner.

5 Conclusion

We have introduced a simple extension of relevant modal logic suitable for expressing a combination of relevant logic with classical epistemic logic, based on the idea that agents reason relevantly while situated in classical possible worlds. While our framework is rooted in previous work, most notably [24], it constitutes a significant simplification. A particular advantage of our framework, on which this paper has focused, is a straightforward algebraic generalisation.

Topics for future research include the development of a comprehensive multi-agent epistemic logic based on our framework with different sorts of group epistemic modalities. Another interesting project is to generalise the present framework in two ways. Firstly, the distinguished situations in W can be governed by a logic weaker than classical logic, and secondly, we can have different agents reasoning according to different logics.

References

[1] Jc Beall, Ross Brady, J. Michael Dunn, Allen P. Hazen, Edwin Mares, Robert K. Meyer, Graham Priest, Greg Restall, David Ripley, John Slaney, and Richard Sylvan. On the ternary relation and conditionality. *Journal of Philosophical Logic*, 41(3):595–612, 2012.

[2] Marta Bílková, Ondrej Majer, Michal Peliš, and Greg Restall. Relevant agents. In L. Beklemishev, V. Goranko, and V. Shehtman, editors, *Advances in Modal Logic, Volume 8*, pages 22–38, London, 2010. College Publications.

[3] Marta Bílková, Ondrej Majer, and Michal Peliš. Epistemic logics for sceptical agents. *Journal of Logic and Computation*, 26(6):1815–1841, 2016.

[4] Katalin Bimbó and J. Michael Dunn. *Generalized Galois Logics. Relational Semantics of Nonclassical Logical Calculi*, volume 188 of *CSLI Lecture Notes*. CSLI Publications, Stanford, CA, 2008.

[5] J. Michael Dunn and Gary M. Hardegree. *Algebraic Methods in Philosophical Logic*, volume 41 of *Oxford Logic Guides*. Oxford University Press, Oxford, 2001.

[6] J. Michael Dunn and Greg Restall. Relevance logic. In D. M. Gabbay and F. Guenthner, editors, *Handbook of Philosophical Logic*, volume 6, pages 1–128. Kluwer, Amsterdam, 2002.

[7] Nicholas Ferenz. First-order relevant reasoners in classical worlds. *The Review of Symbolic Logic*, 17(3):793–818, 2023.

[8] André Fuhrmann. *Relevant Logics, Modal Logics and Theory Change*. PhD Thesis, Australian National University, 1988.

[9] André Fuhrmann. Models for relevant modal logics. *Studia Logica*, 49(4):501–514, 1990.

[10] George Grätzer. *General Lattice Theory*. Birkhäuser, Basel, 2nd edition, 2003.

[11] Hector Levesque. A logic of implicit and explicit belief. In *Proceedings of the Fourth AAAI Conference on Artificial Intelligence*, pages 198–202. AAAI Press, 1984.

[12] Edwin D. Mares. Relevant logic and the theory of information. *Synthese*, 109: 345–360, 1996.

[13] Edwin D. Mares. *Relevant Logic: A Philosophical Interpretation*. Cambridge University Press, Cambridge, 2004.

[14] Robert K. Meyer and Richard Routley. Algebraic analysis of entailment I. *Logique et Analyse*, 15(59/60):407–428, 1972.

[15] Vít Punčochář, Igor Sedlár, and Andrew Tedder. Relevant epistemic logic with public announcements and common knowledge. *Journal of Logic and Computation*, 33(2):436–461, 2023.

[16] Greg Restall. Information flow and relevant logic. In J. Seligman and D. Westerståhl, editors, *Logic, Language and Computation: The 1994 Moraga Proceedings*, pages 463–477. CSLI Publications, Stanford, CA, 1995.

[17] Greg Restall. Negation in relevant logics (how i stopped worrying and learned to love the Routley star). In D. M. Gabbay and H. Wansing, editors, *What is Negation?*, pages 53–76. Kluwer, Dordrecht, 1999.

[18] Greg Restall. *An Introduction to Substrucutral Logics*. Routledge, London, 2000.

[19] Greg Restall. Logics, situations and channels. *Journal of Cognitive Science*, 6: 125–150, 2005.

[20] Richard Routley and Robert K. Meyer. The semantics of entailment — III. *Journal of Philosophical Logic*, 1(2):192–208, 1972.

[21] Richard Routley and Val Routley. The semantics of first degree entailment. *Noûs*, 6(4):335–359, 1972.

[22] Igor Sedlár. Substructural epistemic logics. *Journal of Applied Non-Classical Logics*, 25(3):256–285, 2015.

[23] Igor Sedlár. Epistemic extensions of modal distributive substructural logics. *Journal of Logic and Computation*, 26(6):1787–1813, 2016.

[24] Igor Sedlár and Pietro Vigiani. Relevant reasoners in a classical world. In D. Fernández-Duque, A. Palmigiano, and S. Pinchinat, editors, *Proc. 14th Int. Conference on Advances in Modal Logic (AiML 2022)*, pages 697–719, London, 2022. College Publications.

[25] Igor Sedlár and Pietro Vigiani. Relevant reasoning and implicit beliefs. In H. H. Hansen, A. Scedrov, and R. J. de Queiroz, editors, *Logic, Language, Information, and Computation. WoLLIC 2023*, number 13923 in Lecture Notes in Computer Science, pages 336–350, Cham, 2023. Springer Nature Switzerland.

[26] Igor Sedlár and Pietro Vigiani. Epistemic logics for relevant reasoners. *Journal of Philosophical Logic*, 53(5):1383–1411, 2024.

[27] Shawn Standefer. Tracking reasons with extensions of relevant logics. *Logic Journal of the IGPL*, 27(4):543–569, 2019.

[28] Shawn Standefer, Ted Shear, and Rohan French. Getting some (non-classical) closure with justification logic. *Asian Journal of Philosophy*, 2(2), 2023.

[29] Alasdair Urquhart. Duality for algebras of relevant logics. *Studia Logica*, 56 (1–2):263–276, 1996.

Received 5 March 2025

Universal Necessity and Deep Classicality

Shawn Standefer[*]
Department of Philosophy, National Taiwan University
<standefer@ntu.edu.tw>

Rohan French
Department of Philosophy, UC Davis
<rfrench@ucdavis.edu>

Abstract

The universal conception of necessity says that necessary truth is truth in all possible worlds. This idea is well studied in the context of classical possible worlds models, and there its logic is S5. The universal conception of necessity is less well studied in models for non-classical logics. We will present some preliminary results on universal necessity on models for intuitionistic logic, first-degree entailment, and relevant logics. We will close by discussing a way in which universal necessity is a very classical concept.

2020 *Mathematics Subject Classification.* Primary: 03B45, Secondary: 03B47, 03B20.

1   Introduction

It is well known that there are many different presentations of the classically-based modal logic S5. One way is via any of the common axiomatizations. Another is the use of relational frames $\langle W, S \rangle$ where the modal accessibility relation, S is an equivalence relation. Yet another is the use of relational frames with a universal modal accessibility relation, S, i.e., for all worlds $x, y \in W$, Sxy. In the setting of models on universal frames, the usual truth condition for necessity is equivalent to saying that a formula is true at a world iff it is true at all worlds. Because of this, we

We would like to thank the audience of the 2nd Third workshop, Ren-June Wang, Hitoshi Omori, an anonymous referee, and Katalin Bimbó for feedback on this work.

[*]Standefer's work was supported by by the National Science and Technology Council of Taiwan grant 111-2410-H-002-006-MY3.

can, equally, consider frames without an accessibility relation, which will be useful in considering frames for non-classical logics.

S5 is, in many ways, an interesting logic, and its distinctive features are brought out via the coincidence of these presentations.[1] Once one moves away from frames for classically-based modal logic, however, the situation changes. Standefer [37] showed that in the setting of relevant logics, these three presentations yield at least three distinct logics, with some being incomparable.

The purpose of this paper is to further investigate the necessity operator of universal frames, what we will call *universal necessity*. We will consider universal necessity against the backdrop of three sorts of frames for non-classical logics: intuitionistic logic, first-degree entailment, and the relevant logic B. By studying universal necessity in these settings, we will uncover some commonalities to the logics of universal necessity. We will discuss this commonality in the final section, where we will argue that universal necessity is, in a sense, a deeply classical concept.

2   Intuitionistic Logic

The basic language for this paper is $\{\rightarrow, \wedge, \vee, \neg\}$. There will be a countable set of atoms, At and complex formulas will be constructed in the usual way. We will add a unary connective, $\mathbb{U}$, for universal necessity, which will be the main topic of our investigations below. We will define the biconditional, $A \leftrightarrow B$ as $(A \rightarrow B) \wedge (B \rightarrow A)$. We will define three logics semantically using different classes of frames, so that L will be the set of formulas valid in the appropriate class frames.[2] For a logic L so presented, the logic $\mathsf{L}^{\mathbb{U}}$ will be the extension of L with $\mathbb{U}$, namely, the set of formulas valid in the extended language in the appropriate class of frames.

In this section, we will study the addition of $\mathbb{U}$ to intuitionistic logic, IL. We will study this combination using Kripke frames. Before defining the frames, it is worth noting that there have been studies of S5-ish logics over IL. Ono [28] considers different S5-type axioms in the context of IL, which gives rise to a range of different logics. It follows from results in section 5.3 of Niki & Omori [24] that the present logic $\mathsf{IL}^{\mathbb{U}}$ is equivalent to Ono's L4—the strongest S5-type logic considered in [28].

Definition 1 (Kripke frame for IL). *A Kripke frame for IL is a pair* $\langle W, \leq \rangle$*, where*

---

[1]As an example of the interesting features, S5 has a connection to monadic quantified logics. This connection was noted by Halmos [17] in the context of monadic algebra, and Mints [23] proves the coincidence for the case of classical logic. Bull [11] proves it for the case of intuitionistic logic. Caicedo et al. [12] show it for Dummett's logic LC. Ferenz [16] shows this for certain S5-ish extensions of some relevant logics.

[2]These logics are in the framework FMLA of Humberstone [18].

- W *is non-empty, and*

- $\leq$ *is a reflexive, transitive binary relation on* W.

A Kripke model M *is a triple* $\langle W, \leq, V \rangle$ *where* $\langle W, \leq \rangle$ *is a frame and* V *is a function from* At *to sets of worlds such that if* $x \in V(p)$ *and* $x \leq y$, *then* $y \in V(p)$.

The verification relation $\Vdash$ is defined over the whole language as follows.

- $w \Vdash p$ iff $w \in V(p)$.

- $w \Vdash B \wedge C$ iff $w \Vdash B$ and $w \Vdash C$.

- $w \Vdash B \vee C$ iff $w \Vdash B$ or $w \Vdash C$.

- $w \Vdash B \to C$ iff for all $x \in W$ such that $w \leq x$ and $x \Vdash B$, $x \Vdash C$.

- $w \Vdash \neg B$ iff for all $x \in W$ such that $w \leq x$, $x \nVdash B$.

- $w \Vdash \mathbb{U}B$ iff for all $x \in W$, $x \Vdash B$.

A consequence of the definitions is that the heredity condition extends to arbitrary formulas. We will state this, without proof.

Lemma 2 (Heredity). *For all Kripke models* M, *all* $x, y \in W$, *and all formulas* A, *if* $x \leq y$ *and* $x \Vdash A$, *then* $y \Vdash A$.

There is a special class of formulas that we will examine throughout this paper.

Definition 3 (U-formula). *A formula* A *is a* U-formula *iff it is of the form* $\mathbb{U}B$.

The $\mathbb{U}$-formulas are a special case of a more general class of formulas, namely, the modalized formulas. A formula A is *modalized* iff every atom in A occurs in the scope of $\mathbb{U}$. Modalized formulas are, in many ways, distinctive, and they are generally important in the study of modal logics.[3] Nonetheless, we only need to consider the special case of $\mathbb{U}$-formulas for the results of interest to this paper.

The main feature of universal necessity on which we will rely is one that is common across the different models we will look at.

Lemma 4 (Ubiquity). *In a model* M, *if* $w \Vdash \mathbb{U}A$, *for some* $w \in W$, *then* $u \Vdash \mathbb{U}A$, *for all* $u \in W$.

---

[3]Mints [23] uses modalized formulas in his discussion of the connection between S5 and monadic quantification. Prawitz [30] uses modalized formulas in his natural deduction systems for S5.

Proof. Suppose $w \Vdash \mathbb{U}A$. Then it follows that for all $u \in W$, $u \Vdash A$, so $x \Vdash \mathbb{U}A$, for arbitrary x. $\qquad\square$

In the terminology of Standefer [36, 37], formulas with $\mathbb{U}$ as their main connective are *ubiquitous*, either true at all points in a model or true at no points in a model. $\mathbb{U}A$ either holds everywhere or nowhere. This feature gives rise to a surprising tension with a distinctive feature of intuitionistic logic, the disjunction property.

Definition 5 (Disjunction property). *A logic L has the* disjunction property *iff if $A \vee B$ is valid, then A is valid or B is valid.*

IL enjoys the disjunction property. The disjunction property is an important part of what is meant by saying that IL is a *constructive* logic. When considering extensions of intuitionistic logic with new connectives, the extension enjoying the disjunction property is one indication that the new connective aligns with intuitionistic logic. Adding $\mathbb{U}$ to IL results in violations of the disjunction property. It is easiest to see this using some lemmas.

Lemma 6. *For all Kripke models M, for all $w \in W$, $w \Vdash \neg\mathbb{U}A$ iff $w \nVdash \mathbb{U}A$.*

Proof. The left-to-right direction is immediate.

Next, suppose that $w \nVdash \mathbb{U}A$. By lemma 4, this implies that for all $x \in W$, $x \nVdash \mathbb{U}A$. Therefore, for all $x \in W$ such that $w \leq x$, $x \nVdash \mathbb{U}A$, whence $w \Vdash \neg\mathbb{U}A$, as desired. $\qquad\square$

For the special case of $\mathbb{U}$-formulas, intuitionistic negation collapses to that of classical logic.

Corollary 7. *In the class of Kripke frames for IL, $\mathbb{U}A \vee \neg\mathbb{U}A$ is valid.*

Proof. Let M be a model on a Kripke frame and let $w \in W$. Since $w \Vdash \mathbb{U}A$ or $w \nVdash \mathbb{U}A$ from lemma 6 it follows that $w \Vdash \mathbb{U}A$ or $w \Vdash \neg\mathbb{U}A$. Therefore, $w \Vdash \mathbb{U}A \vee \neg\mathbb{U}A$. $\qquad\square$

Given that it is easy to see that neither $\mathbb{U}p$ nor $\mathbb{U}\neg p$, for example, are valid, it follows that the addition of $\mathbb{U}$ leads to widespread failure of the disjunction property.

We can obtain another corollary of lemma 6, demonstrating a collapse of the intuitionistic implication to the classical material conditional for $\mathbb{U}$-formulas.

Corollary 8. *For all Kripke models M for IL, for all $w \in W$, $w \Vdash (\mathbb{U}A \rightarrow \mathbb{U}B) \leftrightarrow (\neg\mathbb{U}A \vee \mathbb{U}B)$.*

Proof. The left-to-right direction holds for all formulas in IL already, so we will prove the converse.

Suppose that for some $x \in W$, $w \leq x$, $x \Vdash \mathbb{U}A \to \mathbb{U}B$, and $x \nVdash \neg\mathbb{U}A \vee \mathbb{U}B$. Therefore, $x \nVdash \neg\mathbb{U}A$ and $x \nVdash \mathbb{U}B$. By the preceding corollary, $x \nVdash \neg\mathbb{U}A$ implies $x \Vdash \mathbb{U}A$, so it follows that $x \Vdash \mathbb{U}B$. This is a contradiction, so $x \Vdash \neg\mathbb{U}A \vee \mathbb{U}B$, which establishes the desired claim. $\square$

This corollary demonstrates that, for a certain class of formulas, the intuitionistic implication collapses the classical material conditional, namely, the conditional $\mathbb{U}A \to \mathbb{U}B$ defined as $\neg\mathbb{U}A \vee \mathbb{U}B$. While this holds for certain choices of A and B in IL, namely when A and B are theorems, the are no restrictions on A and B in the corollary above.

To sum up this section, we have, in $\mathsf{IL}^{\mathbb{U}}$ violations of key features of IL that are important to proponents of IL. This is a story we will see repeated below.

3 First-Degree Entailment

The logic FDE, or first-degree entailment, is, perhaps, the most well-known four-valued logic.[4] It was isolated as first-degree fragment of the relevant logic E, meaning the set of formulas $A \to B$ where A and B do not contain $\to$, by Anderson and Belnap. It has since been found to be the first-degree fragment of all the standard relevant logics.

FDE has been a focus of study both on its own and in the context of relevant logics. Omori and Wansing [26] provide a good overview of work on FDE. Beall [2, 3] has argued that FDE is the basic subclassical logic. Levesque [19] uses FDE in a logic of awareness, and Standefer et al. [40] argue that FDE should be used for logical closure of justifications.

We will focus on a few specific features of FDE to be presented shortly. We will study the star frames for FDE similar to those of Routley and Routley [34].[5]

Definition 9 (Star frame for FDE). *A star frame for FDE is a pair $\langle K, * \rangle$ where*

- *K is non-empty, and*

- *$a^{**} = a$.*

---

[4]See Dunn [15] or Belnap [4, 5]. Anderson and Belnap [1] cover FDE in chapter 3. The collection Omori and Wansing [27] contains this material, as well as many other papers on FDE.

[5]See also Dunn [15].

A star model M is a triple $\langle K, *, V \rangle$ such that $\langle K, * \rangle$ is a star frame and V is a function from At to sets of worlds. Such a star model is said to be built on the star frame.

In a given model, the verification relation $\Vdash$ is defined over the language without $\rightarrow$ as follows.

- $a \Vdash p$ iff $w \in V(p)$.
- $a \Vdash B \wedge C$ iff $a \Vdash B$ and $a \Vdash C$.
- $a \Vdash B \vee C$ iff $a \Vdash B$ or $a \Vdash C$.
- $a \Vdash \neg B$ iff $a^* \nVdash B$.
- $a \Vdash \mathbb{U}B$ iff for all $b \in K$, $b \Vdash B$.

We are evaluating formulas at worlds only if they lack the implication connective.

Definition 10 (Holding, validity). *For all formulas A, B, $A \rightarrow B$ is holds in a star model iff for all $a \in K$, if $a \Vdash A$, then $a \Vdash B$.*

For all formulas A, B, $A \rightarrow B$ is valid in the class of star frames iff $A \rightarrow B$ holds in any model built on a star frame.

Validity is defined in a restricted form, namely, only for formulas of the form $A \rightarrow B$ where A and B do not contain $\rightarrow$, because we are interested in the valid first-degree formulas.

There are some features of **FDE** that are well-known and generally regarded as virtues of **FDE**, which we will state here without proof.

- **FDE** is paraconsistent, i.e., there is no A such that for all B is $(A \wedge \neg A) \rightarrow B$ valid.

- **FDE** is paracomplete, i.e., there is no A such that for all B is is $B \rightarrow (A \vee \neg A)$ valid.

- **FDE** enjoys variable-sharing, i.e., if $A \rightarrow B$ is valid, then A and B share an atom.

All three of these features are violated in $\mathsf{FDE}^{\mathbb{U}}$. To show this, we note that lemma 4 carries over to star models.

Lemma 11. *In a star model M, if $a \Vdash \mathbb{U}A$, for some $a \in K$, then $b \Vdash \mathbb{U}A$, for all $b \in K$.*

Proof. The proof is essentially the same as that of lemma 4. $\qquad\square$

This lemma has consequences for negation in star models.[6]

Lemma 12. *In a star model M, $a \Vdash \neg \mathbb{U}A$ iff $a \nVdash \mathbb{U}A$.*

Proof. Let M be a star model. Suppose that $a \Vdash \neg \mathbb{U}A$. It follows that $a^* \nVdash \mathbb{U}A$. From lemma 11, this implies $a \nVdash \mathbb{U}A$.
For the converse, suppose $a \nVdash \mathbb{U}A$. By lemma 11, we have $a^* \nVdash \mathbb{U}A$, so $a \Vdash \neg \mathbb{U}A$. $\square$

We can then obtain the following two corollaries.

Corollary 13. *Let M be a star model. Then for all $a \in K$, $a \Vdash \mathbb{U}A \vee \neg \mathbb{U}A$.*

Corollary 14. *Let M be a star model. Then for all $a \in K$, $a \nVdash \mathbb{U}A \wedge \neg \mathbb{U}A$.*

These suffice for the following theorem.

Theorem 15. *For all formulas A and B, both*

- $A \to (\mathbb{U}B \vee \neg \mathbb{U}B)$, *and*

- $(\mathbb{U}B \wedge \neg \mathbb{U}B) \to A$

are valid in the class of all star frames.

Proof. For the first, note that by corollary 13, the consequent holds at all points in all models. There can, then, be no counterexamples. Similarly, for the second, note that by corollary 14, the antecedent fails at all points in all models, so there can, similarly, be no counterexamples. $\square$

Corollary 16. *FDE$^{\mathbb{U}}$ does not enjoy variable-sharing.*

Proof. By the preceding theorem, $p \to (\mathbb{U}q \vee \neg \mathbb{U}q)$ is valid. $\square$

Thus, see that three of the key features of **FDE** fail in the context of **FDE**$^{\mathbb{U}}$. Much as in **IL**$^{\mathbb{U}}$, violations of the highlighted features arise in the context of $\mathbb{U}$-formulas. In the next section, we will turn to relevant logics to further extend our results.

---

[6]A further similarity with the intuitionistic case is illuminated if we work with models closer to those of Routley and Routley [34] which have a designated state $g \in K$ where $g = g^*$. In this setting we can, following Routley and Routley [34, p. 348], define an analogue of classical **S5** strict implication by saying that $A \dashv B$ holds in a star model iff for all $a \in K$ where $a = a^*$, if $a \Vdash A$, then $a \Vdash B$. In this setting lemma 4 shows that we have a collapse between the entailments and the strict entailments between $\mathbb{U}$-formulas, similar to the way in which in **IL**$^{\mathbb{U}}$ we have a collapse between the intuitionistic and classical implication over $\mathbb{U}$-formulas (as recorded in corollary 8).

4 Relevant Logics

In this section, we turn to relevant logics. Relevant logics are substructural logics whose implication connectives are supposed to enforce a strong connection between antecedent and consequent.[7] A key feature of relevant logics is that they enjoy the *variable-sharing* property, which says that if $A \to B$ is valid, then A and B share an atom.[8] There are many relevant logics, but we will focus on the weak relevant logic B. For the most part, the particular choice of which relevant logic to use will not make a difference, for reasons that will become clearer below.

We will study relevant logics using Routley–Meyer ternary relational frames.[9]

Definition 17 (Routley–Meyer frame). *A Routley–Meyer frame is a quadruple* $\langle K, N, R, {}^* \rangle$, *where* $K \neq \emptyset$, $N \subseteq K$, $R \subseteq K^3$, *and* ${}^* \colon K \mapsto K$, *that obeys the following conditions, where* $a \leq b =_{Df} \exists x \in N\, Rxab$,

(B1) $\leq$ *is a partial order,*

(B2) $a^{**} = a$,

(B3) $a \leq b$ *only if* $b^* \leq a^*$,

(B4) *if* $d \leq a$, $e \leq b$, $c \leq f$, *and* $Rabc$, *then* $Rdef$, *and*

Definition 18 (Routley–Meyer model). *A Routley–Meyer model* M *is a quintuple* $\langle K, N, R, {}^*, V \rangle$, *where the first four components make up a frame and* V *is a function from* At *to subsets of* K *obeying the condition that if* $a \in V(p)$ *and* $a \leq b$, *then* $b \in V(p)$. *The model* $M = \langle K, N, R, {}^*, V \rangle$ *is said to built on the frame* $F = \langle K, N, R, {}^* \rangle$.

The verification relation $\Vdash$ is defined over the whole language as follows.

- $a \Vdash p$ iff $a \in V(p)$.

- $a \Vdash B \wedge C$ iff $a \Vdash B$ and $a \Vdash C$.

- $a \Vdash B \vee C$ iff $a \Vdash B$ or $a \Vdash C$.

- $a \Vdash B \to C$ iff for all $b, c \in K$, if $Rabc$ and $b \Vdash B$, then $c \Vdash C$.

- $a \Vdash \neg B$ iff $a^* \nVdash B$.

---

[7]See Dunn and Restall [14], Bimbó [8], Mares [22], Logan [20], or Standefer [39] for an overview of relevant logics.

[8]See Standefer [38] for discussion.

[9]Routley–Meyer models were first introduced by Routley and Meyer [31, 33, 32]. See Bimbó et al. [9] for discussion of the early development of Routley–Meyer models.

- $a \Vdash \mathbb{U}B$ iff for all $b \in K$, $b \Vdash B$.

Given these definitions, we can define holding in a model and validity in a class of frames.

Definition 19. *A formula A holds in a model M iff for all $a \in N$, $a \Vdash A$.*

A formula A is valid in a class of models iff for all frames F in the class, A holds in M, for all M built on F.

The class of all basic Routley–Meyer frames gives us the logic B.

Our interest will be in the logic $\mathsf{B}^{\mathbb{U}}$. As with $\mathsf{IL}^{\mathbb{U}}$, U-formulas have several important properties.

Lemma 20. *In a Routley–Meyer model M, if $a \Vdash \mathbb{U}A$, for some $a \in K$, then $b \Vdash \mathbb{U}A$, for all $b \in K$.*

Proof. The proof is essentially the same as in the IL case. $\qquad\qquad\square$

While heredity is postulated for atoms, it extends to the full language, as in $\mathsf{IL}^{\mathbb{U}}$.

Lemma 21 (Heredity). *For all Routley–Meyer models M, all $a, b \in K$, and all formulas A, if $a \leq b$ and $a \Vdash A$, then $b \Vdash A$.*

Proof. The proof is by induction on the structure of A. The only new case is when A is of the form $\mathbb{U}B$. The case is covered by lemma 20. $\qquad\qquad\square$

Using this lemma, we can prove another lemma, whose proof we will omit as it is standard.

Lemma 22 (Verification). *For all models M, the following are equivalent.*

- *For all $a \in K$, if $a \Vdash A$, then $a \Vdash B$.*

- *$A \to B$ holds in M.*

The verification lemma drastically simplifies the verification of formulas $A \to B$ in a model, especially when A and B do not contain any implications. That suggests a connection to star frames, which bears out. Given a Routley–Meyer frame $\langle K, N, R, * \rangle$ its reduct $\langle K, * \rangle$ is a star frame. Therefore, we can obtain a corollary of lemma 20 concerning negation, as in $\mathsf{FDE}^{\mathbb{U}}$.[10]

---

[10]We are adopting an Australian-plan approach to negation here. The Australian plan has been defended recently by Berto [6] and Berto and Restall [7]. The rival American plan for negation has recently been defended by De and Omori [13] and Omori and De [25]. While in many contexts, the two plans give equivalent results, in the context of universal necessity, there are some differences that can emerge with respect to valid formulas. We would like to thank Ellie Ripley and Hitoshi Omori for discussion of this point.

Corollary 23. *In a Routley–Meyer model M, $a \Vdash \neg \mathbb{U}A$ iff $a \nVdash \mathbb{U}A$.*

As a consequence of the preceding corollary, we obtain the validity of some formulas that can violate variable-sharing.

Corollary 24. *The formulas $(\mathbb{U}A \wedge \neg \mathbb{U}A) \to B$ and $B \to (\mathbb{U}A \vee \neg \mathbb{U}A)$ are valid in the class of all Routley–Meyer frames.*

Proof. The proof is essentially the same as the proof of the similar fact in the previous section. $\qquad\square$

Corollary 25. *$\mathsf{B}^{\mathbb{U}}$ violates variable-sharing.*

There are other violations of variable-sharing in $\mathsf{B}^{\mathbb{U}}$ that are not in $\mathsf{FDE}^{\mathbb{U}}$. Two examples are $\mathbb{U}p \to (\mathbb{U}q \to \mathbb{U}p)$ and $\mathbb{U}q \to (\mathbb{U}p \to \mathbb{U}p)$, which are instances of the paradoxes of implication, paradoxes relevant logics were meant to avoid. As with $\mathsf{IL}^{\mathbb{U}}$, we get a collapse of the implication to the material conditional.

Lemma 26. *For all formulas A and B, $(\neg \mathbb{U}A \vee \mathbb{U}B) \to (\mathbb{U}A \to \mathbb{U}B)$ is valid.*

Proof. Let M be a Routley–Meyer model, and suppose $a \Vdash \neg \mathbb{U}A \vee \mathbb{U}B$. Then, either $a \Vdash \neg \mathbb{U}A$ or $a \Vdash \mathbb{U}B$. Suppose $a \Vdash \neg \mathbb{U}A$. Further, suppose that $Rabc$ and $b \Vdash \mathbb{U}A$. From $a \Vdash \neg \mathbb{U}A$ and lemma 20, it follows that $b \nVdash \mathbb{U}A$, which is a contradiction, so, classically, $c \Vdash \mathbb{U}B$. Next, suppose that $a \Vdash \mathbb{U}B$. Suppose further that $Rabc$ and $b \Vdash \mathbb{U}A$. From lemma 20 and the supposition of $a \Vdash \mathbb{U}B$, it follows that $c \Vdash \mathbb{U}B$. In both cases, we have established $c \Vdash \mathbb{U}B$, which suffices for $a \Vdash \mathbb{U}A \to \mathbb{U}B$. $\qquad\square$

While the implication from material conditional to non-classical implication holds generally in IL, it does not hold generally in any of the standard relevant logics. The converse implication, from relevant implication to material conditional, holds for $\mathbb{U}$-formulas, under an additional supposition.

Lemma 27. *Let $\mathcal{C}$ be the class of Routley–Meyer frames such that for all $a \in K$, there are $b, c \in K$ such that $Rabc$. Then for all formulas A, B, $(\mathbb{U}A \to \mathbb{U}B) \to (\neg \mathbb{U}A \vee \mathbb{U}B)$ is valid in $\mathcal{C}$.*

Proof. Let M be a model built on a Routley–Meyer frame in $\mathcal{C}$. Suppose that $a \Vdash \mathbb{U}A \to \mathbb{U}B$. Suppose, for reductio, that $a \nVdash \neg \mathbb{U}A \vee \mathbb{U}B$. This implies that $a \nVdash \neg \mathbb{U}A$. It follows that $a \Vdash \mathbb{U}A$. From the assumption, there are $b, c \in K$ such that $Rabc$. By lemma 20 and $a \Vdash \mathbb{U}A$, we can conclude $b \Vdash \mathbb{U}A$. This together with the supposition $a \Vdash \mathbb{U}A \to \mathbb{U}B$ entails that $c \Vdash \mathbb{U}B$. By lemma 20 again, $a \Vdash \mathbb{U}B$, whence $a \Vdash \neg \mathbb{U}A \vee \mathbb{U}B$, which is a contradiction. $\qquad\square$

As far as we can tell, the restriction of the lemma is essential for the proof to work. The condition, that for all $a \in K$, there are $b, c \in K$ such that $Rabc$, is known as R-*seriality*.[11] While R-seriality is not built in the Routley–Meyer frames for B, it can be assumed without changing the logic. The class of Routley–Meyer frames obeying R-seriality also generates B. The reason is that, in Henkin-style canonical model completeness proofs for the standard relevant logics, the canonical frame is R-serial. The condition does make a difference once U is in the language, so it is not uncontroversial. The preceding result shows that, for the R-serial frames, for U-formulas, the relevant implication collapses to the classical material conditional, which should be a deeply unappealing result to the relevant logician. The frames for some relevant logics will obey R-seriality, such as those that contain $(A \wedge (A \rightarrow B)) \rightarrow B$, whose frame condition is $\forall x \in K\, Rxxx$. The possibility of avoiding R-seriality only arises for some of the weaker, contraction-free relevant logics.

5 Classicality

We have introduced three different logics, defined using three different kinds of frames. These logics all have distinctive non-classical features that are taken to be virtues by their proponents. We have seen that the result of adding the universal necessity operator U to the language results in violations of these features. In the case of IL and B, the addition of U results in a collapse of the non-classical implication to the classical material conditional, at least for U-formulas.

Many authors in the relevant logic literature have viewed universal necessity as a deeply classical concept. Parks and Byrd [29, p. 180] remark that among relevant logicians there was a "feeling that the Leibnizian view of necessity is incompatible with key tenets of relevance logic." By "Leibnizian view of necessity," they mean universal necessity. Standefer [37] takes the fact that adding U to relevant logics results in violations of variable-sharing to vindicate the suspicion voiced by Parks and Byrd. We want to go further. Universal necessity is incompatible with key tenets of relevant logic, because it smuggles in a kind of covert classicality. To put it loosely, U-formulas behave similarly to formulas of classical logic.

To illustrate what we mean, consider B. Unlike some relevant logics, no classical tautologies in the vocabulary $\{\neg, \vee, \wedge\}$ are valid for B. As we saw in the previous section, $\mathbb{U}A \vee \neg \mathbb{U}A$ is valid for $\mathsf{B}^{\mathbb{U}}$. Due to De Morgan equivalences and the double negation laws that are built into B, it follows that $\neg(\mathbb{U}A \wedge \neg \mathbb{U}A)$ is also valid. Next, note that $\mathsf{B}^{\mathbb{U}}$ is closed under the rule of disjunctive syllogism for U-formulas,

---

[11] R-seriality comes up in the discussion of Halldén completeness for relevant modal logics. See Mares [21] and Seki [35] for more on Halldlén completeness for relevant logics.

- if $\mathbb{U}A$ is valid and $\neg\mathbb{U}A \vee \mathbb{U}B$ is valid, then $\mathbb{U}B$ is valid.

It follows that all substitution instances of classical tautologies in the vocabulary $\{\neg, \wedge, \vee\}$ where the atoms are replaced by $\mathbb{U}$-formulas will be valid. A form of classical logic emerges, even in the austere setting of Routley–Meyer frames for **B**. But, one of the reasons non-classical logicians want to use non-classical logics is because there is something about classical logic that we want to avoid. The classical tautologies reemerging against the backdrop of our favored non-classical logics is not a welcome result.

The problem stems, we think, from the way we model non-classical logics and how that interacts with universal necessity. In frame-based models for non-classical logics, we often obtain non-classical, or generally intensional, logical behavior from a connective by attending to how a subformula is evaluated at potentially distinct points. The approach to negation in star models will serve as an example. A formula, A, and its negation, $\neg A$, can both hold at a point because A fails at the star of the point. When considering the basic vocabulary of non-classical logics, we are restricted to a local view of the model, able to consider only certain points, while there may be other points in the model that are not considered.[12] Again, consider negation in star models. In evaluating $\neg p$ at a point b, we consider only whether p holds at b^*. For that evaluation, assuming $b^* \neq b$, whether p holds at b is immaterial.

When universal necessity enters the picture, one can take a global view on the model. When that happens, the local view is lost. $\mathbb{U}$-formulas hold either at all points or at none. Let us, for a moment, take sets of points as propositions, as is common for models for non-classical logics, and take the set of points where a formula holds to be its proposition. We can see that the only propositions available for $\mathbb{U}$-formulas are the set of all points and the empty set. This limited set of options is very close to the usual two truth-values of classical logic.

We think that this idea is on the right track, and in future work we hope to make this idea both precise and general to prove that against the backdrop of models for many non-classical logics, the addition of $\mathbb{U}$ will result in classical logic reemerging. For now, we will have to make do with the suggestive examples. A consequence of this classicality idea is that we expect that the addition of $\mathbb{U}$ to other non-classical logics will result in violations of the salient non-classical features of the underlying logic.[13]

To close, we will note a curious feature of $\mathbb{U}$. Despite the arguably negative consequences that we have seen of adding $\mathbb{U}$ to a logic defined in terms of frames,

---

[12]The idea of local and global views of a model is common in classically-based modal logic. See, for example, comments about locality by Blackburn et al. [10, pp. 67ff., p. 417].

[13]We thank Hitoshi Omori and Andrzej Indrzejczak for discussion of this point.

there is a silver lining. One can interpret $\mathbb{U}$ in any model on any frame of the classes we considered, since one can do so without adding an accessibility relation to the frame. Alternatively, one can trivially add a universal binary relation. This means that the addition of $\mathbb{U}$ is *conservative* over the base logic, so that if a formula A that does not contain $\mathbb{U}$ is valid in the class of the frames with $\mathbb{U}$ in the language, then A is valid in the class of frames for the base logic. The addition of $\mathbb{U}$, then, does not disturb the base logic, although it does introduce a form of classical logic, which in the eyes of many of the early relevant logicians would seemingly make it a conservative extension in a different sense also.

References

[1] Alan Ross Anderson and Nuel D. Belnap. *Entailment: The Logic of Relevance and Necessity*, volume I. Princeton University Press, Princeton, NJ, 1975.

[2] Jc Beall. There is no logical negation: True, false, both, and neither. *Australasian Journal of Logic*, 14(1): (article no. 1), 2017.

[3] Jc Beall. The simple argument for subclassical logic. *Philosophical Issues*, 28 (1):30–54, 2018.

[4] Nuel Belnap. How a computer should think. In G. Ryle, editor, *Contemporary Aspects of Philosophy*, pages 30–55. Oriel Press, Stocksfield, 1977.

[5] Nuel Belnap. A useful four-valued logic. In J. M. Dunn and G. Epstein, editors, *Modern Uses of Multiple-valued Logic*, pages 8–37. Reidel, Dordrecht, 1977.

[6] Francesco Berto. A modality called 'negation'. *Mind*, 124(495):761–793, 2015.

[7] Francesco Berto and Greg Restall. Negation on the Australian plan. *Journal of Philosophical Logic*, 48(6):1119–1144, 2019.

[8] Katalin Bimbó. Relevance logics. In D. Jacquette, editor, *Philosophy of Logic*, volume 5 of *Handbook of the Philosophy of Science*, pages 723–789. Elsevier, Amsterdam, 2007.

[9] Katalin Bimbó, J. Michael Dunn, and Nicholas Ferenz. Two manuscripts, one by Routley, one by Meyer: The origins of the Routley–Meyer semantics for relevance logics. *Australasian Journal of Logic*, 15(2):171–209, 2018.

[10] Patrick Blackburn, Maarten de Rijke, and Yde Venema. *Modal Logic*. Cambridge University Press, Cambridge, 2002.

[11] Robert A. Bull. MIPC as the formalisation of an intuitionist concept of modality. *Journal of Symbolic Logic*, 31(4):609–616, 1966.

[12] Xavier Caicedo, George Metcalfe, Ricardo Rodríguez, and Olim Tuyt. The one-variable fragment of Corsi logic. In R. Iemhoff, M. Moortgat, and R. de Queiroz, editors, *Logic, Language, Information, and Computation*, pages 70–83, Berlin, 2019. Springer.

[13] Michael De and Hitoshi Omori. There is more to negation than modality. *Journal of Philosophical Logic*, 47(2):281–299, 2018.

[14] J. Michael Dunn and Greg Restall. Relevance logic. In D. M. Gabbay and F. Guenthner, editors, *Handbook of Philosophical Logic*, volume 6, pages 1–128. Kluwer, Amsterdam, 2nd edition, 2002.

[15] Jon Michael Dunn. *The Algebra of Intensional Logics*. Doctoral dissertation, University of Pittsburgh, 1966. (Published as Vol. 2 in the *Logic PhDs* series by College Publications, London, UK, 2019).

[16] Nicholas Ferenz. One variable relevant logics are S5ish. *Journal of Philosophical Logic*, pages 1–23, 2024. Forthcoming.

[17] Paul Richard Halmos. *Algebraic Logic*. Chelsea Publishing Company, New York, NY, 1962. (Reprinted by Dover Publications, Mineola, NY, 2016).

[18] Lloyd Humberstone. *The Connectives*. MIT Press, Cambridge, MA, 2011.

[19] Hector J. Levesque. A logic of implicit and explicit belief. In *Proceedings of AAAI 1984*, pages 198–202, 1984.

[20] Shay Allen Logan. *Relevance Logic*. Elements in Philosophy and Logic. Cambridge University Press, Cambridge, 2024.

[21] Edwin D. Mares. Halldén-completeness and modal relevant logic. *Logique et Analyse*, 46(181):59–76, 2003.

[22] Edwin D. Mares. Relevance logic. In Edward N. Zalta, editor, *The Stanford Encyclopedia of Philosophy*. Metaphysics Research Lab, Stanford University, winter 2020 edition, 2020.

[23] Grigori Mints. *A Short Introduction to Modal Logic*. Number 30 in CSLI Lecture Notes. Center for the Study of Language and Information, Stanford, CA, 1992.

[24] Satoru Niki and Hitoshi Omori. Actuality in Intuitionistic Logic. In N. Olivetti, R. Verbrugge, S. Negri, and G. Sandu, editors, *Advances in Modal Logic, AiML 2020*, volume 13, pages 459–479. College Publications, London, 2020.

[25] Hitoshi Omori and Michael De. Shrieking, shrugging, and the Australian plan. *Notre Dame Journal of Formal Logic*, 63(2), 2022.

[26] Hitoshi Omori and Heinrich Wansing. 40 years of FDE: An introductory overview. *Studia Logica*, 105(6):1021–1049, 2017.

[27] Hitoshi Omori and Heinrich Wansing. *New Essays on Belnap–Dunn Logic*. Springer International Publishing, Switzerland, 2019.

[28] Hiroakira Ono. On some intuitionistic modal logics. *Publications of RIMS Kyoto University*, 13:687–722, 1977.

[29] Zane Parks and Michael Byrd. Relevant implication and Leibnizian necessity. In Jean Norman and Richard Sylvan, editors, *Directions in Relevant Logic*, volume 1 of *Reason and Argument*, pages 179–184. Kluwer, Dordrecht, 1989.

[30] Dag Prawitz. *Natural Deduction: A Proof-theoretical Study*. Stockholm Studies in Philosophy. Almqvist and Wiksell, Stockholm, 1965.

[31] Richard Routley and Robert K. Meyer. The semantics of entailment — II. *Journal of Philosophical Logic*, 1(1):53–73, 1972.

[32] Richard Routley and Robert K. Meyer. The semantics of entailment — III. *Journal of Philosophical Logic*, 1(2):192–208, 1972.

[33] Richard Routley and Robert K. Meyer. The semantics of entailment. In H. Leblanc, editor, *Truth, Syntax, and Modality: Proceedings of the Temple University Conference on Alternative Semantlcs*, pages 199–243. North-Holland, Amsterdam, 1973.

[34] Richard Routley and Valerie Routley. The semantics of first degree entailment. *Noûs*, 6(4):335–359, 1972. doi: 10.2307/2214309.

[35] Takahiro Seki. Halldén completeness for relevant modal logics. *Notre Dame Journal of Formal Logic*, 56(2):333–350, 2015.

[36] Shawn Standefer. What is a relevant connective? *Journal of Philosophical Logic*, 51(4):919–950, 2022.

[37] Shawn Standefer. Varieties of relevant S5. *Logic and Logical Philosophy*, 32(1): 53–80, 2023.

[38] Shawn Standefer. Variable-sharing as relevance. In I. Sedlár, S. Standefer, and A. Tedder, editors, *New Directions in Relevant Logic*, pages 97–117. Springer Nature, Switzerland, 2025.

[39] Shawn Standefer. Relevant logic: Implication, modality, quantification. Manuscript, 202x.

[40] Shawn Standefer, Ted Shear, and Rohan French. Getting some (non-classical) closure with justification logic. *Asian Journal of Philosophy*, 2(2):1–25, 2023.

Received 4 November 2024

A Relevant Framework for Barriers to Entailment

Yale Weiss

The Philosophy Program and Saul Kripke Center,
The Graduate Center, CUNY, New York, USA
`<yweiss@gradcenter.cuny.edu>`

Abstract

In her recent book, Russell (2023) examines various so-called "barriers to entailment," including Hume's law, roughly the thesis that an 'ought' cannot be derived from an 'is.' Hume's law bears an obvious resemblance to the proscription on fallacies of modality in relevance logic, which has traditionally formally been captured by the so-called Ackermann property. In the context of relevant modal logic, this property might be articulated thus: No conditional whose antecedent is box-free and whose consequent is box-prefixed is valid (for the connection, interpret box deontically). While the deontic significance of Ackermann-like properties has been observed before, Russell's new book suggests a more broad-scoped formal investigation of the relationship between barrier theses of various kinds and corresponding Ackermann-like properties. In this paper, I undertake such an investigation by elaborating a general relevant bimodal logical framework in which several of the barriers Russell examines can be given formal expression. I then consider various Ackermann-like properties corresponding to these barriers and prove that certain systems satisfy them. Finally, I respond to some objections Russell makes against the use of relevance logic to formulate Hume's law and related barriers.

2020 *Mathematics Subject Classification.* Primary: 03B47, Secondary: 03B44, 03B45.

An early version of this paper was presented on April 13, 2024 at the Impromptu CUNY Logic Workshop held at the CUNY Graduate Center in New York, USA. Mature versions were presented on May 13, 2024 at the 2nd Third Workshop held at the University of Alberta in Edmonton, Canada, and on September 27, 2024 at the Logic Colloquium of the University of Connecticut in Storrs, USA. I am grateful to the attendees of these talks for helpful feedback. Further thanks are due to two anonymous referees for the journal, Katalin Bimbó, Lloyd Humberstone, and Shawn Standefer.

1   Introduction

In her recent book [43], Russell examines various so-called *barriers to entailment*, including Hume's law, roughly the thesis that an 'ought' cannot be derived from an 'is' (or, more generally, that no collection of descriptive claims entails any normative claim). The law bears an obvious resemblance to the proscription on fallacies of modality traditionally used to motivate relevant systems of entailment especially (see, e.g., Anderson and Belnap [5, §5.2]). This proscription has formally been captured by the so-called *Ackermann property* (see, e.g., Ackermann [1, §6] and Anderson and Belnap [5, §§5.2, 8.12]), which in the context of relevant modal logic might be articulated thus (cf. Meyer [33, p. 476]): No conditional of the form $\varphi \to \Box\psi$ is valid, where φ is $\Box$-free (for the connection, read $\Box$ deontically).[1] The deontic significance of Ackermann-like properties has been observed before, for example, by Mares [29]. Nevertheless, Russell's new book suggests a more broad-scoped formal investigation of the relationship between barrier theses of various kinds and corresponding Ackermann-like properties.

The primary purpose of this paper is to pursue such a formal investigation. To that end, in Section 2, I present a fairly general relevant bimodal framework in which several (though not all) of the barriers Russell [43] is concerned with can be given formal expression. Routley–Meyer semantics is canvassed for a variety of (bi)modal extensions of **R** and soundness and completeness theorems are sketched in each case.

In Sections 3, 4, and 5, this framework is applied to present relevant systems of modal logic, deontic logic, and tense logic, all of which satisfy versions of the Ackermann property corresponding to particular barrier theses studied by Russell [43]. The exact formulation of the Ackermann property depends on the system and language under consideration, but proofs are in each case given using matrices.[2] Pertinent historical and philosophical issues are discussed as appropriate.

Finally, in Section 6, I close by examining some of the critical remarks Russell [43, pp. 36–41] makes against the use of relevance logic to formulate barrier theses. Against Russell, I contend that relevance logic, and the (bi)modal relevant framework propounded in this paper in particular, is a very natural setting in which to express barrier theses of various kinds, and prove that they obtain.

The foregoing notwithstanding, let me add some remarks about what this paper is not. This paper is not intended as a book review or critical notice of [43]. My engagement with Russell's critical comments will be selective, and my engagement

---

[1]The Ackermann property was first examined (by Ackermann [1, §6]) as a desideratum to be imposed on a primitive entailment connective. Anderson and Belnap [5, §5.2] spell out the property in connection with $\mathbf{E}_\to$ thus: $\varphi \to (\psi \to \theta)$ is never a theorem when φ is a propositional variable.

[2]Many (though not all) of these matrices were found using Slaney's program MaGIC [49].

with her positive proposals will be minimal. A more thorough comparison of the relative merits of the proposals made here with her own would be desirable, but such a project is left for future work.

2   A Relevant Bimodal Framework

Fix a countable set of propositional variables $\Pi = \{p_0, p_1, \ldots\}$. The set of formulae, Φ, is built up in the usual way from Π and the connectives $\neg, \vee, \wedge, \rightarrow, \square$, and possibly, $\blacksquare$. (I mostly leave implicit what the language under discussion is.) The usual definition $\Diamond\varphi := \neg\square\neg\varphi$ (and, if applicable, $\blacklozenge\varphi := \neg\blacksquare\neg\varphi$) is adopted.[3] I write $p, q, \ldots$ for arbitrary propositional variables, and $\varphi, \psi, \ldots$ for arbitrary formulae.

Throughout, I take $\mathbf{R}$ as the background relevant logic. (I have no special love for $\mathbf{R}$; however, it is well-known, and many interesting results that can be proved for $\mathbf{R}$ carry over, mutatis mutandis, to weaker systems.) For concreteness, let $\mathbf{R}$ be axiomatized thus:[4]

$$\text{(I)} \qquad \varphi \rightarrow \varphi$$

$$\text{(B')} \qquad (\varphi \rightarrow \psi) \rightarrow ((\psi \rightarrow \theta) \rightarrow (\varphi \rightarrow \theta))$$

$$\text{(C)} \qquad (\varphi \rightarrow (\psi \rightarrow \theta)) \rightarrow (\psi \rightarrow (\varphi \rightarrow \theta))$$

$$\text{(W)} \qquad (\varphi \rightarrow (\varphi \rightarrow \psi)) \rightarrow (\varphi \rightarrow \psi)$$

$$\text{($\wedge$E1)} \qquad (\varphi \wedge \psi) \rightarrow \varphi$$

$$\text{($\wedge$E2)} \qquad (\varphi \wedge \psi) \rightarrow \psi$$

$$\text{($\wedge$I)} \qquad ((\varphi \rightarrow \psi) \wedge (\varphi \rightarrow \theta)) \rightarrow (\varphi \rightarrow (\psi \wedge \theta))$$

$$\text{($\vee$I1)} \qquad \varphi \rightarrow (\varphi \vee \psi)$$

$$\text{($\vee$I2)} \qquad \psi \rightarrow (\varphi \vee \psi)$$

$$\text{($\vee$E)} \qquad ((\varphi \rightarrow \theta) \wedge (\psi \rightarrow \theta)) \rightarrow ((\varphi \vee \psi) \rightarrow \theta)$$

$$\text{($\wedge\vee$)} \qquad (\varphi \wedge (\psi \vee \theta)) \rightarrow ((\varphi \wedge \psi) \vee (\varphi \wedge \theta))$$

$$\text{(DNE)} \qquad \neg\neg\varphi \rightarrow \varphi$$

$$\text{(CP)} \qquad (\varphi \rightarrow \neg\psi) \rightarrow (\psi \rightarrow \neg\varphi)$$

---

[3]Incidentally, because relevant negation is nonclassical, this has the downstream effect that the truth condition for $\Diamond$ (and $\blacklozenge$) is somewhat nonstandard (cf. Fuhrmann [18, pp. 505–506]).

[4]This is the axiomatization given by Anderson and Belnap [5, p. 341], except that I have used C in lieu of their R3 (Assertion) and $\wedge\vee$ in lieu of their R11; the equivalence of these axiomatizations of $\mathbf{R}$ is well-known.

(MP) $\dfrac{\varphi,\ \varphi \to \psi}{\psi}$

(ADJ) $\dfrac{\varphi,\ \psi}{\varphi \wedge \psi}$

I turn now to (bi)modal extensions of $\mathbf{R}$. My approach throughout largely follows Fuhrmann [17, 18] who, however, was only concerned with monomodal relevant logic.[5] To obtain the basic regular (bi)modal extension of $\mathbf{R}$, $\mathbf{R}^{\square}$ ($\mathbf{R}^{\blacksquare}$), add the following:

(RM) $\dfrac{\varphi \to \psi}{\boxdot\varphi \to \boxdot\psi}$, where $\boxdot$ is $\square$ (or $\blacksquare$)

($\wedge$C) $(\boxdot\varphi \wedge \boxdot\psi) \to \boxdot(\varphi \wedge \psi)$, where $\boxdot$ is $\square$ (or $\blacksquare$)

$\mathbf{R}^{\square}$ is the system $R.C$ from Fuhrmann [18]. Further systems of interest are obtained by adding selections of the following axiom schemata and rules (in general, my interest will be in bimodal logics where different principles are adopted for each of the modal connectives, whence the redundancy):

($\square$K) $\square(\varphi \to \psi) \to (\square\varphi \to \square\psi)$

($\blacksquare$K) $\blacksquare(\varphi \to \psi) \to (\blacksquare\varphi \to \blacksquare\psi)$

($\square$D) $\square\neg\varphi \to \neg\square\varphi$

($\blacksquare$D) $\blacksquare\neg\varphi \to \neg\blacksquare\varphi$

($\square$T) $\square\varphi \to \varphi$

($\blacksquare$T) $\blacksquare\varphi \to \varphi$

---

[5]There is relatively little discussion of relevant bimodal (or multimodal) logic to be found in the literature. Seki [46, 47, 48] invariably treats $\square$ and $\Diamond$ as primitive connectives with independent accessibility relations (cf. Routley [39, pp. 273–276]), and so in a sense gives a bimodal treatment of what is usually thought of as a monomodal logic. (This is an established approach in modal logic over logics with a nonclassical negation; see, e.g., the intuitionistic system $\mathbf{HK}\square\Diamond$ from Božić and Došen [9].) Wansing [54] presents a bimodal relevant epistemic logic. Cheng has sketched purely syntactic developments of versions of relevant tense logic and multimodal tense-deontic logic in a series of short papers [11, 12, 13]. Standefer [50, §6.4] devotes some discussion to relevant logics containing combinations of an actuality operator, an alethic modal operator, and a fixedly operator. The work in this section is more substantially anticipated by Routley [39, §2], which I was unfamiliar with when I first drafted this piece. While Routley articulates a general relevant multimodal framework and even discusses a (somewhat peculiar) bimodal relevant tense logic [39, p. 276], his approach and my own differ in at least some important respects nonetheless (e.g., in generally treating diamonds as primitives). Thanks to Shawn Standefer for a number of valuable reference pointers here.

$(\Box 4)$ $\quad\quad \Box\varphi \to \Box\Box\varphi$

$(\blacksquare 4)$ $\quad\quad \blacksquare\varphi \to \blacksquare\blacksquare\varphi$

$(\Box B)$ $\quad\quad \varphi \to \Box\Diamond\varphi$

$(\blacksquare B)$ $\quad\quad \varphi \to \blacksquare\blacklozenge\varphi$

$(\blacklozenge\Box)$ $\quad\quad \blacklozenge\Box\varphi \to \varphi$

$(\Diamond\blacksquare)$ $\quad\quad \Diamond\blacksquare\varphi \to \varphi$

$(\Box\blacksquare D)$ $\quad\quad \Box\neg\varphi \to \neg\blacksquare\varphi$

$(\Box NEC)$ $\quad\quad \dfrac{\varphi}{\Box\varphi}$

$(\blacksquare NEC)$ $\quad\quad \dfrac{\varphi}{\blacksquare\varphi}$

The smallest extension of $\mathbf{R}^{\Box}$ ($\mathbf{R}^{\Box}_{\blacksquare}$) by axioms/rules from some subset $\mathcal{C}$ of the list above will be written $\mathbf{R}^{\Box}(\mathcal{C})$ ($\mathbf{R}^{\Box}_{\blacksquare}(\mathcal{C})$). Where no particular such extension is intended, I will usually just write $\mathbf{L}$. I adopt the standard useful fiction that a logic $\mathbf{L}$ is a set of formulae and identify theoremhood for $\mathbf{L}$ ($\vdash_{\mathbf{L}}$) with membership. Bootstrapping, $\mathbf{L}$-derivability is defined as follows.

Definition 1. A set Δ is $\mathbf{L}$-*derivable* from a set Γ (in symbols, $\Gamma \vdash_{\mathbf{L}} \Delta$) if and only if there are some $\gamma_1, \ldots, \gamma_m \in \Gamma$ and $\delta_1, \ldots, \delta_n \in \Delta$ such that $\vdash_{\mathbf{L}} \bigwedge \gamma_i \to \bigvee \delta_j$.

I turn now to matters semantic. The basic framework I will avail myself to is the (unreduced) Routley–Meyer semantics (see, e.g., Routley et al. [41, Ch. 4]).

Definition 2 (Frame). A *frame* for $\mathbf{R}$ is a structure $\mathfrak{F} = \langle W, N, R, {}^{*} \rangle$, where $\emptyset \neq N \subseteq W$, $R \subseteq W^3$, and $^{*}: W \to W$, and these components satisfy the following postulates:[6]

p1. $a \leq a$;

p2. $a \leq b$ and $Rbcd$ imply $Racd$;

p3. $a = a^{**}$;

p4. $Rabc$ implies $Rac^{*}b^{*}$;

p5. $R^2(ab)cd$ implies $R^2b(ac)d$;

p6. $R^2(ab)cd$ implies $R^2(ac)bd$;

---

[6]Abbreviations: $a \leq b := \exists x(x \in N \wedge Rxab)$, $R^2(ab)cd := \exists x(Rabx \wedge Rxcd)$, and $R^2a(bc)d := \exists x(Raxd \wedge Rbcx)$.

p7. *Rabc* implies $R^2(ab)bc$.

Generalizing Fuhrmann [17, 18], to get a frame for $\mathbf{R}_{\blacksquare}^{\square}$, two new components are added to Definition 2. (In all of what follows, for $\mathbf{R}^{\square}$ and its monomodal extensions, just leave out the material specific to $\blacksquare$.)

Definition 3 (Modal frame). A *modal frame* for $\mathbf{R}_{\blacksquare}^{\square}$, $\mathfrak{F} = \langle W, N, R, {}^{*}, S_{\square}, S_{\blacksquare} \rangle$, is an $\mathbf{R}$ frame with two new binary relations $S_{\square}, S_{\blacksquare} \subseteq W^2$ which satisfy the further condition:[7]

p8. $a \leq b$ implies $S_{\boxdot}(b) \subseteq S_{\boxdot}(a)$, where $\boxdot \in \{\square, \blacksquare\}$.

Frames for extensions $\mathbf{L}$ of $\mathbf{R}^{\square}$ ($\mathbf{R}_{\blacksquare}^{\square}$) are obtained by imposing constraints which correspond to the associated axiom schemata and rules (such frames are called *fit for* $\mathbf{L}$).[8]

	Axiom/Rule	Frame Condition
p9	$\square$K	$\exists x(Rabx \wedge S_{\square}xc) \Rightarrow \exists y \exists z(S_{\square}ay \wedge S_{\square}bz \wedge Ryzc)$
p10	$\blacksquare$K	$\exists x(Rabx \wedge S_{\blacksquare}xc) \Rightarrow \exists y \exists z(S_{\blacksquare}ay \wedge S_{\blacksquare}bz \wedge Ryzc)$
p11	$\square$D	$\exists x(S_{\square}ax^* \wedge S_{\square}a^*x)$
p12	$\blacksquare$D	$\exists x(S_{\blacksquare}ax^* \wedge S_{\blacksquare}a^*x)$
p13	$\square$T	$S_{\square}aa$
p14	$\blacksquare$T	$S_{\blacksquare}aa$
p15	$\square$4	$S_{\square}ab \wedge S_{\square}bc \Rightarrow S_{\square}ac$
p16	$\blacksquare$4	$S_{\blacksquare}ab \wedge S_{\blacksquare}bc \Rightarrow S_{\blacksquare}ac$
p17	$\square$B	$S_{\square}ab \Rightarrow S_{\square}b^*a^*$
p18	$\blacksquare$B	$S_{\blacksquare}ab \Rightarrow S_{\blacksquare}b^*a^*$
p19	$\blacklozenge\square$	$S_{\blacksquare}ab \Rightarrow S_{\square}b^*a^*$
p20	$\lozenge\blacksquare$	$S_{\square}ab \Rightarrow S_{\blacksquare}b^*a^*$
p21	$\square\blacksquare$D	$\exists x(S_{\blacksquare}a^*x \wedge S_{\square}ax^*)$
p22	$\square$NEC	$a \in N \wedge S_{\square}ab \Rightarrow b \in N$
p23	$\blacksquare$NEC	$a \in N \wedge S_{\blacksquare}ab \Rightarrow b \in N$

Definition 4 (Model). A *model* fit for a system $\mathbf{L}$ extending $\mathbf{R}_{\blacksquare}^{\square}$ is a structure $\mathfrak{M} = \langle \mathfrak{F}, V \rangle$, where $\mathfrak{F} = \langle W, N, R, {}^{*}, S_{\square}, S_{\blacksquare} \rangle$ is a frame (Definition 3) fit for $\mathbf{L}$ and $V \colon \Pi \to \mathcal{P}(W)$ subject to the heredity condition: $x \in V(p)$ and $x \leq y$ imply $y \in V(p)$.

---

[7]Abbreviation: $S_{\boxdot}(a) := \{b \in W : S_{\boxdot}ab\}$, where $\boxdot \in \{\square, \blacksquare\}$.

[8]Many of these correspondences are noted by Fuhrmann [18, pp. 507–508] (cf. Routley [39, pp. 275–276]).

Given a model $\mathfrak{M} = \langle W, N, R, {}^{*}, S_{\square}, S_{\blacksquare}, V \rangle$ and $w \in W$, the relation $\vDash^{\mathfrak{M}}_{w}$ is defined by the following conditions:[9]

p. $\vDash^{\mathfrak{M}}_{w} p$ iff $w \in V(p)$;

$\neg$. $\vDash^{\mathfrak{M}}_{w} \neg\varphi$ iff $\nvDash^{\mathfrak{M}}_{w^{*}} \varphi$;

$\wedge$. $\vDash^{\mathfrak{M}}_{w} \varphi \wedge \psi$ iff $\vDash^{\mathfrak{M}}_{w} \varphi$ and $\vDash^{\mathfrak{M}}_{w} \psi$;

$\vee$. $\vDash^{\mathfrak{M}}_{w} \varphi \vee \psi$ iff $\vDash^{\mathfrak{M}}_{w} \varphi$ or $\vDash^{\mathfrak{M}}_{w} \psi$;

$\rightarrow$. $\vDash^{\mathfrak{M}}_{w} \varphi \rightarrow \psi$ iff, for all $x, y \in W$, if $Rwxy$ and $\vDash^{\mathfrak{M}}_{x} \varphi$, then $\vDash^{\mathfrak{M}}_{y} \psi$;

$\boxdot$. $\vDash^{\mathfrak{M}}_{w} \boxdot\varphi$ iff $S_{\square}(w) \subseteq [\varphi]^{\mathfrak{M}}$, where $\boxdot \in \{\square, \blacksquare\}$.

The following lemma, used in the argument for Theorem 1, has a routine proof which I omit.

Lemma 1 (Heredity). *For any model* $\mathfrak{M} = \langle W, N, R, {}^{*}, S_{\square}, S_{\blacksquare}, V \rangle$ *and* $a, b \in W$, *if* $a \leq b$ *and* $\vDash^{\mathfrak{M}}_{a} \varphi$, *then* $\vDash^{\mathfrak{M}}_{b} \varphi$.

Definition 5 (Validity). A formula φ is valid over a model $\mathfrak{M}$ (in symbols, $\vDash^{\mathfrak{M}} \varphi$) if for all $w \in N$, $\vDash^{\mathfrak{M}}_{w} \varphi$. φ is *valid* in the system $\mathbf{L}$ ($\vDash_{\mathbf{L}} \varphi$) if for all models $\mathfrak{M}$ fit for $\mathbf{L}$, $\vDash^{\mathfrak{M}} \varphi$.

Theorem 1 (Soundness). *For any of the systems* $\boldsymbol{L}$, *if* $\vdash_{\boldsymbol{L}} \varphi$, *then* $\vDash_{\boldsymbol{L}} \varphi$.

Proof. Just the usual checking of cases. All of the nonmodal and monomodal cases can be found in the literature (see, e.g., Routley et al. [41] and Fuhrmann [17, 18]). I quickly run through the mixed-modality cases here.

Ad $\blacklozenge\square$: If $\mathbf{L}$ contains $\blacklozenge\square$ and $\nvdash_{\mathbf{L}} \blacklozenge\square\varphi \rightarrow \varphi$, then there is a model $\mathfrak{M} = \langle W, N, R, {}^{*}, S_{\square}, S_{\blacksquare}, V \rangle$ fit for $\mathbf{L}$ and $a, b \in W$ such that $a \leq b$, $\vDash^{\mathfrak{M}}_{a} \blacklozenge\square\varphi$ (i.e., $\vDash^{\mathfrak{M}}_{a} \neg\blacksquare\neg\square\varphi$), and $\nvDash^{\mathfrak{M}}_{b} \varphi$. Then $\vDash^{\mathfrak{M}}_{a} \neg\blacksquare\neg\square\varphi$ implies $\nvDash^{\mathfrak{M}}_{a^{*}} \blacksquare\neg\square\varphi$ implies $\nvDash^{\mathfrak{M}}_{c} \neg\square\varphi$ for some c such that $S_{\blacksquare}a^{*}c$, which implies $\vDash^{\mathfrak{M}}_{c^{*}} \square\varphi$. By p3 and p19, $S_{\blacksquare}a^{*}c$ implies $S_{\square}c^{*}a$, whence $\vDash^{\mathfrak{M}}_{c^{*}} \square\varphi$ implies $\vDash^{\mathfrak{M}}_{a} \varphi$; by Lemma 1, $\vDash^{\mathfrak{M}}_{a} \varphi$ implies $\vDash^{\mathfrak{M}}_{b} \varphi$, a contradiction. (The case of $\lozenge\blacksquare$ is entirely analogous to this case and is omitted.)

Ad $\square\blacksquare$D: If $\mathbf{L}$ contains $\square\blacksquare$D and $\nvdash_{\mathbf{L}} \square\neg\varphi \rightarrow \neg\blacksquare\varphi$, then there is a model $\mathfrak{M} = \langle W, N, R, {}^{*}, S_{\square}, S_{\blacksquare}, V \rangle$ fit for $\mathbf{L}$ and $a, b \in W$ such that $a \leq b$, $\vDash^{\mathfrak{M}}_{a} \square\neg\varphi$, and $\nvDash^{\mathfrak{M}}_{b} \neg\blacksquare\varphi$, whence $\vDash^{\mathfrak{M}}_{b^{*}} \blacksquare\varphi$. By p21, there is some c such that $S_{\blacksquare}b^{*}c$ and $S_{\square}bc^{*}$. Thus, $\vDash^{\mathfrak{M}}_{c} \varphi$, and by p8 and $a \leq b$, $S_{\square}ac^{*}$, whence $\vDash^{\mathfrak{M}}_{c^{*}} \neg\varphi$, and so $\nvDash^{\mathfrak{M}}_{c} \varphi$, a contradiction. $\square$

---

[9]Abbreviation: $[\varphi]^{\mathfrak{M}} := \{w \in W : \vDash^{\mathfrak{M}}_{w} \varphi\}$.

For completeness, a number of preliminary definitions and results concerning theory building are required (cf. Routley et al. [41, pp. 306–307]). Fix a system **L**. An **L**-*theory* is a set of formulae Γ closed under ADJ and such that if $\varphi \in \Gamma$ and $\vdash_{\mathbf{L}} \varphi \to \psi$, then $\psi \in \Gamma$. An **L**-theory Γ is *regular* if $\mathbf{L} \subseteq \Gamma$, and *prime* if whenever $\varphi \vee \psi \in \Gamma$, then either $\varphi \in \Gamma$ or $\psi \in \Gamma$. A pair $\langle \Gamma, \Delta \rangle$ is **L**-*maximal* iff 1) $\Gamma \nvdash_{\mathbf{L}} \Delta$ and 2) $\Gamma \cup \Delta = \Phi$.

The following standard results (and others besides), rehearsed in detail in Routley et al. [41, pp. 307–312], carry over wholesale (the key result is Lemma 2, from which much of the rest ultimately follow).

Lemma 2 (Pair extension). *If $\Gamma \nvdash_L \Delta$, then there is an **L**-maximal pair $\langle \Gamma', \Delta' \rangle$ such that $\Gamma \subseteq \Gamma'$ and $\Delta \subseteq \Delta'$.*

Corollary 1 (Priming). *If Γ is an **L**-theory and Δ is a set of formulae disjoint from Γ and such that for any $\varphi, \psi \in \Delta$ (distinct), $\varphi \vee \psi \in \Delta$, then there is a prime **L**-theory Γ' such that $\Gamma \subseteq \Gamma'$ and $\Gamma' \cap \Delta = \emptyset$.*

Corollary 2 (Witness). *If $\nvdash_L \varphi$, then there is a prime regular **L**-theory Π such that $\varphi \notin \Pi$.*

Given sets of formulae Γ, Δ, and Σ, define $R^c\Gamma\Delta\Sigma$ iff $\{\psi \colon \exists \varphi \in \Delta(\varphi \to \psi \in \Gamma)\} \subseteq \Sigma$.

Corollary 3. *For **L**-theories Γ, Δ, Σ, where Σ is prime, if $R^c\Gamma\Delta\Sigma$, then there is a prime **L**-theory Π such that $\Delta \subseteq \Pi$ and $R^c\Gamma\Pi\Sigma$.*

Corollary 4. *For **L**-theories Γ, Δ, Σ, where Σ is prime, if $R^c\Gamma\Delta\Sigma$, then there is a prime **L**-theory Π such that $\Gamma \subseteq \Pi$ and $R^c\Pi\Delta\Sigma$.*

Definition 6 (Canonical model). The canonical model for **L** is the structure $\mathfrak{M}^c = \langle W^c, N^c, R^c, *^c, S^c_\square, S^c_\blacksquare, V^c \rangle$ defined as follows:

1. W^c is the set of all prime **L**-theories;

2. N^c is the set of all regular prime **L**-theories;

3. $R^c\Gamma\Delta\Sigma$ iff $\{\psi \colon \exists \varphi \in \Delta(\varphi \to \psi \in \Gamma)\} \subseteq \Sigma$ (as above);[10]

4. $\Gamma^{*c} = \{\varphi \colon \neg\varphi \notin \Gamma\}$;

---

[10]Strictly speaking, as defined above, $R^c \subseteq \mathcal{P}(\Phi)^3$, whereas in compliance with Definition 2, $R^c \subseteq (W^c)^3$ here. It proves convenient and causes no real confusion to employ a wider notion at various points (similarly for $S^c_\square, S^c_\blacksquare$).

5. $S_\Box^c \Gamma \Delta$ iff $\{\varphi : \Box\varphi \in \Gamma\} \subseteq \Delta$;

6. $S_\blacksquare^c \Gamma \Delta$ iff $\{\varphi : \blacksquare\varphi \in \Gamma\} \subseteq \Delta$;

7. $V^c(p) = \{\Gamma : p \in \Gamma\}$.

Observe that in the canonical model for $\mathbf{L}$, $\mathfrak{M}^c = \langle W^c, N^c, R^c, {}^{*c}, S_\Box^c, S_\blacksquare^c, V^c \rangle$, $\leq^c = \subseteq$.

Lemma 3. *The canonical model for $\mathbf{L}$, $\mathfrak{M}^c$, is a model fit for $\mathbf{L}$.*

Proof. I ignore the nonmodal cases, and cover the (representative) cases of p8, p9, p20, p21, and p22. (For some of the arguments given below, cf. Fuhrmann [17, pp. 48–50] and Routley [39, pp. 277–278].)

Ad p8: Suppose $\Gamma \subseteq \Delta$, $S_\Box^c \Delta \Sigma$, and $\Box\varphi \in \Gamma$, to show that $\varphi \in \Sigma$; this will suffice to show $S_\Box^c \Gamma \Sigma$. Then $\Box\varphi \in \Gamma \subseteq \Delta$ and $\{\psi : \Box\psi \in \Delta\} \subseteq \Sigma$, so in particular, $\varphi \in \Sigma$.

Ad p9: Suppose $\mathbf{L}$ contains $\Box$K and that $R^c \Gamma \Delta \Sigma$ and $S_\Box^c \Sigma \Pi$; it must be shown that there are prime $\mathbf{L}$-theories Λ and Ξ such that $S_\Box^c \Gamma \Lambda$, $S_\Box^c \Delta \Xi$, and $R^c \Lambda \Xi \Pi$. Put $\Lambda' := \{\varphi : \Box\varphi \in \Gamma\}$ and $\Xi' := \{\varphi : \Box\varphi \in \Delta\}$. It is clear that Λ' and Ξ' are $\mathbf{L}$-theories such that $S_\Box^c \Gamma \Lambda'$ and $S_\Box^c \Delta \Xi'$. Suppose $\varphi \in \Xi'$ and $\varphi \to \psi \in \Lambda'$; then $\Box(\varphi \to \psi) \in \Gamma$, whence by $\Box$K, $\Box\varphi \to \Box\psi \in \Gamma$. Since also $\Box\varphi \in \Delta$, $\Box\psi \in \Sigma$ by the hypothesis that $R^c \Gamma \Delta \Sigma$, and so $\psi \in \Pi$ by the hypothesis that $S_\Box^c \Sigma \Pi$. Therefore, $R^c \Lambda' \Xi' \Pi$. By Corollaries 3 and 4, there are prime $\mathbf{L}$-theories Λ and Ξ such that $\Lambda' \subseteq \Lambda$ and $\Xi' \subseteq \Xi$ (whence, $S_\Box^c \Gamma \Lambda$ and $S_\Box^c \Delta \Xi$), and $R^c \Lambda \Xi \Pi$, as desired.

Ad p20: Suppose $\mathbf{L}$ contains $\Diamond\blacksquare$, that $S_\Box^c \Gamma \Delta$, and that $\blacksquare\varphi \in \Delta^{*c}$. Then $\neg\blacksquare\varphi \notin \Delta$, whence $\Box\neg\blacksquare\varphi \notin \Gamma$, whence $\neg\Box\neg\blacksquare\varphi \in \Gamma^{*c}$. But $\mathbf{L}$ contains $\Diamond\blacksquare$ and Γ^{*c} is an $\mathbf{L}$-theory, whence $\varphi \in \Gamma^{*c}$, as desired.

Ad p21: Suppose $\mathbf{L}$ contains $\Box\blacksquare$D and fix an arbitrary prime $\mathbf{L}$-theory Γ. It is to be shown that there is some prime $\mathbf{L}$-theory Δ such that $S_\blacksquare^c \Gamma^{*c} \Delta$ and $S_\Box^c \Gamma \Delta^{*c}$. Define $\Delta' := \{\varphi : \blacksquare\varphi \in \Gamma^{*c}\}$; clearly, Δ' is an $\mathbf{L}$-theory and $S_\blacksquare^c \Gamma^{*c} \Delta'$. Moreover, suppose $\Box\varphi \in \Gamma$; then since $\mathbf{L}$ contains $\Box\blacksquare$D, $\neg\blacksquare\neg\varphi \in \Gamma$, whence $\blacksquare\neg\varphi \notin \Gamma^{*c}$, whence $\neg\varphi \notin \Delta'$. Now, define $\Sigma := \{\varphi : \Box\neg\varphi \in \Gamma\}$. Observe that $\Delta' \cap \Sigma = \emptyset$; for if $\varphi \in \Sigma$, then $\Box\neg\varphi \in \Gamma$, whence by the foregoing, $\neg\neg\varphi \notin \Delta'$, and so $\varphi \notin \Delta'$. Furthermore, Σ is closed under disjunction. Therefore, by Corollary 1, there is a prime $\mathbf{L}$-theory Δ such that $\Delta' \subseteq \Delta$ and $\Delta \cap \Sigma = \emptyset$. Then clearly $S_\blacksquare^c \Gamma^{*c} \Delta$. On the other hand, suppose that $\Box\varphi \in \Gamma$; then $\Box\neg\neg\varphi \in \Gamma$, from which it follows that $\neg\varphi \in \Sigma$, $\neg\varphi \notin \Delta$, and finally, that $\varphi \in \Delta^{*c}$, as desired.

Ad p22: Suppose $\mathbf{L}$ contains $\Box$NEC and that $\Gamma \in N^c$ and $S_\Box^c \Gamma \Delta$. If $\vdash_\mathbf{L} \varphi$, then ex hypothesi, $\vdash_\mathbf{L} \Box\varphi$, whence $\Box\varphi \in \Gamma$ and so $\varphi \in \Delta$. But then Δ is regular, that is, $\Delta \in N^c$. $\qquad\square$

Lemma 4 (Truth lemma). *In the canonical model for L, $\mathfrak{M}^c$, the following biconditional obtains: $\vDash_\Gamma^{\mathfrak{M}^c} \varphi$ iff $\varphi \in \Gamma$.*

Proof. The basis case holds by Definition 6. The only induction case I examine is that concerning $\square$.

Ad $\square$: Suppose $\square\varphi \in \Gamma$ and $S_\square^c \Gamma\Delta$, that is, $\{\theta : \square\theta \in \Gamma\} \subseteq \Delta$; then $\varphi \in \Delta$, so $\vDash_\Delta^{\mathfrak{M}^c} \varphi$ by the induction hypothesis, which suffices. Conversely, suppose $\square\varphi \notin \Gamma$; then observe that $\Delta := \{\theta : \square\theta \in \Gamma\}$ is an L-theory: $\alpha, \beta \in \Delta$ imply $\square\alpha, \square\beta \in \Gamma$ implies $\square(\alpha \wedge \beta) \in \Gamma$ ($\wedge$C) implies $\alpha \wedge \beta \in \Delta$; if $\alpha \in \Delta$ and $\vdash_L \alpha \to \beta$, then since $\square\alpha \in \Gamma$ and $\vdash_L \square\alpha \to \square\beta$ (RM), $\square\beta \in \Gamma$, whence $\beta \in \Delta$. Now apply Corollary 1 to the pair $\langle \Delta, \{\varphi\}\rangle$; then there is a prime L-theory $\Delta' \supseteq \Delta$ such that $\varphi \notin \Delta'$. Clearly, $S_\square^c \Gamma\Delta'$ and $\nvDash_{\Delta'}^{\mathfrak{M}^c} \varphi$, which suffices. $\square$

Theorem 2 (Completeness). *For any of the systems L, if $\vDash_L \varphi$, then $\vdash_L \varphi$.*

Proof. Suppose $\nvdash_L \varphi$; by Corollary 2, there is a prime regular L-theory Γ such that $\varphi \notin \Gamma$. By Definition 6, $\Gamma \in N^c$ in the canonical model $\mathfrak{M}^c$. By Lemma 3, $\mathfrak{M}^c$ is fit for L. By Lemma 4, $\nvDash_\Gamma^{\mathfrak{M}^c} \varphi$. Therefore, $\nvDash_L \varphi$, as desired. $\square$

3   The Modal Barrier

The first barrier thesis I examine belongs to alethic modal logic proper, that is, the logic of what is *necessarily so*, or of what *must be*. (Throughout this section, I focus on monomodal logic.) The intuitive idea is that there exists a barrier between what is merely the case, on the one hand, and what must necessarily be the case, on the other hand. Russell glosses the idea thus:

> According to the *is/must* barrier, no matter what we learn about the way the world is, nothing follows logically about how it must be: no *must* from an *is*. [43, p. 119]

Although Russell [43] does not remark on it, exactly such an intuition as this was what Anderson and Belnap availed themselves to early on in order to motivate a theory of entailment free of the fallacies of modality:[11]

> Modal fallacies arise when it is claimed that entailments follow from, or are entailed by, *contingent* propositions [...] Consider $A \to .A \to A$. Though "snow is white" and "that snow is white entails that snow is

---

[11]In response to objections such as those of Routley and Routley [40], Anderson and Belnap [5, §§5.2.1, 22.1.2] would later be somewhat more circumspect in characterizing the fallacies of modality. More on this below.

white" are both true—the latter necessarily so—it seems implausible that "snow is white" should *entail* that it entails itself. [4, pp. 44–45]

Recall that relevance logic in the Anderson–Belnap tradition was conceived (sinlessly) with the task of rectifying two types of fallacies; those of relevance and those of modality.[12] The ensuing history of the field has shifted the focus almost entirely onto the fallacies of relevance, and the repudiation of such chimeras as $(\varphi \wedge \neg \varphi) \to \psi$, but Anderson and Belnap were at least as (if not more) concerned with the latter category.

How plausible is the is/must barrier? First, it only appears plausible, even in a prima facie way, in the direction from 'is' to 'must;' few would contest the T-axiom ($\Box$T) for metaphysical or even much weaker shades of necessity.[13] But even the prima facie plausible direction seems problematic on further reflection. Routley and Routley [40] offer many counterexamples to what they call DC (the Distributivity of Contingency). One particularly acute failure which they note is embodied by $\Box$B:

$$(\Box \text{B}) \quad \varphi \to \Box \Diamond \varphi$$

If the correct logic of what must be includes $\Box$B, then it would seem that each and every 'is' yields a 'must.'[14]

While it is therefore far from obvious that one should even want the is/must barrier, I will, nevertheless, now show that one can have it, or a suitable form of the Ackermann property corresponding to it, even in a fairly strong relevant modal logic (which, however, will naturally not contain $\Box$B).

Fix a background logic **L** extending $\mathbf{R}^{\Box}$. If $\vdash_{\mathbf{L}} \varphi \to \psi$, then the conditional holds as a matter of logic, and it is natural enough to describe this state of affairs as φ entails ψ. A *barrier to entailment* or *Ackermann* theorem will then be a result to the effect that $\nvdash_{\mathbf{L}} \varphi \to \psi$ whenever φ and ψ meet certain conditions (in which case, **L** will be said to enjoy the barrier or Ackermann property under consideration). In the modal case, it is to be shown that $\nvdash_{\mathbf{L}} \varphi \to \psi$ whenever φ is extensional (an 'is') and ψ is necessary (a 'must'). To give this content, a taxonomy of formulae must

---

[12]See, for example, Anderson and Belnap [4, pp. 42–50], Anderson and Belnap [5, §5], and Meyer [33, p. 472].

[13]Incidentally, Anderson and Belnap [5, §8.12, pp. 95–96] describe a converse of the Ackermann property, and remark that, "One might wish a system to have this feature if one had the opinion, as we do not, that a non-necessitative could never follow from a necessitative." But a bidirectional barrier is plausible in other cases, for example, in the deontic case discussed below (see Section 4; cf. Weiss [57, p. 396, n. 7]).

[14]Of course, not everyone thinks that $\Box$B ought to be part of the logic of what must be (consult, e.g., Salmon [44]).

be pinned down.[15]

Intuitively speaking, φ is *necessary* if it is **L**-equivalent to some formula $\Box\psi$,[16] and it is *extensional* if it has no modal content whatsoever (i.e., it is $\Box$-free). I find this taxonomy natural enough, but it does have the rather unattractive feature that the class of necessary formulae is identified by a properly logical property while the class of extensional formulae is characterized by a syntactic property. However, if the result is proved for necessary formulae characterized in this way, it will also hold (trivially) for necessary formulae characterized in the most natural syntactic fashion (as formulae *of the form* $\Box\psi$).[17] On the other hand, I cannot see an easy way to formulate a 'properly logical' characterization of the extensional formulae which would be amenable to proving a form of the Ackermann property. So, the definitions given at the start of this paragraph are the definitions that I adopt.

Proposition 1. $\nvdash_L \varphi \to \psi$ *whenever* φ *is extensional and* ψ *is necessary, for* $L \subseteq R^\Box(\Box K, \Box T, \Box 4, \Box NEC)$.

Proof. Let φ be extensional and ψ be necessary; then $\vdash_\mathbf{L} \varphi \to \psi$ if and only if $\vdash_\mathbf{L} \varphi \to \Box\theta$, for some θ. That $\nvdash_{\mathbf{R}^\Box(\Box K, \Box T, \Box 4, \Box NEC)} \varphi \to \Box\theta$ (for extensional φ and any θ) is a result of Meyer's proved using matrices from [33, p. 476].[18] $\qquad\Box$

An immediate consequence of Proposition 1 is that the foregoing taxonomy is trichotomous (cf. Russell [43, pp. 26–27]);[19] fixing $\mathbf{R}^\Box(\Box K, \Box T, \Box 4, \Box NEC)$ for concreteness, p is extensional, $\Box p$ is necessary, and $q \vee \Box p$ is neither (if $q \vee \Box p$ were equivalent to some $\Box\theta$, then it would follow that $\vdash_{\mathbf{R}^\Box(\Box K, \Box T, \Box 4, \Box NEC)} q \to \Box\theta$, which is impossible). That no formula can be both extensional and necessary is also immediate from the proposition (if there were such a formula φ, it would be the case that $\nvdash_{\mathbf{R}^\Box(\Box K, \Box T, \Box 4, \Box NEC)} \varphi \to \varphi$, contradicting axiom I).

---

[15]Russell [43, pp. 4–7] (cf. [42, pp. 627–628]) helpfully lays out different approaches to setting up a taxonomy or classification of expressions (as normative vs. descriptive, etc.), including a syntax-driven approach—using 𝕿𝖍𝖊 𝕷𝖎𝖘𝖙—favored in some form or other and with suitable caveats by, for example, Prior [36, p. 89] and Jackson [24, pp. 89–90], and the model-theoretic approach she favors [38, 42, 43]. My own approach will be, in a sense, syntactic, though I hope not crudely so.

[16]Anderson and Belnap [5, p. 36] call such φ "necessitives," reserving "necessary" for something else.

[17]The syntactic characterization of necessary formulae has the untoward consequence that $\Box\varphi \wedge \Box\psi$ is not a 'must,' despite being equivalent (modulo **L**) to $\Box(\varphi \wedge \psi)$.

[18]This requires a slight qualification: Meyer [33] uses a formulation of **R** in which $\neg$ is not primitive (he instead uses a constant f). But the negation table is easily reconstructed (cf. Mares [29, p. 8]).

[19]Thanks to Ed Mares for asking about this.

Thus, the is/must barrier is satisfied in relevant modal logics up to and including a relevant analogue of **S4**.[20] It should be pointed out explicitly that classical **S4** does not enjoy this barrier (e.g., $\vdash_{\mathbf{S4}} p \to \Box(q \vee \neg q)$).

4 Hume's Law

I turn now to Hume's law, which is a barrier thesis belonging to deontic logic, the logic of what ought to be. Hume's law, in slogan form, is that you can't derive an 'ought' from an 'is.'[21] Like the modal barrier from Section 3, it has already received some attention in the relevance logic literature. Mares [29, 31], in particular, has been keen to connect it to relevance logic's rejection of the fallacies of modality and the insistence on Ackermann-like properties.

Mares's work provides a convenient jumping-off point for the project of this section. While he (correctly) connects Hume's law and the fallacies of modality, his approach in [29] is a bit idiosyncratic in that he follows the Andersonian reduction of deontic logic to alethic modal logic.[22] Whatever the philosophical merits of that approach, it does not fit neatly into the framework elaborated in Section 2. Furthermore, [31] is principally intended as a philosophical, rather than as a technical, piece. So, there is work yet to be done.

Below, in Subsections 4.1 and 4.2, I examine two ways of formulating formal Ackermann-like properties corresponding to Hume's law; the ways are, in effect, distinguished by whether the relevant deontic logic in question is monomodal or bimodal. (The rough motivation for going bimodal is to capture certain connections between alethic possibility and deontic necessity, notably, a suitable form of 'ought implies can;' cf. Schurz [45, p. 37].)

Before getting to the technical business, one last remark is in order. Russell [43, pp. 1–2] suggests Hume's law is one of the most controversial barrier theses—it's

---

[20]Meyer [33] refers to $\mathbf{R}^{\Box}(\Box\mathrm{K},\Box\mathrm{T},\Box 4,\Box\mathrm{NEC})$ as **NR**. Is **NR** *the* relevant analogue of classical **S4**? In a certain technical sense, the answer turns out to be 'no' (see, e.g., Mares and Meyer [32]), but this is not important for any of my purposes.

[21]Here is what Hume had to say on the matter. After observing that much philosophical discourse on morality proceeds from constructions concerning 'is' and 'is not' to constructions concerning 'ought' and 'ought not,' he writes, "For as this *ought*, or *ought not*, expresses some new relation or affirmation, 'tis necessary that it shou'd be observ'd and explain'd; and at the same time that a reason should be given, for what seems altogether inconceivable, how this new relation can be a deduction from others, which are entirely different from it" [22, §3.1.1, p. 469].

[22]Anderson [2] proposed to reduce deontic logic to alethic modal logic by adding a propositional constant representing a bad state of affairs, and interpreting statements to the effect that φ ought to hold as statements to the effect that φ failing to hold would entail the bad state. For more discussion, see also Prior [35, Appendix D], Anderson [3], and Kanger [25, pp. 53–54].

the sort of thing philosophers have even ended up disagreeing with themselves on (see, especially, Prior [36, p. 88]). I myself am inclined to view it favorably, in no small part because I am favorably inclined towards relevance logic, and many of the arguments against Hume's law trade on principles that are obviously relevantly fallacious (more on this in Section 6). But even if the reader is unconvinced of its desirability, its *attainability* within a suitable relevant framework should suffice to motivate what follows.

4.1 Monomodal Deontic Logic

Throughout this subsection, I restrict my attention to the monomodal language (no ■). $\Box\varphi$ is interpreted (impersonally) as 'φ ought to be so;' derivatively, $\Diamond\varphi$ is interpreted as 'φ is permissible.' Strong modal logics do not make for plausible deontic logics, and the two systems I focus on in what follows are relatively weak extensions of $\mathbf{R}^\Box$. These systems are $\mathbf{R}^\Box(\Box\mathrm{K})$ and $\mathbf{R}^\Box(\Box\mathrm{K},\Box\mathrm{D})$ (aliases **OR.1** and **DR.1**; see Goble [19]).[23]

Taking these in reverse order, I will first specify a form of Hume's law for $\mathbf{R}^\Box(\Box\mathrm{K},\Box\mathrm{D})$ and prove that it obtains. I will also show that $\mathbf{R}^\Box(\Box\mathrm{K},\Box\mathrm{D})$ is non-normal; in particular, $\Box$NEC is not admissible in it. Both of these results carry over to the weaker system $\mathbf{R}^\Box(\Box\mathrm{K})$. However, I will then proceed to show that a more stringent form of Hume's law is enjoyed by the weaker system, as well as a desirable converse of it.

On the paradigm of what was done in Section 3, I am interested in proving a result to the effect that $\nvdash_\mathbf{L} \varphi \to \psi$, whenever φ is extensional (an 'is') and ψ is obligatory (an 'ought'), where $\mathbf{L} \subseteq \mathbf{R}^\Box(\Box\mathrm{K},\Box\mathrm{D})$. The same characterization as from Section 3 may as well be adopted, and it may be stipulated that φ is *obligatory* if it is **L**-equivalent to some formula $\Box\psi$, and it is *extensional* if it is $\Box$-free.[24] Since

---

[23]$\Box$D is arguably what is most characteristic of classical deontic logic (see, e.g., von Wright [52, p. 13], Lemmon [27, §5], and Lemmon [28, p. 40]). Goble [19, §1] argues against requiring $\Box$D, at least if one is using a relevant background logic, for which reason the weakest relevant deontic systems he examines omit it. I need not take any position on the matter.

[24]I mention one prima facie peculiar consequence of these definitions: $\Box\varphi \vee \Box\psi$ is not generally obligatory. Suppose it were; then, by definition, $\Box\varphi \vee \Box\psi$ is equivalent—over, say, $\mathbf{R}^\Box(\Box\mathrm{K},\Box\mathrm{D})$—to some $\Box\theta$. Then clearly $\vdash_{\mathbf{R}^\Box(\Box\mathrm{K},\Box\mathrm{D})} \Box\varphi \to \Box\theta$, $\vdash_{\mathbf{R}^\Box(\Box\mathrm{K},\Box\mathrm{D})} \Box\psi \to \Box\theta$, and $\vdash_{\mathbf{R}^\Box(\Box\mathrm{K},\Box\mathrm{D})} \Box\theta \to \Box(\varphi\vee\psi)$. A simple syntactic argument shows $\dfrac{\Box\alpha \to \Box\beta}{\alpha \to \beta}$ to be admissible in $\mathbf{R}^\Box(\Box\mathrm{K},\Box\mathrm{D})$ (cf. Chellas [10, pp. 124–125]). Consequently, $\vdash_{\mathbf{R}^\Box(\Box\mathrm{K},\Box\mathrm{D})} \varphi \to \theta$, $\vdash_{\mathbf{R}^\Box(\Box\mathrm{K},\Box\mathrm{D})} \psi \to \theta$, and $\vdash_{\mathbf{R}^\Box(\Box\mathrm{K},\Box\mathrm{D})} \theta \to \varphi \vee \psi$, whence $\vdash_{\mathbf{R}^\Box(\Box\mathrm{K},\Box\mathrm{D})} \theta \leftrightarrow (\varphi \vee \psi)$, and so $\vdash_{\mathbf{R}^\Box(\Box\mathrm{K},\Box\mathrm{D})} \Box\theta \leftrightarrow \Box(\varphi \vee \psi)$. Thus, by transitivity, $\Box\varphi\vee\Box\psi$ is equivalent over $\mathbf{R}^\Box(\Box\mathrm{K},\Box\mathrm{D})$ to $\Box(\varphi\vee\psi)$, an equivalence which fails for many substitution instances. I believe a more nuanced taxonomy could accommodate such disjunctions (cf. Anderson and Belnap [5, §22.1.2]), but the technical details are formidable and beyond the scope of this paper.

$\mathbf{R}^\square(\square K, \square D)$ is clearly a subsystem of $\mathbf{R}^\square(\square K, \square T, \square 4, \square NEC)$, the desired result is really a corollary of Proposition 1.

Nevertheless, I will give a proof of the result using different matrices, based on **RM3**, because the same matrices can also be used to establish other results of interest (cf. Mares [29, pp. 11–13]). The matrices are as follows (2 and 1 are designated):

$\rightarrow$	2	1	0
2	2	0	0
1	2	1	0
0	2	2	2

$\wedge$	2	1	0
2	2	1	0
1	1	1	0
0	0	0	0

$\vee$	2	1	0
2	2	2	2
1	2	1	1
0	2	1	0

	$\neg$
2	0
1	1
0	2

	$\square$
2	0
1	0
0	0

Proposition 2. $\nvdash_L \varphi \rightarrow \psi$ whenever φ is extensional and ψ is obligatory, for $L \subseteq \mathbf{R}^\square(\square K, \square D)$.[25]

Proof. As in Proposition 1, the problem reduces to showing that $\nvdash_{\mathbf{L}} \varphi \rightarrow \square\theta$, whenever φ is $\square$-free. It is easy to verify that all theorems of $\mathbf{R}^\square(\square K, \square D)$ come out designated in the above matrices. Now, consider the valuation ν assigning 1 to all variables; inspection indicates $\nu(\varphi) = 1$ but, clearly, $\nu(\square\theta) = 0$, whence the conditional takes an undesignated value. $\square$

Proposition 3. *No formula of the form* $\square\varphi$ *is a theorem of* $\mathbf{R}^\square(\square K, \square D)$.

Proof. Using the same matrices and any valuation whatsoever, it is clear that $\nu(\square\varphi) = 0$. $\square$

Before the advent of Kripke semantics, deontic logicians showed a marked preference for nonnormal systems such as **D2** (see Lemmon [27, p. 185]). $\square NEC$ seems to have been generally considered to be deontically implausible (see, e.g., Prior [34, pp. 221–222] and Lemmon [27, p. 185]). Indeed, it is deontically implausible. Why should *logic* deliver that there are any obligations whatsoever? Fortunately, as Proposition 3 shows, $\mathbf{R}^\square(\square K, \square D)$ is not blemished in this way.

Thanks to Rohan French for pressing me to comment on this issue.

[25]Incidentally, Proposition 2 shows that $p, \neg p \nvdash_{\mathbf{L}} \square q$ (because $\nvdash_{\mathbf{L}} (p \wedge \neg p) \rightarrow \square q$), thereby ruling out a certain counterexample considered by Russell [42, p. 626] on grounds that actually have nothing to do with variable sharing (more on this in Section 6).

Corollary 5. $\Box NEC$ *is inadmissible in* $\mathbf{R}^{\Box}(\Box K, \Box D)$.

Proof. Immediate from Proposition 3 (cf. Mares [29, p. 13]). $\qquad\Box$

I turn now to the weaker system $\mathbf{R}^{\Box}(\Box K)$. I continue to classify formulae as extensional if they are $\Box$-free, but now I classify a formula φ as *deontic* if it's equivalent to either $\Box\psi$ or $\Diamond\psi$ (for some ψ). A natural extension of Hume's line of thought would hold that no deontic propositions (not just obligations) are entailed by extensional propositions.[26] This result demonstrably holds for $\mathbf{R}^{\Box}(\Box K)$.

Proposition 4. $\nvdash_{\mathbf{L}} \varphi \to \psi$ *whenever* φ *is extensional and* ψ *is deontic, for* $\mathbf{L} \subseteq \mathbf{R}^{\Box}(\Box K)$.

Proof. If $\vdash_{\mathbf{L}} \psi \leftrightarrow \Box\theta$ (for some θ), then $\varphi \to \Box\theta$ (and so $\varphi \to \psi$) can be dispatched using Proposition 2 and its associated matrices. On the other hand, if $\vdash_{\mathbf{L}} \psi \leftrightarrow \Diamond\theta$ (for some θ), then use the same **RM3** matrices except the following matrix for $\Box$:

	$\Box$
2	2
1	2
0	2

It can be verified that every theorem of $\mathbf{R}^{\Box}(\Box K)$ takes a designated value in these matrices. (Incidentally, $\Box D$ does *not*.) By inspection, if 1 is assigned to every variable, $\nu(\varphi) = 1$ and $\nu(\Diamond\theta) = 0$; therefore, $\nu(\varphi \to \Diamond\theta) = 0$, as desired. $\qquad\Box$

An attractive converse of Hume's law can also be proved for systems contained in $\mathbf{R}^{\Box}(\Box K)$, where extensional and deontic formulae are characterized in the same way as above.

Proposition 5. $\nvdash_{\mathbf{L}} \varphi \to \psi$ *whenever* φ *is deontic and* ψ *is extensional, for* $\mathbf{L} \subseteq \mathbf{R}^{\Box}(\Box K)$.

Proof. The argument is symmetric to that given in the proof of Proposition 4. That is, if $\vdash_{\mathbf{L}} \varphi \leftrightarrow \Box\theta$ (for some θ), the pertinent conditional ($\Box\theta \to \psi$) can be dispatched using the **RM3** matrices and the constant 2 $\Box$-matrix from Proposition 4. If, instead, $\vdash_{\mathbf{L}} \varphi \leftrightarrow \Diamond\theta$ (for some θ), the pertinent conditional ($\Diamond\theta \to \psi$) can be dispatched using the **RM3** matrices and the constant 0 $\Box$-matrix from earlier. $\qquad\Box$

---

[26] A parallel line would not be plausible in the alethic case from Section 3. In the presence of $\Box T$—which seems largely uncontroversial—one would expect validities such as $p \to \Diamond p$.

Propositions 4 and 5 together show that there is a thoroughgoing logical separation of extensional and deontic propositions in $\mathbf{R}^{\square}(\square\mathrm{K})$. $\mathbf{R}^{\square}(\square\mathrm{K})$ therefore avoids forms of the naturalistic fallacy as well as what might be called the "wishful thinking" fallacy. As to whether $\mathbf{R}^{\square}(\square\mathrm{K},\square\mathrm{D})$ enjoys these same properties, I cannot say; the proofs I have do not extend to it, but I do not have counterexamples either.

4.2 Bimodal Deontic Logic

I turn now to developing a relevant bimodal deontic logic. As in Subsection 4.1, $\square$ (and, derivatively, $\Diamond$) is used to interpret the deontic modalities. I now also include $\blacksquare$ (and, derivatively, $\blacklozenge$) in the language, and use it to interpret the alethic modalities. The key connecting thesis is 'ought implies can,' or:[27]

$$(\square\blacksquare\mathrm{D}) \quad \square\neg\varphi \to \neg\blacksquare\varphi$$

I here consider an extension of $\mathbf{R}^{\square}_{\blacksquare}$ whose (putative) deontic fragment corresponds to the stronger system examined in the previous subsection, and whose (putative) alethic fragment is **S4**-ish: $\mathbf{R}^{\square}_{\blacksquare}(\square\mathrm{K},\square\mathrm{D},\blacksquare\mathrm{K},\blacksquare\mathrm{T},\blacksquare4,\blacksquare\mathrm{NEC},\square\blacksquare\mathrm{D})$.

It is to be noted straightaway that $\square\blacksquare\mathrm{D}$ presents problems for framing a barrier based on what might be thought of as the naïve taxonomy: *Nondeontic* (in lieu of *extensional*) formulae are those which are $\square$-free, whereas a formula is *deontic* if it is (logically equivalent to something) of the form $\square\varphi$ or $\Diamond\varphi$. For it is clear that even in a system as weak as $\mathbf{R}^{\square}_{\blacksquare}(\square\blacksquare\mathrm{D})$, one can derive the deontic from the nondeontic as water from a stone (as witnessed by the theorem $\blacksquare\varphi \to \Diamond\varphi$).

So, I will restrict myself to a less ambitious taxonomy. A formula φ is *nondeontic* if it is $\square$-free, whereas it is *obligatory* if it is equivalent to a formula of the form $\square\psi$ (for some ψ).

Proposition 6. $\nvdash_L \varphi \to \psi$ *whenever φ is nondeontic and ψ is obligatory, for* $L \subseteq \mathbf{R}^{\square}_{\blacksquare}(\square K,\square D,\blacksquare K,\blacksquare T,\blacksquare 4,\blacksquare NEC,\square\blacksquare D).$[28]

Proof. As in earlier cases, the problem comes to showing that $\nvdash_{\mathbf{L}} \varphi \to \square\theta$ whenever φ is $\square$-free. Once again, I use the **RM3** matrices for the nonmodal connectives, and the following matrices for the modal connectives:

---

[27]The informal thesis is sometimes known as 'Kant's law,' although the status and interpretation of the principle in Kant is not unambiguous (for a recent discussion, consult Kohl [26]).

[28]Schurz [45], interestingly, formulates a bimodal alethic deontic predicate logic and proves a sort of version of Hume's law for it. The formulation of Hume's law he gives might be described as 'relevant-adjacent,' though it is not based on relevance logic (see, especially, [45, pp. 92–93, n. 3]).

	□			■
2	0		2	1
1	0		1	1
0	0		0	0

It can be verified that all the theorems of $\mathbf{R}_{\blacksquare}^{\square}(\square K,\square D,\blacksquare K,\blacksquare T,\blacksquare 4,\blacksquare NEC,\square\blacksquare D)$ come out designated. Now, assigning 1 to all variables, it is clear that $\nu(\varphi) = 1$ while $\nu(\square\theta) = 0$, whence the pertinent conditional fails. $\square$

As in the case of $\mathbf{R}^{\square}(\square K,\square D)$, it is clear that $\square NEC$ is not admissible in $\mathbf{R}_{\blacksquare}^{\square}(\square K,\square D,\blacksquare K,\blacksquare T,\blacksquare 4,\blacksquare NEC,\square\blacksquare D)$, and the system has no theorems of the form $\square\varphi$. (The argument simply reuses the matrices from the proof of Proposition 6.)

5   The Time Barrier

The last barrier I examine—the past/future barrier—belongs to tense logic. Russell [43, p. 85] glosses it thus:

> *no set of premises about the past entails a conclusion about the future* [...] Or, in slogan form: No *will* from a *was*.

The unidirectional gloss notwithstanding, the barrier in question seems to be intuitively bidirectional, and in what follows I formulate a system for which statements about the future and statements about the past are, to a large extent, logically separated from one another.

As Russell [43, p. 84] observes, this barrier can also be found in Hume.[29] Prior [37, p. 57] dismisses it, at least in part because, bluntly interpreted, it conflicts with certain intuitive temporal bridge principles (more on this anon). My own position is that, in a suitably restricted form, the barrier is acceptable after all.

Tense logic is inherently bimodal with one set of modalities concerning the future and another set concerning the past. Throughout this section, I use the language with both modal operators, though for ease of exposition and to bring the presentation into conformity with the more typical conventions of tense logic, I redecorate the boxes and diamonds à la Prior [35, 37]:

---

[29] Hume writes, "all inferences from experience suppose, as their foundation, that the future will resemble the past [...] If there be any suspicion that the course of nature may change, and that the past may be no rule for the future, all experience becomes useless, and can give rise to no inference or conclusion. It is impossible, therefore, that any arguments from experience can prove this resemblance of the past to the future; since all these arguments are founded on the supposition of that resemblance" [23, §4.2, pp. 37–38].

1. $\mathcal{H}\varphi$ ('φ has always been the case') replaces $\Box\varphi$;

2. $\mathcal{P}\varphi$ ('φ was at one time the case') replaces $\Diamond\varphi$;

3. $\mathcal{G}\varphi$ ('φ is always going to be the case') replaces $\blacksquare\varphi$;

4. $\mathcal{F}\varphi$ ('φ will at some time be the case') replaces $\blacklozenge\varphi$.

The foregoing notwithstanding, I (confusingly) continue to refer to axioms according to the conventions of Section 2. (In fact, it will not be very confusing, because I adopt essentially the same axioms for each modal operator.)

Since there is little previous technical work on relevant tense logic and tense logic is, in any case, not the focus of the present study, I will restrict my attention to a fairly weak system, which is sort of a relevant analogue of the basic classical tense logic $\mathbf{K}_t$.[30] This is the system $\mathbf{R}_\blacksquare^\Box(\Box\mathrm{K},\Box\mathrm{NEC},\blacksquare\mathrm{K},\blacksquare\mathrm{NEC},\blacklozenge\Box,\Diamond\blacksquare)$, that is, roughly, $\mathbf{K}$ for both $\mathcal{H}$ and $\mathcal{G}$, plus the bridge axioms:

($\blacklozenge\Box$) $\mathcal{F}\mathcal{H}\varphi \to \varphi$

($\Diamond\blacksquare$) $\mathcal{P}\mathcal{G}\varphi \to \varphi$

$\blacklozenge\Box$ expresses that if φ will at some time have always been the case, then φ is now the case; and $\Diamond\blacksquare$ expresses that if φ was at some time always going to be the case, then φ is now the case.

As $\Box\mathrm{B}$ creates obstacles for the alethic barrier considered in Section 3, so $\blacklozenge\Box$ and $\Diamond\blacksquare$ create obstacles for $\mathbf{R}_\blacksquare^\Box(\Box\mathrm{K},\Box\mathrm{NEC},\blacksquare\mathrm{K},\blacksquare\mathrm{NEC},\blacklozenge\Box,\Diamond\blacksquare)$. To be more concrete, per $\Diamond\blacksquare$, from at least some statements ostensibly about the past (e.g., $\mathcal{P}\mathcal{G}\mathcal{F}p$) some statements ostensibly about the future (e.g., $\mathcal{F}p$) *do* follow. Nevertheless, in the bimodal setting under consideration, these obstacles can be overcome if the barrier, and corresponding formula taxonomy, is implemented with sufficient subtlety.

My proposal is this. A formula φ is classified as *pure futuristic* if it is logically equivalent to $\mathcal{G}\psi$ or $\mathcal{F}\psi$ for some ψ which is $\mathcal{H}$-free (e.g., $\mathcal{F}\mathcal{G}p$); and it is classified as *simple futuristic* if it is logically equivalent to $\mathcal{G}\psi$ for some ψ which is itself $\mathcal{G}$-free (e.g., $\mathcal{G}p$, but also $\mathcal{G}\mathcal{P}p$; not $\mathcal{F}p$). Analogously, a formula φ is classified as *pure past* if it is logically equivalent to $\mathcal{H}\psi$ or $\mathcal{P}\psi$ for some ψ which is $\mathcal{G}$-free (e.g., $\mathcal{H}p \wedge \mathcal{H}q$); and it is classified as *simple past* if it is logically equivalent to $\mathcal{H}\psi$ for some ψ which is itself $\mathcal{H}$-free (e.g., $\mathcal{H}\mathcal{F}p$; not $\mathcal{P}p$).

The results to be proved make use of the following matrices found using MaGIC (the designated values are $n \geq 2$):

---

[30]For $\mathbf{K}_t$, see, for example, Prior [37, p. 176] and Blackburn et al. [8, p. 205].

$\to$	4	3	2	1	0
4	4	0	0	0	0
3	4	3	1	1	0
2	4	3	2	1	0
1	4	3	3	3	0
0	4	4	4	4	4

$\wedge$	4	3	2	1	0
4	4	3	2	1	0
3	3	3	2	1	0
2	2	2	2	1	0
1	1	1	1	1	0
0	0	0	0	0	0

$\vee$	4	3	2	1	0
4	4	4	4	4	4
3	4	3	3	3	3
2	4	3	2	2	2
1	4	3	2	1	1
0	4	3	2	1	0

	$\neg$
4	0
3	1
2	2
1	3
0	4

	$\mathcal{H}$
4	4
3	3
2	3
1	0
0	0

	$\mathcal{G}$
4	4
3	2
2	2
1	2
0	0

Proposition 7. $\nvdash_L \varphi \to \psi$ *whenever* ψ *is pure futuristic and* φ *is simple past, for* $L \subseteq \boldsymbol{R}^{\square}_{\blacksquare}(\square K, \square NEC, \blacksquare K, \blacksquare NEC, \blacklozenge\square, \lozenge\blacksquare)$.

Proof. First, observe that every theorem of $\boldsymbol{R}^{\square}_{\blacksquare}(\square K, \square NEC, \blacksquare K, \blacksquare NEC, \blacklozenge\square, \lozenge\blacksquare)$ is designated in the relevant matrices. Now, the problem falls into two halves: Showing that $\mathcal{H}\alpha \to \mathcal{G}\beta$ always fails and that $\mathcal{H}\alpha \to \mathcal{F}\beta$ always fails (where α, β are $\mathcal{H}$-free). The same valuation works for both, namely, assign 2 to every variable. It is clear that $\nu(\mathcal{H}\alpha) = 3$ but $\nu(\mathcal{G}\beta) = \nu(\mathcal{F}\beta) = 2$, whence the conditionals fail. $\square$

Proposition 8. $\nvdash_L \varphi \to \psi$ *whenever* ψ *is pure past and* φ *is simple futuristic, for* $L \subseteq \boldsymbol{R}^{\square}_{\blacksquare}(\square K, \square NEC, \blacksquare K, \blacksquare NEC, \blacklozenge\square, \lozenge\blacksquare)$.

Proof. The argument is essentially the same as that for Proposition 7, but the matrices for $\mathcal{H}$ and $\mathcal{G}$ given above are swapped, and the relevant subformulae are required to be $\mathcal{G}$-free. $\square$

Propositions 7 and 8 jointly yield a decent bidirectional barrier between certain kinds of statements about the past, on the one hand, and certain kinds of statements about the future, on the other. They preclude from theoremhood $\mathcal{H}\varphi \to \mathcal{G}\varphi$ (strong forwards induction), $\mathcal{H}\varphi \to \mathcal{F}\varphi$ (weak forwards induction), $\mathcal{G}\varphi \to \mathcal{H}\varphi$ (strong backwards induction), $\mathcal{G}\varphi \to \mathcal{P}\varphi$ (weak backwards induction), and a host of related aberrations. But they are silent about such suspicious characters as $\mathcal{P}\varphi \to \mathcal{F}\varphi$ (forwards recurrence) and $\mathcal{F}\varphi \to \mathcal{P}\varphi$ (backwards recurrence).[31]

---

[31] Actually, over $\boldsymbol{R}^{\square}_{\blacksquare}(\square K, \square NEC, \blacksquare K, \blacksquare NEC, \blacklozenge\square, \lozenge\blacksquare)$, all of forwards recurrence, strong backwards induction, backwards recurrence, and strong forwards induction are equivalent (cf. Prior [37, pp. 63–64]). While all are thereby ruled out as theorems, Propositions 7 and 8 cannot be appealed to (directly) in every case since the *taxonomic applicability conditions* are not met in every case. Thanks to Katalin Bimbó for drawing my attention to this point.

6   Concluding Polemical Remarks

In this paper, I have elaborated a relevant bimodal framework in which many particular systems can be implemented and shown to satisfy Ackermann-like properties corresponding to various barrier theses. I focused on three particular barriers, which were drawn from Russell [43]: The modal barrier (is/must), Hume's law (is/ought), and the time barrier (was/will). The framework developed above does not cover two other barriers Russell [43] is concerned with—the particular/universal barrier and the indexical barrier—but this is more indicative of a certain measure of slothfulness on my part than of any essential limitation.[32]

In fact, work has already been done on relevant quantified modal logic (not to mention *mere* relevant quantificational logic), and work has also been done on some (putative) indexicals in relevance logic.[33] I naïvely presume that the framework developed above could be extended to formulate systems—or even formulate one single relevant quantified multimodal system—implementing versions of all of the barriers Russell examines.

Therefore, it would seem (at least prima facie) that relevance logic is a natural home for examining and implementing all of the barriers Russell [43] is interested in. Alas, this is not the approach she favors, though not for lack of consideration. In fact, after surveying a range of purported counterexamples to Hume's law, Russell [43, pp. 36–41] explicitly airs the question of whether adopting a relevant (or paraconsistent) logic to formulate barriers such as Hume's law is advisable, and comes out against the idea.[34] I conclude this paper by examining and responding to Russell's comments on this matter.

---

[32]Russell [43, p. 1] glosses the particular/universal and indexical barriers thus: "no universal claims from particular ones [...] no indexical claims from claims which are not indexical."

[33]For relevant quantified modal logic, see, for example, Ferenz [15]. For some discussion of an actuality operator in relevance logic, see Standefer [50].

[34]Let me point out here that the matrices given throughout this paper, while paraconsistent, are not relevant in the sense that none of the finite many-valued logics determined by them satisfy the variable sharing property; in each, $\neg(q \to q) \to (p \to p)$, for example, comes out designated. One might therefore be tempted to conclude that this paper's real thesis is that a *paraconsistent framework* ought to be adopted for implementing the barriers. However, this would be mistaken. Paraconsistency is not doing the interesting work here; paraconsistency is not sufficient, in general, for the Ackermann property (consider once more **RM3** and the constant 2-function interpreting $\Box$; the resulting logic is paraconsistent, but every conditional of the form $\varphi \to \Box\psi$ is designated). I suppose I concede that relevance in the sense of the variable sharing property is also not doing the interesting work here, but as relevance logic traditionally drew significant motivation from the Ackermann property, I do not see that it is inappropriate to describe a framework mainly motivated by it as relevant. In any case, the logics I am really interested in are not these finite many-valued logics, which validate all sorts of problematic theses (e.g., Dugundji-like formulae), though they are, nevertheless, useful as tools. Thanks to Shay Logan for pressing this issue.

By way of situating Russell's comments, a short detour through Prior's dilemma (from [36]; cf. Russell [43, pp. 24–27]) will prove instructive. The dilemma is roughly as follows. Let p be some unambiguous descriptive claim (e.g., 'Cats are mammals') and let $\Box q$ be some unambiguous normative claim (e.g., 'Cats ought to be petted'). Is $p \vee \Box q$ normative or descriptive?[35] If it is normative, then something normative $(p \vee \Box q)$ follows from something descriptive (p) via Addition $(p \vdash p \vee \Box q)$; but if it is descriptive, then something normative $(\Box q)$ follows from some descriptive claims $(\neg p, p \vee \Box q)$ via Disjunctive Syllogism $(\neg p, p \vee \Box q \vdash \Box q)$.[36] Therefore, it would appear, Hume's law is not correct.

The friend of relevance logic will not be moved. Disjunctive Syllogism fails to be valid in all standard relevance logics, so there is, so to speak, no dilemma at all. Mares [31, p. 286] writes:

> Prior's argument too should be rejected by relevantists because it appeals to Disjunctive Syllogism for extensional disjunction (henceforth 'EDS'). The rejection of EDS has been treated by non-relevant logicians and philosophers as a serious problem for relevant logic. In fact, I think it is a virtue. The Lewis, and especially the Prior, arguments show that it is a virtue. Because it allows the derivation of irrelevances (such as *ex falso* and the Prior inference), EDS is an unsafe inference form.

As Russell [43, pp. 36–38] observes, many of the proposed counterexamples to Hume's law turn on principles (such as Disjunctive Syllogism, or EDS) which are not relevantly valid. Moreover, in a vein that is consonant with the final sentence quoted from Mares [31] above, she suggests that the relevantist might reject such counterexamples *because* they commit fallacies of relevance:

> There is a link between traditional motivations for relevant logic and a tempting idea about the reason the *is/ought* barrier exists. One motivation for relevant logic is the idea that the premises of a valid argument should be *relevant* to its conclusion, and an informal way of understanding this is as requiring that the conclusion concern the same subject matter as the premises. It is sometimes suggested, with regard to the *is/ought* barrier, that the problem with "of a sudden" drawing conclusions containing *ought*s from premises that contain only *is*es and *is not*s, is that *what ought to be* is a different subject matter from *what is*. [43, p. 38]

---

[35]One might be tempted to respond 'neither.' More on that below (cf. Russell [43, pp. 26–27]).

[36]In the general case, the argument assumes that the class of descriptive expressions is closed under negation. In the particular case under consideration, this is clearly uncontroversial.

The problem with this line of thought, as Section 3 will have made clear, is that the "traditional motivations for relevant logic" include not only the rejection of fallacies of relevance, but also the rejection of fallacies of modality. The latter are not to be explained by the former—they just are taken to be, intuitively, fallacious. While some putative counterexamples are both fallacies of relevance and fallacies of modality—for instance, $(p \wedge \neg p) \to \Box q$—it is clear that the type of fallacy operative in most of the (formal) counterexamples is in fact the modal variety.[37]

For this reason, several of Russell's subsequent remarks against the use of relevance logic to formulate barrier theses are wide of the mark. Her points in [43, p. 38] that the "line-up between the subject-matter motivation and relevant logic is not as clean as it might seem" and that there "are problems with the subject-matter explanation of Hume's Law" don't address the rejection of the fallacies of modality at all.

Russell [43, pp. 38–39] also claims, taking a page from Humberstone [20, pp. 133–135], that relevance logic will not allow for all of the counterexamples to be avoided. The envisioned counterexample, using the same p and $\Box q$ from above, is something like $\neg p \to \neg(p \wedge \Box q)$. Certainly, this is a theorem of $\mathbf{R}^{\Box}$, $\mathbf{R}^{\Box}_{\blacksquare}$, and all of their extensions, but is it a genuine counterexample?

The line of thought seems to be that since $\Box q$ is entailed by $p \wedge \Box q$, by Hume's law, the latter must itself be normative, and therefore its negation must be as well. This argument is really too quick, though. It's not clear that $p \wedge \Box q$ is normative (it plainly fails to be deontic relative to $\mathbf{R}^{\Box}(\Box \mathrm{K})$ by Proposition 5), nor is it clear that Hume's law, suitably understood, should require that it be so. Why isn't it compatible with Hume's law that normative conclusions can be drawn from premises which are neither normative nor descriptive, as one might be tempted to classify $p \wedge \Box q$?[38]

Russell [43, p. 38] further asserts that relevant consequence is too complicated to capture the intuitive content of Hume's law and of the other barriers. I certainly

---

[37]In particular, even in Prior's dilemma the operative fallacy can be taken to be one of modality as much as one of relevance. Fix some $\mathbf{R}^{\Box} \subseteq \mathbf{L} \subseteq \mathbf{R}^{\Box}(\Box \mathrm{K}, \Box \mathrm{D})$. Then, $\neg p, p \vee \Box q \vdash_{\mathbf{L}} \Box q$ if and only if $\vdash_{\mathbf{L}} (\neg p \wedge (p \vee \Box q)) \to \Box q$ if and only if $\vdash_{\mathbf{L}} ((\neg p \wedge p) \vee (\neg p \wedge \Box q)) \to \Box q$ if and only if $\vdash_{\mathbf{L}} (\neg p \wedge p) \to \Box q$ and $\vdash_{\mathbf{L}} (\neg p \wedge \Box q) \to \Box q$; but $\nvdash_{\mathbf{L}} (\neg p \wedge p) \to \Box q$ by Proposition 2.

[38]Russell [43, pp. 27–28] considers a problem, attributed to Vranas [53], to the effect that classifying expressions like $p \wedge \Box q$ as 'neither' leads to formulations of Hume's law that are too weak. It is beyond the scope of the present work to engage with these issues in any detail, but it seems to me that the problems raised depend on either taking the conditional as material implication (as no relevantist would) or interpreting statements that ought to be formalized using a primitive conditional obligation connective as composites of other connectives (on why one should not do this, see, e.g., Chisholm [14] and Chellas [10, p. 201]). On these matters, see also Humberstone [21, pp. 1379ff.].

would not wish to suggest that ordinary reasoners and speakers have the model theory of Section 2 in mind, or that they themselves could formulate technical results such as those given in Propositions 2 and 6—nor could they, of course, formulate Russell's own results. I think, nevertheless, that there is a clear intuitive connection between what philosophers like Hume have said, and the rigorous formulations of proscriptions on modal fallacies given above. So I do take myself to have met the desideratum on intuitiveness given by Russell [43, p. 41].

The only objection of Russell's that remains is that there is no need to go relevant, since "we can prove each of the barriers in its stronger, classical form" [43, p. 38]. To properly evaluate this claim, I would need to engage at much greater length with Russell's positive proposal than can be done here. But, for the sake of saying something, let me just remark that it is not really clear that strength is a virtue, and even if one could get by with classical logic for the purposes of formulating barrier theses, another argument would be required to show that it is optimal among the possible solution frameworks.

Let me close by raising an objection of my own. In the current proposal, there is not really any substantial explanation of why the barriers hold. To say, for example, that you can't get a necessity from a contingency because that's a fallacy of modality and relevance logic rejects the fallacies of modality just does not seem to clarify very much.

What one would like is for the failure of the fallacies of modality to fall out of some deeper logical fact—ideally, in my view, one concerning the interpretation of the different pieces of the semantic apparatus presented in Section 2.[39] The Routley–Meyer semantics and its variations leave much to be desired in terms of naturalness,[40] and I am doubtful that any particularly illuminating interpretation of the machinery developed above will appear quickly. But the situation is not, I now think,[41] entirely hopeless, and I will content myself to leave the matter there.

---

[39] One would similarly like for the failure of the fallacies of relevance—captured formally by the variable sharing property (see, e.g., Belnap [7] and Anderson and Belnap [5, §§5.1, 22.1.3])—to fall out of some deeper semantic fact. Elsewhere, I have shown that there is indeed a pleasing correspondence between the formal relevance property and the natural semantic notion of exact verification (cf. Fine [16, p. 558]) within the (positive) semilattice semantics [55, 56].

[40] For some valiant attempts to elucidate aspects of the Routley–Meyer semantics, see, for example, Mares [30] and Beall et al. [6].

[41] In the past, I have been considerably more hostile towards the Routley–Meyer semantics than I am now. Nevertheless, I continue to strongly favor the operational semantics of Urquhart [51] and variations thereof.

References

[1] Wilhelm Ackermann. Begründung einer strengen Implikation. *Journal of Symbolic Logic*, 21(2):113–128, 1956.

[2] Alan Ross Anderson. A reduction of deontic logic to alethic modal logic. *Mind*, 67(265):100–103, 1958.

[3] Alan Ross Anderson. Some nasty problems in the formal logic of ethics. *Noûs*, 1(4):345–360, 1967.

[4] Alan Ross Anderson and Nuel D. Belnap. The pure calculus of entailment. *Journal of Symbolic Logic*, 27(1):19–52, 1962.

[5] Alan Ross Anderson and Nuel D. Belnap. *Entailment: The Logic of Relevance and Necessity*, volume I. Princeton University Press, Princeton, 1975.

[6] Jc Beall, Ross Brady, J. Michael Dunn, A. P. Hazen, Edwin Mares, Robert K. Meyer, Graham Priest, Greg Restall, David Ripley, John Slaney, and Richard Sylvan. On the ternary relation and conditionality. *Journal of Philosophical Logic*, 41(3):595–612, 2012.

[7] Nuel D. Belnap. Entailment and relevance. *Journal of Symbolic Logic*, 25(2): 144–146, 1960.

[8] Patrick Blackburn, Maarten de Rijke, and Yde Venema. *Modal Logic*. Cambridge University Press, Cambridge, 2001.

[9] Milan Božić and Kosta Došen. Models for normal intuitionistic modal logics. *Studia Logica*, 43(3):217–245, 1984.

[10] Brian F. Chellas. *Modal Logic: An Introduction*. Cambridge University Press, Cambridge, 1980.

[11] Jingde Cheng. Temporal relevant logic as the logic basis for reasoning about dynamics of concurrent systems. In *IEEE International Conference on Systems, Man, and Cybernetics (SMC '98)*, pages 794–799. IEEE, 1998.

[12] Jingde Cheng. Temporal deontic relevant logic as the logical basis for decision making based on anticipatory reasoning. In *2006 IEEE International Conference on Systems, Man and Cybernetics*, pages 1036–1041. IEEE, 2006.

[13] Jingde Cheng. A temporal relevant logic approach to modeling and reasoning about epistemic processes. In *2009 Fifth International Conference on Semantics, Knowledge and Grid*, pages 18–25. IEEE, 2009.

[14] Roderick Chisholm. Contrary-to-duty imperatives and deontic logic. *Analysis*, 24(2):33–36, 1963.

[15] Nicholas Ferenz. Quantified modal relevant logics. *Review of Symbolic Logic*, 16(1):210–240, 2023.

[16] Kit Fine. Truthmaker semantics. In B. Hale, C. Wright, and A. Miller, editors, *A Companion to the Philosophy of Language*, pages 556–577. John Wiley & Sons, West Sussex, 2nd edition, 2017.

[17] André Fuhrmann. *Relevant Logics, Modal Logics and Theory Change*. PhD thesis, Australian National University, Canberra, 1988.

[18] André Fuhrmann. Models for relevant modal logics. *Studia Logica*, 49(4):501–514, 1990.

[19] Lou Goble. Deontic logic with relevance. In P. McNamara and H. Prakken, editors, *Norms, Logics and Information Systems: New Studies in Deontic Logic and Computer Science*, pages 331–345. IOS Press, Amsterdam, 1999.

[20] Lloyd Humberstone. A study in philosophical taxonomy. *Philosophical Studies*, 83(2):121–169, 1996.

[21] Lloyd Humberstone. Recent thought on *is* and *ought*: Connections, confluences and rediscoveries. *IFCoLog Journal of Logics and their Applications*, 6(7):1373–1446, 2019.

[22] David Hume. *A Treatise of Human Nature*. Clarendon Press, Oxford, 1739/1888.

[23] David Hume. *Enquiries Concerning the Human Understanding and Concerning the Principles of Morals*. Clarendon Press, Oxford, 2nd edition, 1777/1902.

[24] Frank Jackson. Defining the autonomy of ethics. *Philosophical Review*, 83(1):88–96, 1974.

[25] Stig Kanger. New foundations for ethical theory. In R. Hilpinen, editor, *Deontic Logic: Introductory and Systematic Readings*, pages 36–58. D. Reidel, Dordrecht, 1971.

[26] Markus Kohl. Kant and 'ought implies can'. *Philosophical Quarterly*, 65(261): 690–710, 2015.

[27] Edward J. Lemmon. New foundations for Lewis modal systems. *Journal of Symbolic Logic*, 22(2):176–186, 1957.

[28] Edward J. Lemmon. Deontic logic and the logic of imperatives. *Logique et Analyse*, 8(29):39–71, 1965.

[29] Edwin D. Mares. Andersonian deontic logic. *Theoria*, 58(1):3–20, 1992.

[30] Edwin D. Mares. *Relevant Logic: A Philosophical Interpretation*. Cambridge University Press, Cambridge, 2004.

[31] Edwin D. Mares. Supervenience and the autonomy of ethics: Yet another way in which relevant logic is superior to classical logic. In C. R. Pigden, editor, *Hume on Is and Ought*, pages 272–289. Palgrave Macmillan, Basingstoke, 2010.

[32] Edwin D. Mares and Robert K. Meyer. The semantics of **R4**. *Journal of Philosophical Logic*, 22(1):95–110, 1993.

[33] Robert K. Meyer. Entailment and relevant implication. *Logique et Analyse*, 11 (44):472–479, 1968.

[34] Arthur N. Prior. *Formal Logic*. Oxford University Press, Oxford, 2nd edition, 1955/1962.

[35] Arthur N. Prior. *Time and Modality*. Clarendon Press, Oxford, 1957.

[36] Arthur N. Prior. The autonomy of ethics. In P. T. Geach and A. J. P. Kenny, editors, *A. N. Prior: Papers in Logic and Ethics*, pages 88–96. University of Massachusetts Press, Amherst, 1960/1976.

[37] Arthur N. Prior. *Past, Present and Future*. Clarendon Press, Oxford, 1967.

[38] Greg Restall and Gillian K. Russell. Barriers to implication. In C. R. Pigden, editor, *Hume on Is and Ought*, pages 243–259. Palgrave Macmillan, Basingstoke, 2010.

[39] Richard Routley. Philosophical and linguistic inroads: Multiply intensional relevant logics. In R. Routley and J. Norman, editors, *Directions in Relevant Logic*, pages 269–304. Kluwer Academic Publishers, Dordrecht, 1989.

[40] Richard Routley and Val Routley. A fallacy of modality. *Noûs*, 3(2):129–153, 1969.

[41] Richard Routley, Robert K. Meyer, Val Plumwood, and Ross T. Brady. *Relevant Logics and their Rivals: The Basic Philosophical and Semantical Theory*, volume 1. Ridgeview Publishing Company, Atascadero, California, 1982.

[42] Gillian K. Russell. How to prove Hume's law. *Journal of Philosophical Logic*, 51(3):603–632, 2022.

[43] Gillian K. Russell. *Barriers to Entailment: Hume's Law & other limits on logical consequence*. Oxford University Press, Oxford, 2023.

[44] Nathan Salmon. The logic of what might have been. *Philosophical Review*, 98 (1):3–34, 1989.

[45] Gerhard Schurz. How far can Hume's is-ought thesis be generalized? *Journal of Philosophical Logic*, 20(1):37–95, 1991.

[46] Takahiro Seki. A Sahlqvist theorem for relevant modal logics. *Studia Logica*, 73(3):383–411, 2003.

[47] Takahiro Seki. The γ-admissibility of relevant modal logics I—the method of normal models. *Studia Logica*, 97(2):199–231, 2011.

[48] Takahiro Seki. Some metacomplete relevant modal logics. *Studia Logica*, 101 (5):1115–1141, 2013.

[49] John Slaney. *MaGIC, Matrix Generator for Implication Connectives: Release 2.1 Notes and Guide*. Automated Reasoning Project, ANU, 1995.

[50] Shawn Standefer. Actual issues for relevant logics. *Ergo*, 7(8):241–276, 2020.

[51] Alasdair Urquhart. Semantics for relevant logics. *Journal of Symbolic Logic*, 37 (1):159–169, 1972.

[52] Georg H. von Wright. Deontic logic. *Mind*, 60(237):1–15, 1951.

[53] Peter Vranas. Comments on 'Barriers to Implication'. In C. R. Pigden, editor, *Hume on Is and Ought*, pages 260–267. Palgrave Macmillan, Basingstoke, 2010.

[54] Heinrich Wansing. Diamonds are a philosopher's best friends: The knowability paradox and modal epistemic relevance logic. *Journal of Philosophical Logic*, 31(6):591–612, 2002.

[55] Yale Weiss. A note on the relevance of semilattice relevance logic. *Australasian Journal of Logic*, 16(6):177–185, 2019.

[56] Yale Weiss. A reinterpretation of the semilattice semantics with applications. *Logica Universalis*, 15(2):171–191, 2021.

[57] Yale Weiss. New(ish) foundations for theories of entailment. In Y. Weiss and R. Birman, editors, *Saul Kripke on Modal Logic*, volume 30 of *Outstanding Contributions to Logic*, pages 389–408. Springer, Cham, 2024.

Received 10 October 2024

On Strong and Weak Logics
for Paraconsistent Computability

Zach Weber

Philosophy Programme, University of Otago, Dunedin, New Zealand
<zach.weber@otago.ac.nz>

Fernando Cano-Jorge[*]

Philosophy Programme, University of Otago, Dunedin, New Zealand
Department of Philosophy, University of Canterbury, Christchurch, New Zealand
Facultad de Filosofía, Universidad Panamericana, Ciudad de México, México
<fernando.cano-jorge@otago.ac.nz>

Abstract

One tradition in relevant and paraconsistent logics has been to develop systems intended for applications to arithmetic and computability theory. The aspiration, as in Meyer [38] and others, is to recover enough working mathematics for real computation, but without the limitative results of Turing, Gödel, etc.; or more cautiously, as in Dunn [22], to respect *relevance* and with that be insulated against the possibility of a genuine inconsistency. We distill these goals into GUIDING QUESTIONS, and study the options for logics within a range of relevant systems. We focus on strong truth functional logics **RM3** and **PAC** [6] and their expansions, with application to inconsistent arithmetics [61, 62]. We argue that this approach, while having many virtues, does not fully answer our guiding questions. This points to weak relevant logics like Routley/Sylvan's **DKQ** [54], Brady's **MCQ** [14], and Logan and Boccuni's **DL2Q**^{t,fc} [31]. The recurring theme is that paraconsistent computability struggles with *functionality* [17, 41, 43]. A method for advancing on the 'function problem' is sketched with Kleene's theorem as a worked example.

2020 *Mathematics Subject Classification.* Primary: 03B47, Secondary: 03B53, 03B80.

Thanks to audiences at Logic Day in Dunedin 2023 and the 2nd Third Workshop at the University of Alberta, 2024, and to Jack Copeland. Thanks to the referees for helpful comments that improved the paper, and to the editor of this special issue.

[*]Research supported by the Marsden Fund, Royal Society of New Zealand; the PAPIIT project IN406225; and the CONAHCYT project CBF2023-2024-55.

1 Introduction: Logics Strong and Weak

For a long time, research in paraconsistency has sought a logic (or logics) for *computability*. Motivations range from wanting to compute better with inconsistent information, as in Belnap [7], or "wagering" carefully about the possibility of inconsistency as in Dunn [22], to the more speculative hopes of surpassing the limitative results of Turing, Gödel, etc., as in Meyer [38], Brady [11], and others.[1] This would be to recover enough working mathematics for 'real' computation, and to go beyond the classical Church–Turing barrier—and maybe, as in Routley/Sylvan and Priest [50], compute better with inconsistent truths.

Whatever the motivations, insofar as computation is about operations from numbers to numbers, in practice this project means looking for paraconsistent *arithmetic* (or arithmetics) that can support calculating but evade classical collapse. And this in turn means considering what happens with arithmetic under different paraconsistent logics.[2] Here, we consider a spectrum of suitable relevant and closely related logics, ranging from logics so strong they aren't relevant at all, like **RM3** and **PAC**, to logics so relevant that they are extremely weak, like **DK** and **DL**, and corresponding arithmetics. In the background, we hear Hilbert's quixotic question [28]: Is *ignorabimus* inevitable? Or, per impossible, could everything be *decidable*? Of course, this has been definitively answered classically in the negative, but the non-classicist may still wonder.[3]

Less dreamily, more operationally, around every corner lurks the *function problem for (inconsistent) paraconsistency*, most recently raised in [17], which is as follows. Computation is often glossed in terms of recursive functions, which do not sit well with inconsistency. For a recursive relation A one usually expects a computable function f such that

$$f_A(n) = \begin{cases} 1 & \text{if } A(n) \\ 0 & \text{if } \neg A(n). \end{cases}$$

---

[1] Asenjo explicitly connects his pioneering work to avoiding Gödel [4].

[2] One may wonder why we take this low-level approach to computability, i.e., looking at computability via arithmetic, instead of a high-level approach using some modal logic for computation, like dynamic logics. The latter assumes some form of imperative programming language in the background and some properties of algorithms which we do not take for granted in a paraconsistent setting. Since the different paraconsistent arithmetics we examine here are not as well known as classical or intuitionistic arithmetics, and since all vary in their properties, we think it is best at this stage of research to look at the most fundamental theory underlying computation, i.e., arithmetic, in order to examine the prospects of paraconsistent computability. For some other approaches in related areas, see [27, 18, 29].

[3] For discussion of paraconsistent computability, see also [32, 60].

But if some A is inconsistent,

$$A(n) \wedge \neg A(n)$$

then this leads by transitivity to the absurdity $0 = f_A(n) = 1$, which is bad—cf. [41, 43].[4]

Since inconsistency does not agree well with functionality, it is tempting to consider doing without functions, if it were somehow possible (and see §5.4). But one (presumably) needs some working notion of function in order to calculate—that is, to compute. As Priest outlines the situation,

> First-order arithmetic could be formulated just as well with no function symbols, but ... the inconsistent models would have no interesting structure, as far as I can see. ... At the other extreme, we could formulate arithmetic with many more function symbols. This would make [finding any inconsistent models] much more difficult [48, p. 1528].

We look for ways this problem might be addressed, measured against standard results like the Rabin–Scott theorem about non-deterministic finite automata [52], and a related limiting result by Agudelo and Carnielli [1] for strong paraconsistent logics. At the end we will suggest a way that the apparent impasse can be turned to an advantage, namely, the problem can be used to differentiate computable *functions* from properly paraconsistent *relations*.

It turns out that the choice of logic, *per se*, in most cases makes little difference to the function problem. Rather it is the difference between approaches—cautiously *paraconsistent*, and ultimately conservative, or *inconsistent* and so ultimately revisionary—that marks the fork in the path.

2 Paraconsistent Arithmetic: Some Guiding Questions

The quest for a paraconsistent computability theory, specially one which is immune to some classical limitative results, goes hand in hand with the quest for an appropriate theory of arithmetic. For if arithmetic itself is seriously limited, it is only natural that this in turn restricts the class of computable relations. We should then elaborate what an appropriate theory of arithmetic means in this context.

Gödel's incompleteness theorems stand as the most important limitative results about arithmetical theories. But these come with various assumptions, most notably the negation consistency of the theory. It is then natural to ask, as e.g., Meyer [35]

---

[4]A referee suggests defining $f_A(n) = 0$ when $A(n)$ is untrue, rather than false (thus distinguishing a true negation (falsity) from not being true). This points to larger questions, about the logic of the metatheory and relatedly the *just true* problem for paraconsistent logics—see [44], and further discussion in §4.4 below.

and Priest [47] do, whether a paraconsistent logic, in being able to accommodate possible inconsistencies, would sidestep Gödel's results. For, indeed, we have that:

G1. If a theory of arithmetic is negation consistent and can represent all computable relations, then that theory is incomplete.

G2. Moreover, such a theory cannot prove its own (absolute) consistency.

While a paraconsistent theory of computability will certainly demand a theory of arithmetic that can represent all computable relations, it will not in general assume that arithmetic is negation consistent.[5] The whole idea behind going paraconsistent in the underlying logic of our theory is to be prepared to handle inconsistencies without triviality. So natural GUIDING QUESTIONS to ask, reasoning in the other direction, are:

Q1. Is there a theory of arithmetic T that is negation inconsistent, non-trivial, and capable of representing all computable relations, such that T is complete?

Q2. Can such T prove its own absolute consistency, i.e., its non-triviality?

While Gödel's theorems decisively show that Hilbert's dream of a consistent and complete system of mathematics, expressively rich enough to show its own consistency through finitary means, has no hope whatsoever, one may still ask whether features like completeness and non-triviality can be had by going inconsistent. Notably, **G1** says that if a theory of arithmetic is complete and can represent all computable relations, then the theory is inconsistent.

And a similar line of thought can be followed regarding decidability—another salient goal of Hilbert's programme. Leaning on Turing's work, we have:

T1. If a theory of arithmetic is negation consistent, and can represent all computable relations, and is decidable, then the halting problem is effectively solvable.

T2. The halting problem is effectively unsolvable.

But the argument for **T2** famously rests on the inadmissibility of contradictions in the underlying logic of our theory. So again, reasoning dually, we ask:

---

[5] According to tradition, if T is a consistent theory in the language of arithmetic, we say that a k-place function f is *representable* in T if there is a formula $A(x_1, \ldots, x_k, y)$ such that, whenever $f(a_1, \ldots, a_k) = b$, then $\vdash_T \forall y(A(\overline{a_1}, \ldots, \overline{a_k}, y) \leftrightarrow y = \overline{b})$—where the $\overline{a_i}$ and $\overline{b}$ are numerals standing for the a_i and b, and $\leftrightarrow$ is a biconditional to be specified by choice of logic. Similarly, a k-place relation R is representable in T if there is a formula $A(x_1, \ldots, x_k)$ such that, whenever $R(a_1, \ldots, a_k)$ holds, $\vdash_T A(\overline{a_1}, \ldots, \overline{a_k})$. See [58, Ch. 12].

Q3. Is there a theory T as in Q1–Q2 where it can be shown that the halting function is computable? And if so, is T decidable then?

Certainly, these considerations would require that the provability relation of arithmetic is itself a representable, computable relation in the theory. We know that self-reference by Gödelian constructions gives arithmetic the possibility of talking about itself. But this is related to another important limitative result, due to Löb (see §5.1 below):

L1. If a theory of arithmetic can prove its own soundness, then it is trivial.

A theory T answering to the GUIDING QUESTIONS we have been considering so far should indeed be a sound theory and, moreover, be capable of expressing such a fact within itself through finitary means [49, Ch. 3, 17]. While the limitative results of Gödel and Turing directly depend on the negation consistency of a theory of arithmetic, Löb's theorem, like Curry's paradox, crucially depends (given other assumptions) on a different feature of the logic underlying our theory of arithmetic, namely, that Contraction,

$$\vdash A \to (A \to B) \;\therefore\; \vdash A \to B$$

is logically valid. So we ask:

Q4. Is there a theory T as in Q1–Q3 whose underlying logic is not only paraconsistent but also Contraction-free? If so, can T prove its own soundness without this leading to triviality?

These GUIDING QUESTIONS make it look like we are after a non-trivial theory of arithmetic that represents all computable relations and which is negation inconsistent, complete, decidable and capable of proving its own soundness. But such a theory might only be a mythological creature, and indeed the arithmetics surveyed in this paper show that no investigation has yet found it. The lesson of limitative results is that *something* has to give; we will see that there are pros and cons for different candidate paraconsistent arithmetics.

These pros and cons hinge largely on the logic that underlies a given arithmetic. Indeed, this logic must be paraconsistent and perhaps even Contraction-free, but at the same time it must also be strong enough to carry out valid reasoning in arithmetical proofs and describe the structure of the number line; the arithmetic must be, in some sense, recognizable as *arithmetic*.[6] This might make it impossible to have all these features in a single system. This is why we will analyze systems of logic which we deem strong enough to fulfill some or all of the properties involved in Q1–Q4. This is to ask a final, more practical GUIDING QUESTION:

Q5. Is there a theory T as in Q1–Q4 that is 'sufficiently rich' to carry out calculation and mathematical reasoning?

Much as constructive mathematics differs from classical in the allowable methods, and leads to differences in which proofs (and so theorems) are valid, so here we note where e.g., reductio is available or not, and the effects this may have on what is deemed uncomputable.

In looking at applications to arithmetic and computability, it would be too quick to think that these goals are achievable as a result of the choice of logic alone. More is needed. When we consider the requisite that all computable relations are representable in the theory, we must have some answer to the question 'What is a computable relation in a paraconsistent setting?', and the answer to this question will necessarily depart from the standard, classical response insofar as contradictions must be accommodated in the theory—as shown by the function problem. To illustrate this point, by the end of the paper we will sketch a start at redesigning concepts and methods from computability theory so as to escape the function problem and glimpse how a paraconsistent computability theory might take shape.

3 Relevant Arithmetics

3.1 $\mathbf{R}^\sharp$ and Some of its Extensions

A nice place to begin our expedition is first-order relevant logic **RQ**. Not only is **RQ** one of the strong systems in the spectrum of relevant logics—which are paraconsistent—but it is also considered to be THE logic of relevance by Anderson and Belnap. Moreover, Meyer developed relevant arithmetic $\mathbf{R}^\sharp$ [37] using this system and made enough progress in his investigations to give some answers to our QUESTIONS Q1 and Q2. The axiomatization of **RQ** and a number of relevant logics that will be used throughout this paper (like **RM** and **RM3**) can be found in Appendix 1; cf. §3.3, 4.2, Appendix 2.

---

[6]In the context of this paper, this somewhat informal and imprecise condition needs to be understood without a classical bias. For example, if one asserts that anything worth calling arithmetic must be undecidable, this would clearly be begging the question. The constraint is rather that if *all* we wanted was decidability, a one-element trivial structure would suffice. What more *is* needed, is an open philosophical question; see [33].

The relevant arithmetics $\mathbf{R}^\sharp$ and $\mathbf{RM}^\sharp$, formulated in the standard arithmetical language $\mathscr{L}$, are obtained by adding the following arithmetical axioms and rules to $\mathbf{RQ}$ and $\mathbf{RMQ}$, respectively:

(A1) $\forall x \, \forall y \, (x = y \leftrightarrow x' = y')$

(A2) $\forall x \, \forall y \, \forall z \, (x = y \rightarrow (x = z \rightarrow y = z))$

(A3) $\forall x \, (x' \neq 0)$

(A4) $\forall x \, (x + 0 = x)$

(A5) $\forall x \, \forall y \, (x + y' = (x + y)')$

(A6) $\forall x \, (x \times 0 = 0)$

(A7) $\forall x \, \forall y \, (x \times y' = (x \times y) + x)$

(AR) $\vdash P0, \vdash \forall x \, (Px \rightarrow Px') \, \therefore \, \vdash \forall x \, (Px)$

$\mathbf{R}^{\sharp\sharp}$ and $\mathbf{RM}^{\sharp\sharp}$ are obtained by respectively adding to $\mathbf{R}^\sharp$ and $\mathbf{RM}^\sharp$ the following infinitary rule:

(Ω) $\vdash P0, \vdash P0', \ldots, \vdash P0^{(n)} \, \therefore \, \vdash \forall x \, (Px)$

Meyer was able to show that, though $\mathbf{R}^\sharp$ is not negation complete, since e.g., neither $0 = 2 \rightarrow 0 = 1$ nor $\neg(0 = 2 \rightarrow 0 = 1)$ are provable in $\mathbf{R}^\sharp$, it is however absolutely consistent, i.e., non-trivial, since $0 = 1$ is not provable in $\mathbf{R}^\sharp$ and this can be shown inside $\mathbf{R}^\sharp$ itself through finitistic means using $\mathbf{RM3}$ matrices [38]. Here we have a paraconsistent arithmetic that escapes one of Gödel's results and which resembles classical Peano arithmetic quite closely insofar as $\mathbf{R}^\sharp$ can represent all computable relations and is also undecidable.

Regarding negation consistency, Meyer tried to show that rule γ, i.e., $\vdash A, \vdash \neg A \vee B \, \therefore \, \vdash B$, was admissible in $\mathbf{R}^\sharp$ —in analogy with proofs of the admissibility of Cut in Gentzen systems. He was able to show that there is an exact translation of the theorems of $\mathbf{P}^\sharp$ (classical Peano arithmetic) into $\mathbf{R}^\sharp$, assuming that γ is admissible in $\mathbf{R}^\sharp$. A finitary proof of the admissibility of γ not only would mean that $\mathbf{R}^\sharp$ was negation consistent, but also that so was $\mathbf{P}^\sharp$—but by Gödel's second incompleteness theorem, no such proof was to be found.

In fact, Meyer thought that $\mathbf{R}^\sharp$ was Peano complete, i.e., that $\mathbf{P}^\sharp \subseteq \mathbf{R}^\sharp$—see remarks in [40, p. 917], for instance. However, this is not the case. Friedman and Meyer [26] showed that there is a strictly positive (negation-free) theorem, the quadratic residue formula, which is not provable in $\mathbf{R}^\sharp$ though it is, of course, provable in $\mathbf{P}^\sharp$. This also means that γ is not admissible in $\mathbf{R}^\sharp$. Perhaps this is quite surprising, given the fact that both the logic $\mathbf{RQ}$ and the arithmetic $\mathbf{R}^{\sharp\sharp}$ admit γ [39], [36]. Hence, Friedman and Meyer invite us to find out whether there is such a thing as $\mathbf{R}^{\sharp 1/2}$, a system between $\mathbf{R}^\sharp$ and $\mathbf{R}^{\sharp\sharp}$—which is as of yet unknown; see [30] for a recent approach.

3.2 Relevant Robinson's Arithmetic

One of the difficulties in proving that γ is admissible in $\mathbf{R}^\sharp$ was the induction schema. This lead Dunn to explore Robinson arithmetic $\mathbf{Q_R}$ based on $\mathbf{RQ}$ in [22]. Recall that Robinson arithmetic is given by the Peano axioms without the induction schema (AR), plus the axiom

$$\forall x \, (x \neq 0 \to \exists y \, (x = y'))$$

(phrased this way rather than the more usual $\forall x \, (x = 0 \vee \exists y \, (x = y'))$ so as to avoid need of disjunctive syllogism).

Momentarily happy with his finding a proof of the admissibility of γ for $\mathbf{Q_R}$, Dunn soon realized that, for any formula A, $\vdash_{\mathbf{Q_R}} A$ iff $\vdash_{\mathbf{Q}} A$, i.e., that relevant Robinson arithmetic collapses with classical Robinson arithmetic —which makes the admissibility of γ a trivial fact, for γ, or material detachment, is surely admissible in $\mathbf{Q}$. Hence, $\mathbf{Q_R}$ is a sound theory that can represent all recursive relations, whence it is incomplete, undecidable and cannot prove its own consistency.

Oddly enough, this collapse only happens when arithmetic includes zero. Relevant Robinson arithmetic on the *positive* numbers, denoted $\mathbf{Q_R}(1)$, is different from its classical counterpart $\mathbf{Q}(1)$ since $\nvdash_{\mathbf{Q_R}(1)} \forall x \, \forall y \, \forall z \, (x = y \to z = z)$ [22, §3]. Moreover, neither $\mathbf{Q_R}(1) \subseteq \mathbf{Q_R}$ nor $\mathbf{Q_R} \subseteq \mathbf{R}^\sharp$.

3.3 An Interlude on RMQ

The logic $\mathbf{RM}$ is $\mathbf{R}$ plus the Mingle axiom, $A \to (A \to A)$. This axiom was introduced by McCall in the form $(A \to B) \to ((A \to B) \to (A \to B))$; and it is equivalent to the converse of Contraction, $(A \to B) \to (A \to (A \to B))$.[7] $\mathbf{RM}$ is quasi-relevant: if $\vdash_{\mathbf{RM}} A \to B$ then either A and B share a variable, or else $\vdash_{\mathbf{RM}} \neg A$ and $\vdash_{\mathbf{RM}} B$; see [53].

But as Anderson and Belnap notice, the relevance of $\mathbf{RM}$ is highly dubious, since Linearity (a.k.a. the Chain Property), $(A \to B) \vee (B \to A)$, is provable in

---

[7]According to Dunn, McCall suggested the mingle axiom, $A \to (A \to A)$, to be added to $\mathbf{E}$, and Anderson and Belnap misattribute their own conjecture about restricted mingle $(A \to B) \to ((A \to B) \to (A \to B))$ (giving it to McCall) in [2]. See [10] for a detailed discussion.

this logic (see [2, p. 429] and Appendix 2). Standefer [59] argues that, to the letter, this does not violate variable sharing and so is satisfactory from a purely relevantist standpoint. Perhaps so, but it remains highly counterintuitive and certainly not in the spirit of relevant implication.

Ideological objections aside, **RM** is paraconsistent and is not without its charms. In a 'consumer's report' style note, Dunn lists nice features like decidability [23].[8] The present note can be read as another report on related logics, specifically trialed on paraconsistent arithmetic and computability theory. The verdict is mixed.

One possible risk of using **RM** is Safety, $(A \wedge \neg A) \to (B \vee \neg B)$, which is derivable from the **RM**-theorem $\neg(A \to A) \to (B \to B)$. Tedder [63] warns against Safety, in the following terms. A *regular* theory has every theorem of **RM** as a member; a *funky* theory has the negation of a theorem of **RM** as a member. Because of Safety, all funky theories are regular, and all non-regular theories are consistent. Thus,

> **RM** dictates that a theory may be about something other than logic, or it may be inconsistent, but it may not be both. But why think that the theory of, say, an inconsistent model of arithmetic must also contain the complete theory of logic?

This is a philosophical objection to Safety, which perhaps could be brushed aside—what harm would there be in an inconsistent model also containing the complete theory of logic?—but does suggest that, at the same time as proving too much, these strong logics also are too constrained.

It is to the pros and cons of such logics for arithmetic that we now turn.

4 Arithmetic in Maximally Paraconsistent Logics

When it comes to strong logics for paraconsistent computability theory, there is nothing stronger than *maximally* paraconsistent logics. Avron and collaborators have advanced and refined this concept throughout several papers, e.g., [6, 3]. Most recently, maximal paraconsistency is taken to be the requirement from a paraconsistent logic **L** to retain as much of classical logic as possible, while still allowing non-trivial inconsistent theories. This in turn can be interpreted in two ways: (1) as *absolute* maximal paraconsistency, which means that any extension of **L** (without changing the language) is not paraconsistent; and (2) as maximal *relative to classical logic*, meaning that any further extension of **L** is exactly classical logic.

Some notable examples of maximally paraconsistent logics (in the absolute sense) are Sette's $\mathbf{P}^1$, Priest's **LP**, D'Ottaviano and da Costa's **J3**, all the **LFI** systems of

---

[8]Dunn [20] hoped that **R** would be the intersection of the **RM** systems. But **R** is not decidable [65], so this didn't work. Besides, the **RM** systems are not extensions of **RM**; rather they are intermediate logics between **R** and **RM**.

Carnielli and Marcos, Avron's **PAC** (also known as Baten's **CLuNs**), and Anderson and Belnap's **RM3**. Not all of these logics have been used as a basis for arithmetic. In what follows we will look at some attempts in this direction, with a particular attention to inconsistent models of arithmetic.

4.1 Inconsistent Models of Arithmetic

Paraconsistent arithmetics have been studied mostly in terms of their models. We assume the reader is familiar with these ideas and just mention a few memory joggers.

Dunn [21] initiated a general way of obtaining a three-valued structure from a two-valued classical one. Models for inconsistent arithmetic can be *heap* models, as in Priest [48],

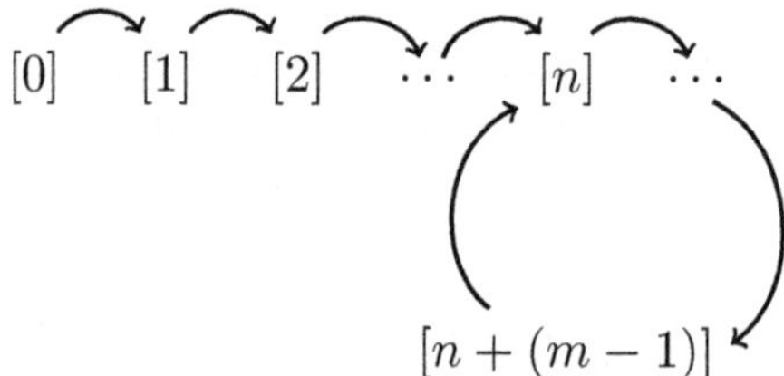

In such a model, there is a consistent initial segment followed by a least inconsistent number, which then is and is not identical to all numbers after it. Such models are amenable to arithmetic based on the logic **LP** [47], as Priest has explored in detail.

Or models can be *cyclic*, as in Meyer and Mortensen [40],

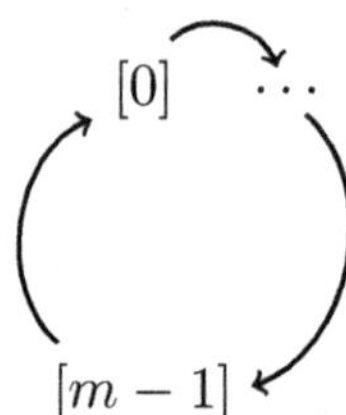

This can be worked with as modular arithmetic—where being congruent modulo m is interpreted as identity. Such models are amenable to arithmetic based on **RM3**.

Both types of models are, notably, finite.[9] This makes properties like decidability much more tractable—Hilbert's call for a decision procedure by finitary methods taken very literally! Let us look at a couple of examples illustrating how inconsistent arithmetic looks in each kind of model.

---

[9]Though not necessarily; see [46]. The general structure of these models is now well understood. See [48, p. 1523], [46].

4.2 Arithmetic Based on RM3

As mentioned in Appendix 1, **RM3** can be obtained from **RM** by adding to it $A \vee (A \to B)$ as an axiom. In contrast with **RM** and its sublogics, **RM3** has a characteristic matrix:

$\neg$	
t	f
b	b
f	t

$\wedge$	t	b	f
t	t	b	f
b	b	b	f
f	f	f	f

$\vee$	t	b	f
t	t	t	t
b	t	b	b
f	t	b	f

$\to_{\mathbf{RM3}}$	t	b	f
t	t	f	f
b	t	b	f
f	t	t	t

The designated values are t and b, and logical consequence preserves designated values from premises to conclusion. Of course, having a characteristic matrix means that **RM3** is not a relevant logic. Nevertheless, **RM3** serves as a 'laboratory of relevance' when it comes to relevant arithmetic.

In [40] we find a general model-theoretic construction for several relevant arithmetics which are extensions of $\mathbf{R}^{\sharp}$ by taking modular arithmetics (of modulus $i \geqslant 2$) with **RM3** as their base logic. An $RM3^i$-*model* is an ordered pair $\langle D^i, I \rangle$, where D^i are the integers modulo i, and I is a function which assigns to the terms, operators and formulas of $\mathscr{L}$ the following values:

- $I(x) \in D^i$ for individual variables x;

- $I(0) = 0$;

- $I(+), I(\times), I(')$ are the operations $+$, $\times$, $'$ of arithmetic modulo i, respectively;

- $I(t_1 + t_2) = I(+)(I(t_1), I(t_2))$, $I(t_1 \times t_2) = I(\times)(I(t_1), I(t_2))$, and $I(t_1') = I(')(I(t_1))$ for any terms t_1 and t_2;

- I assigns the values t, b or f to open or closed formulas as follows:

  - For atomic formulas $t_1 = t_2$, where t_1 and t_2 are terms, $I(t_1 = t_2) = $ b iff $I(t_1) = I(t_2)$ (mod i); otherwise $I(t_1 = t_2) = $ f.

  - $I(\neg A)$, $I(A \wedge B)$, $I(A \vee B)$ and $I(A \to B)$ are determined by the tables for **RM3** given above.

  - $I(\forall x\, A) = \mathrm{glb}\{y \colon I^*(A) = y\}$ for every x-variant I^* of I.

  - A sentence A is $RM3^i$-*true under interpretation* I iff $I(A)$ is designated; A is *true in the* $RM3^i$-*model* iff A is $RM3^i$-true under all interpretations I. The *arithmetic* $\mathbf{RM3}^i$ is the set of sentences true in the $RM3^i$-model.

Remark. The $\mathbf{RM3}^i$ can themselves be considered arithmetics, not just models of arithmetic, due to the following considerations. The $\mathbf{RM3}^i$ are axiomatized by adding to $\mathbf{RM}^\sharp$ the following axioms: $0 = 0^{(i)}$ and, for every j such that $2 \leqslant j < i$, $0 = 0^{(j)} \leftrightarrow 0 = 0'$. This arithmetic is called $\mathbf{RM3}^{i\sharp}$ and, as Meyer and Mortensen showed [40, Prop. 8], the theorems of $\mathbf{RM3}^{i\sharp}$ are exactly the truths of $\mathbf{RM3}^i$.

It follows immediately that the $\mathbf{RM3}^i$ are inconsistent, since, for any modulus $i \geqslant 2$ and any term t, $I(t = t) = \mathsf{b} = I(t \neq t)$. Moreover, the $\mathbf{RM3}^i$ are absolutely consistent [40, Prop. 2], given that $0 = 1$ is not provable in any of them—consider, for instance $i = 2$. And since $\mathbf{RQ}$ is contained in $\mathbf{RM3Q}$, all $RM3^i$-models are models of $\mathbf{R}^\sharp$, from which the absolute consistency of $\mathbf{R}^\sharp$ also follows [40, Prop. 1]. In fact, $\mathbf{RM3}^i$ also contains $\mathbf{R}^{\sharp\sharp}$, $\mathbf{RM}^\sharp$ and $\mathbf{RM}^{\sharp\sharp}$ [40, Prop. 5].

With respect to our GUIDING QUESTIONS, other interesting properties of the $\mathbf{RM3}^i$ are completeness and primeness [40, Prop. 6], decidability [40, Prop. 7], and ω-completeness and ω-inconsistency [40, Props. 2, 4]. Thus, $\mathbf{RM3}^i$ provides an example of an inconsistent arithmetic which answers positively to Q1 and Q2. Regarding Q3, it is unknown whether the halting function is a representable, computable relation, though $\mathbf{RM3}^i$ certainly represents all primitive recursive functions (since $\mathbf{R}^\sharp$ does) and is, in fact, a decidable theory of arithmetic. Alas, the conditions in Q4 are not met, since Contraction is valid in $\mathbf{RM3}$, which we will complain about in §5.1.

Contraction aside, all this sounds really nice, given the GUIDING QUESTIONS we have raised. For the relevant logician, though, the $\mathbf{RM3}^i$ come with some bad news as well: Relevance is lost. Indeed, some irrelevant implications, like Safety, $(A \wedge \neg A) \to (B \vee \neg B)$, are valid in $\mathbf{RM3}$. Moreover, being a finitely valued logic, Dugundji formulas [2, p. 426] hold. For a three valued logic, for any four A_0, A_1, A_2, A_3,

$$\vdash \bigvee_{0 \leqslant j < k \leqslant 3} A_j \leftrightarrow A_k.$$

This shows that desirable properties of paraconsistent arithmetics, as displayed throughout Q1–Q4, are partially attainable but logical relevance might be lost in the process of securing these.

There is a worry, too, that paraconsistency itself is at risk in such a setup. The strength of the $\mathbf{RM3}$ conditional, plus the expressive resources of arithmetic, return something very close to classical negation. If one defines $\bot$ as $0 = 1$, then $\sim A \overset{df}{=} A \to \bot$ is a definable *explosive* negation, $A, \sim A \vdash_{\mathbf{RM3}^i} B$ and, by the instance $A \vee (A \to \bot)$ of the characteristic $\mathbf{RM3}$ axiom $A \vee (A \to B)$, this negation is also *exhaustive*, i.e., $A \vee \sim A$ for all A. This is not classical negation, since $\sim\sim A \to A$ fails when $I(A) = \mathsf{b}$, but exclusion and exhaustion properties certainly give this negation a classical flavor, and would be sufficient for a strengthened liar $L \leftrightarrow {\sim}L$ to do damage. See §4.4.

And regarding computability theory, the function problem is not avoided by $\mathbf{RM3}^i$, since any characteristic function $\chi_A(x)$ of some set A, defined through cases of the form $x = c$ and $x \neq c$ (for some constant c in the domain of χ_A), will lead to $0 = 1$ given that $c = c$ and $c \neq c$ both hold in $\mathbf{RM3}^i$. What this shows, as indicated before, is that a paraconsistent computability theory defined over an inconsistent arithmetic needs to depart from standard methods and definitions of classical computability theory on pain of triviality. We return to this issue in §6. So arithmetic in $\mathbf{RM3}$ has significant drawbacks.

4.3 Arithmetic Based on PAC

Avron's $\mathbf{PAC}$ [5], [6, §3.2–§3.2.2] has exactly the same matrix as $\mathbf{RM3}$, except when it comes to its conditional:

$\rightarrow_{\mathbf{PAC}}$	t	b	f
t	t	b	f
b	t	b	f
f	t	t	t

Whereas $\rightarrow_{\mathbf{RM3}}$ contraposes and does not weaken, $\rightarrow_{\mathbf{PAC}}$ weakens but does not contrapose. The $\rightarrow_{\mathbf{PAC}}$ conditional has the nice property for the logic $\mathbf{LP}$ [47] that $\Gamma, A \vDash_{\mathbf{LP}} B$ iff $\Gamma \vDash_{\mathbf{PAC}} A \rightarrow_{\mathbf{PAC}} B$. Interestingly, $\rightarrow_{\mathbf{RM3}}$ is definable in $\mathbf{PAC}$ as

$$A \rightarrow_{\mathbf{RM3}} B \stackrel{df}{=} (A \rightarrow_{\mathbf{PAC}} B) \wedge (\neg B \rightarrow_{\mathbf{PAC}} \neg A)$$

and $\rightarrow_{\mathbf{PAC}}$ is definable in $\mathbf{RM3}$ as

$$A \rightarrow_{\mathbf{PAC}} B \stackrel{df}{=} (A \rightarrow_{\mathbf{RM3}} B) \vee B$$

Hence, $\mathbf{PAC}$ and $\mathbf{RM3}$ are definitionally equivalent systems.

In [61], Tedder axiomatizes Peano arithmetic in $\mathbf{PAC}$—he calls this logic $\mathbf{A3}$ and the arithmetic based on it is dubbed there $\mathbf{A3}^\sharp$. Just like $\mathbf{R}^\sharp$, $\mathbf{A3}^\sharp$ is properly contained in $\mathbf{P}^\sharp$ [61, p. 534]. Unsurprisingly, given the definitional equivalence of $\mathbf{PAC}$ and $\mathbf{RM3}$, $\mathbf{A3}^\sharp$ has finite models which are non-trivial, inconsistent and decidable; moreover, these properties hold good when one uses either cyclic or heap models along with either $\mathbf{PAC}$ or $\mathbf{RM3}$ [62, Prop. 2.11].

Tedder [62] also studies two versions of Robinson arithmetic in $\mathbf{PAC}$, with a focus on decidability. There are many details, e.g., about which sort of models (cyclic or heap) are under consideration. Let us just give a flavor of the results.

1. Using **PAC**, the characteristic axiom of Robinson arithmetic, $\forall x \, (x \neq 0 \rightarrow \exists y \, (x = y'))$, i.e., that every number is either zero or a successor, is satisfiable in cyclic models. However,

$$(*) \qquad \forall x \, (x \leqslant n \rightarrow (x = 0 \vee \cdots \vee x = n))$$

   is *not*. Tedder shows that **Q** in **PAC** $+ (*)$, *plus* $(*)$'s contrapositive, is undecidable. The contrapositive must be added by hand, so to speak, since $\rightarrow_{\textbf{PAC}}$ does not contrapose.

2. Analogously, **Q** in **RM3** is undecidable if we add special instances of the weakening axiom

$$(**) \qquad A(n) \rightarrow (n \leqslant m \rightarrow A(n))$$

   where A is any Δ_0 formula. Here we are adding infinitely many axioms, since A is schematic.

The undecidability of such theories might actually be an attractive feature. If some arithmetics in **RM3/PAC** are undecidable (or close to it), that might be a good thing since the 'arithmetic' described is more likely to be close to the conventional picture. However, the choice between cyclic or heap models might come with its own problems. Consider the successor axiom

$$(***) \qquad x' = y' \rightarrow x = y$$

This states that the successor function is injective, but we might regard it as saying that the successor function is an injection from $\mathbb{N}$ to a proper subset of itself, which is Dedekind's definition of being infinite. This looks problematic for heap models. Let $\mathfrak{n}$ be the least number such that $\mathfrak{n} = \mathfrak{n}'$. Then for the $m < \mathfrak{n}$ such that $m' = \mathfrak{n}$, we have $m' = \mathfrak{n}'$, so by $(***)$ $m = \mathfrak{n}$. This means the initial segment of the model that is supposed to be before the collapse nevertheless gets 'sucked in', so to speak, leading to the inconsistent identification of all numbers. This seems bad. And it does not even help to rephrase and say that the axiom only applies to numbers before $\mathfrak{n}$, i.e.,

$$(x < \mathfrak{n} \wedge y < \mathfrak{n}) \rightarrow (x' = y' \rightarrow x = y)$$

because $\mathfrak{n} < \mathfrak{n}$, so the antecedent is satisfied by $\mathfrak{n}$! This is an echo of the function problem in the order relation, in one of the simplest recursive notions.

4.4 Consistency Operators and the LFIs

Logics of formal inconsistency, like **LFI1** [16], can be thought of as **LP** plus a primitive connective $\circ$, a consistency operator, with the truth table

$$
\begin{array}{c|c}
A & \circ A \\
\hline
\text{t} & \text{t} \\
\text{b} & \text{f} \\
\text{f} & \text{t}
\end{array}
$$

With $\neg$ as **LP** negation, one can then define a second operator, $\bullet$, which is a dual 'non-classicality' or *inconsistency* marker, $\bullet A \overset{df}{=} \neg \circ A$. Indeed, adding $\circ$ to the language is very powerful; one may further define classical negation as $\sim A \overset{df}{=} \neg A \wedge \circ A$, a bottom constant $\bot \overset{df}{=} A \wedge \sim A$, and detachable conditional $A \to_{\textbf{LFI1}} B \overset{df}{=} \sim A \vee B$. Crucially, the $\circ$ and $\bullet$ operators are mutually exclusive ($\circ A, \bullet A \vDash B$) and cannot themselves be subject to non-classicality [45].

Having (effectively) full classical logic, especially a consistency operator, might help with the function problem. Coming back to the $\mathfrak{n}$ conundrum, if a consistency operator is at hand, we may have

$$
(x < \mathfrak{n} \wedge y < \mathfrak{n} \wedge \circ(x < \mathfrak{n}) \wedge \circ(y < \mathfrak{n})) \to (x' = y' \to x = y)
$$

In general, if $\mathbb{N}^\circ$ is the 'consistent' fragment of arithmetic, maybe we can work with functions in it. Problematic collapses may be averted by restricting to consistent cases, as in:

$$
f_A(n) = \begin{cases} 1 & \text{if } A(n) \wedge \circ(A(n)) \\ 0 & \text{if } \neg A(n) \wedge \circ(\neg A(n)) \end{cases}
$$

This is the intuition Priest suggests at various points: for heap models, consistent reasoning is legitimate in the consistent initial segment. So there is evidence that from the strong logic standpoint, the function problem may be dealt with (if not solved, since there is no decision procedure for consistency $\circ$). This is enough reason to peer a bit farther down the **LFI1** path.

4.4.1 Paraconsistent Turing Machines in LFI1

There is an inchoate sense that a paraconsistent machine should be able to 'do two things at the same time'. Less fancifully, and quite conventionally, one might conjecture that a *non-deterministic* Turing machine has the potential to out-compute a deterministic Turing machine. This turns out not to be so. For finite automata, we have:

Theorem (Rabin and Scott 1959). *Anything computable by a non-deterministic finite automaton is also computable by a deterministic finite automaton [52].*

More generally (though through a different proof strategy), we have:

Theorem. *For every non-deterministic Turing machine there is an equivalent deterministic Turing machine [57, p. 178].*

Could the paraconsistent case be different? Perhaps, but not if the paraconsistent logic is **LFI1** (or any of the equivalent logics we've been studying at the top of the strength ranking). Agudelo and Carnielli show that paraconsistent Turing Machines axiomatized in **LFI1** (ParTMs) do not out-compute their classical counterparts.

Theorem (Agudelo and Carnielli 2010). *Any ParTM can be simulated by a (classical) Turing Machine [1].*

Perhaps, this is *good news*. Agudelo and Carlielli explicitly state that they are not looking to deviate from classical theory too much. For example, if a ParTM could compute, say, the halting function, then this would imply by Church's theorem that first-order logic is decidable, and for Agudelo and Carnielli such a conclusion would be far too radical. Perhaps, a logic that stays as close to classicality as possible, with concomitant results that are as close to classical computability as possible, is desirable, insofar as it remains within the bounds of accepted science.

4.4.2   The Downsides of Consistency Operators

As with the Rabin–Scott theorem, the strong approach has all the costs and benefits of standard, conservative arithmetic and computability. Alternatively, if one is motivated to work on paraconsistent computability theory because of a genuine expectation of inconsistency, or at least a genuine uncertainty about inconsistency, then there are real problems here.

Most saliently, the suggestion that we can solve the function problem by assuming consistency in consistent situations is subject to the ever-present fact that we do not in fact have any guarantee of consistency in almost any situation. Put more technically, a consistency operator $\circ$ must itself be consistent for this to work:

$$(\circ A \wedge \neg \circ A) \rightarrow \bot$$

But why suddenly presume the consistency of consistency? Any qualms about classicality that were meant to be solved by going paraconsistent would be very likely to recur one level up. For example, revenge liar-type sentences of the form

$$L : \neg L \wedge \circ(L)$$

(saying 'this sentence is false and consistently so') are prone to explode. This means the prospects are dim for dealing any better with diagonal arguments to incompleteness in arithmetic based in **LFI1** (or in an equivalent theory). A Gödel sentence that says 'this sentence is not provable, and consistently so' will evade decision. From a new-seeking non-classical direction, all this looks too classical; see [45].

5 Arithmetic in Deep Relevant Logics

Having looked at the strongest possible logics for paraconsistent computability theory, we now turn to logics weaker than **RQ** which may also accommodate (inconsistent) theories of arithmetic.

5.1 The Need for Contraction-free Logics

The alert reader will also have already long-noted that strong logics like **R**, **RM** and **RM3** support Contraction, $\vdash A \to (A \to B) \therefore \vdash A \to B$. In the case of arithmetic, this is not a catastrophe, since Curry-paradox-inducing axioms like a naïve truth schema are not assumed to be around. But contraction is still a cost, since it will support the (pernicious) reasoning behind Löb's theorem: If we have $\Box$ representing provability in arithmetical theories, and provability minimally satisfies

- $\Box(A \to B) \to (\Box A \to \Box B)$

- $\Box A \to \Box\Box A$

- $\vdash A \therefore \vdash \Box A$

then one can show that $\vdash \Box A \to A$ implies $\vdash A$, which is Löb's theorem, and which seems wrong. Questionable steps include Permutation, $\vdash A \to (B \to C) \therefore \vdash B \to (A \to C)$, and Contraction [58, p. 230]. A logic that validates such steps will be forced to say that the provability predicate representable in arithmetic is not the real, informal proof relation that we really use over arithmetic. It is not provable (in the second sense) that all sentences provable (in the first sense) are true, even though it *is* true that all provable sentences (in the second sense) *are* true [49, Ch. 17]. Incompleteness!

Again, this makes the relevant arithmetician no *worse* off than their classical counterparts. Löb's theorem says that arithmetic cannot express its own soundness, which the world has been coming to terms with since Gödel. It does close off, however, the possibility of finding a new way around.

5.2 Ultralogic to the Rescue?

When it comes to relevant, Contraction-free logics, the family of *deep* relevant logics contains the more interesting systems—though of course one has Contraction-free systems which are not depth relevant, like **RW**. Among these, Routley's ultralogic, **DK** [54, p. 48], is a quite strong system—yet weaker than the most well-known systems containing **TW**. Moreover, Routley advanced **DK** as a candidate for a universal logic, i.e., one with which we may reason safely (paradox-free) in any domain of science and philosophy. As such, **DK** was already considered by Routley as a possible base for arithmetic.

Routley's Arithmetic **DKA** [54, §9] is not just the result of adding the Peano axioms to **DKQ**. He has some qualms with the fact that Meyer's $\mathbf{R}^\sharp$ proves theorems like $\forall x\,\forall y\,(x = x \leftrightarrow y = y)$ or the $\forall x\,\forall y\,\forall z\,(x = y \rightarrow z = z)$ we noted when reporting on relevant Robinson arithmetic. His view is that such theorems fail to be relevant implications, for how could $3 = 5$ be relevant (sufficient) to $9 = 9$? He blames $x = y \rightarrow (y = z \rightarrow x = z)$, axiom (A2) from $\mathbf{R}^\sharp$, for the validity of such irrelevant-looking principles, whence he replaces it with $(x = y \wedge y = z) \rightarrow x = z$ and $x = y \rightarrow y = x$. Moreover, he defines $1 \overset{df}{=} 0'$ and $\tau \overset{df}{=} (1 = 1)$ and also modifies the successor axioms thus: $(x = y \wedge \tau) \rightarrow x' = y'$ and $(x' = y' \wedge \tau) \rightarrow x = y$.

Mortensen [42] disputes Routley mainly on the grounds that a true implication need not be a relevant implication. He considers that the relevance of a logic is a matter of its logical theorems satisfying criteria like variable sharing, but these conditions should not be demanded from non-logical theorems of arithmetical theories. Whether Mortensen is right in this point is a matter we do not discuss here but which we recommend to fellow relevant logicians, since it is quite important for any intended applications of relevant logics.

Nevertheless, if one wishes to take Routley's criticism of $x = x \leftrightarrow y = y$ and $x = y \rightarrow z = z$ very seriously, there is a relevantist way to respond to him. As suggested in [25], the implication relation in those principles is not entirely irrelevant even though the Variable Sharing Principle is violated. Estrada-González and Tapia-Navarro advance the notion of weak q-relevance [25, p. 515]: if $A \rightarrow B$ is a theorem, then A and B share weak q-content—where the weak q-content of a formula is defined as the set of its terms and (relevant) predicates. Thus, the identity predicate is shared q-content in the principles criticized by Routley.[10]

---

[10] Besides being a tool to respond to Routley on arithmetical principles with dubious validity as entailments, weak q-relevance is also an interesting response to the objection against relevant mathematics in general, to wit, that some forms of weakening, ruled out in relevant logics, are needed to develop sufficiently strong mathematical theories worth of consideration. See [25, p. 516] for a restricted formulation of weakening which a relevant logician might endorse.

Not enough attention—almost no attention—has been given to **DKA** or other arithmetics based on **DKQ**, owing perhaps in part to weak logics' historical performance against GUIDING QUESTION Q5: they tend to be hard to use.[11] As Routley points out in [54, p. 68], he does not know whether **DKA** can represent all primitive recursive functions.[12] Insofar as **DKA** admits inconsistent models, interesting results might be found. We invite readers to follow this path.

5.3 The Logic of Meaning Containment

A much more developed Contraction-free system is that of Brady's $\mathbf{MC}^\sharp$ [13], based on his logic $\mathbf{MCQ}^-$, a weaker formulation of his logic **MC** of meaning containment [14]. This logic is also weaker than Routley's **DK**, since **MC** lacks LEM, $A \vee \neg A$, and Distribution, $(A \wedge (B \vee C)) \to ((A \wedge B) \vee (A \wedge C))$—though the latter holds in rule form. Moreover, $\mathbf{MCQ}^-$ further removes the quantifier principles $\forall x\,(A \to B) \to (A \to \forall x\,B)$ and $\forall x\,(A \vee B) \to (A \vee \forall x\,B)$, and their corresponding rule forms as well, and instead adds the weaker $A \to \forall x\,A$. The removal of LEM and these quantifier principles gives $\mathbf{MCQ}^-$ an intuitionistic flavor, which may be considered a desirable feature given its intended application in mathematical theories.

Subtle but important modifications to the arithmetical axioms and rules included in $\mathbf{MCQ}^-$ are also made. For instance, identity principles are similar to those of **DKA**, and the successor principles are given in rule form as $\vdash x' = y' \therefore \vdash x = y$ and $\vdash x = y \therefore \vdash x' = y'$ and their contrapositives. Most notably, a classicality axiom for equations, $x = y \vee x \neq y$, is needed given the lack of LEM.

Brady is then able to prove through metavaluations that $\mathbf{MC}^\sharp$ is negation consistent and can also represent all primitive recursive functions [13, §8] [12, §6]. Of course, then, by Gödel's first incompleteness theorem, $\mathbf{MC}^\sharp$ is incomplete. What is quite outstanding of $\mathbf{MC}^\sharp$ is that general recursion can also be accommodated in it [13, §8]. To introduce expressions of the form $\mu x A x$, i.e., the least x such that Ax holds, Brady adds the following least number principles for a formula A:

- $Aa \vee \neg Aa$

- $\vdash \exists x\,Ax \therefore \vdash A\mu x A x$

---

[11]"Several logicians (including Brady, Meyer, Mortensen, Priest, and Routley) have attempted to reconstruct various fragments of classical reasoning in this way. While the results are not definitive, they are not terribly encouraging" [49, p. 221]. Cf. §5.4.

[12]Given Brady's results for $\mathbf{MC}^\sharp$, which we report in the next subsection, one may suspect that arithmetics based on **DKQ** (perhaps, not exactly **DKA** due to its particular axiomatization) can represent not only all primitive recursive functions but also general recursive functions given that Brady's $\mathbf{MCQ}^-$ is contained in **DKQ**.

- $\vdash \exists x\, Ax, \vdash m < \mu x Ax \therefore \vdash \neg Am$

where the expression $A\mu x Ax$ means that A is true of the least x such that Ax.

Finally, regarding Gödel's second incompleteness theorem, Brady [13, p. 469] remarks that

> [...] arithmetic, set up using the above metavaluational proof process and incorporating primitive and general recursion, is simply consistent. This proof is finitary, but that does not mean that Gödel's Second Theorem is contradicted, as the classical component of the logic is restricted by not including the rule $\forall x\,(A \vee B) \Rightarrow A \vee \forall x\, B$, which then prevents the LEM from automatically extending to the two quantifiers.

The careful technical work of Brady provides a nice example of a relevant arithmetic that can carry out classical computability theory, which is a remarkable result. Alas, due to its negation consistency, any hopes of sidestepping the classical limitations of arithmetic and computability theory are futile. But this, of course, was not Brady's goal; but it keeps us looking elsewhere.

5.4 Second-order Dialectical Logic

This brings us to the deepest circle, where a weak logic is used to build foundations. Logan and Boccuni [31] consider how to resuscitate arithmetic from set theory, based on the logic **DL**.[13] They use second-order **DL**, plus Frege's Basic Law V:

$$\forall F\, \forall G\, ((\{F\} = \{G\} \wedge t) \leftrightarrow \forall x\, (Fx \leftrightarrow Gx))$$

(with $\{\}$ a functional abstraction operator). Membership is definable,

$$x \in y \overset{df}{=} \exists F\, (y = \{F\} \wedge Fx)$$

so first-order relations are available. They show that this package can be used to derive the Peano axioms, as we will briefly observe.

In reviewing this impressive achievement, we also note that several of our complaints in the previous sections were about strong logics committing fallacies of relevance. How does a depth-relevant logic like **DL** fare? This is a place where GUIDING QUESTION Q5 arises; deep relevant logics are designed to be weak, and they are. How is this dealt with? Well, constant t plays an essential role, with axiom

$$\vdash t$$

---

[13]Cf. [55]. Routley studied both set theory and arithmetic in the closely related logic **DKQ** [54], but without attempting to derive the axioms of arithmetic from set theory.

and rule

$$\vdash A \therefore \vdash t \to A.$$

Define $0 \overset{df}{=} \{x\colon \neg(x = x)\}$, define successor zermelodically, $s(x) \overset{df}{=} \{x\}$, and define the natural numbers, in the spirit of Frege but with much more attention to the *Sinn* than the *Bedeutung*:

$$\mathbb{N}(x) \overset{df}{=} \forall F \left(\forall y \left(Fy \to (Fs(y) \wedge t)\right) \to (F0 \to Fx)\right).$$

This is a clever permutation of a more usual statement (that the numbers are the intersection of all inductive structures) that turns out to be more usable in a somewhat rigid logic. For example, the axiom

$$\forall x \, s(x) \neq 0$$

follows using $A \to A$, $(A \to \neg A) \to \neg A$, and, crucially, t. Other axioms, like the previously problematic successor injection, $s(x) = s(y) \to x = y$, reduce to the straightforward application of the definition $\{x\} \overset{df}{=} \{z\colon z = x\}$.

From a non-contractive standpoint, the challenging axiom is

$$(\mathbb{N}(x) \wedge t) \to \mathbb{N}(s(x))$$

For example, using light affine set theory, Terui [64] needs an extra modality to get this through; otherwise all one can show in his setting is that if x is a number *and* x is a number, then its successor is, too. Logan and Boccuni solve this with judicious use of t, and lattice conjunction.

Obtaining the axioms, recall, tells us not much more than that the axioms are obtained. We do not know what arithmetic follows from the Peano Axioms in this logic. So far, "an analogue of a piece of arithmetic is available" [31, p. 19]. In this direction lies slow, hard, honest toil. A very intriguing possibility here, meanwhile, is a solution to—or avoidance of—the function problem. The only function symbol in the language is the variable binding term forming operator. Addition is done using a three place relation:

$$S(x, y, z) \overset{df}{=} \forall F \left(\forall a \, \forall b \left(F(a, b) \to (F(s(a), s(b)) \wedge t)\right) \to (F(0, x) \to F(y, z))\right)$$

From this it can be shown that the sum of any two numbers exists—up to contraction, at least.

Foundational work in depth-relevant logic promises to be significantly different than standard classical theory, free of many problems observed above, and with the potential for entirely new results. It is also significantly more difficult to work on, because the deeper the logic, the farther we need to climb. This remains a tantalising speculative exercise.

6 A Lesson from the Function Problem

Having now amassed a good deal of conflicting knowledge about the options for relevant-ish logics for computability, let us look ahead at how any such logics will fare in mathematical practice. We have seen strong logics that will underwrite incomplete arithmetics, and weaker logics that may yet do better but have not yet. In all cases, though, the discussion assumes that key notions like *recursive* and *representation* are fixed, more or less as classically. It is becoming clear that if a non-classical result (like a decidable arithmetic) is what one wants, then a presupposed classical starting point is not what one needs. What might a genuinely paraconsistent *computability theory* look like?

We'll consider a simple example—something you find on the first pages of a computability theory textbook (like [24]). We will go through some simple informal reasoning in mathematical English, based on new definitions intended for paraconsistency, remaining neutral about which of our logics might best support it. The idea being illustrated is that apparent barriers to progress in this area can be taken as markers for where a redesign, rather than recapture, is needed.

Kleene's theorem says that decidable sets are exactly the semi-decidable sets with semi-decidable complements:

Theorem (Kleene's theorem). *S is decidable iff S and $S^{\complement}$ are semi-decidable.*

What here? The idea is to return to the function problem, and use it as a basis for a certain kind of reductio proof. The strategy is to *decide* functionality, or not, as a special property case by case.

Let's presume we have some very basic set theory, and recall that a relation $R \subseteq X \times Y$ is

- *total* iff $\forall x(x \in X$ implies $\exists y(y \in Y \wedge \langle x, y \rangle \in R))$;

- *partial* iff $\exists x(x \in X \wedge \neg \exists y(y \in Y \wedge \langle x, y \rangle \in R))$;

- *overcomplete* iff $\exists x \in X \exists y, z \in Y(y \neq z \wedge \langle x, y \rangle \in R \wedge \langle x, z \rangle \in R)$;

- a *function* iff $\forall x \in X \forall y, z \in Y(\langle x, y \rangle \in R \wedge \langle x, z \rangle \in R$ implies $y = z)$.

A relation can be total *and* partial. An overcomplete relation may be total or partial. If an overcomplete relation is a function, then $z \neq z$ for some $z \in Y$.

In search of some properly paraconsistent relations, consider:

Definition 1. *A set $X \subseteq \mathbb{N}$ is*

- pre-decidable *iff there is a computable relation $f \subseteq \mathbb{N} \times \{0,1\}$ such that*

$$\textit{if } ((x \in X \wedge y = 1) \vee (x \notin X \wedge y = 0)) \textit{ then } \langle x, y \rangle \in f;$$

- decidable *iff it is pre-decidable and f is a function;*

- pre-semi-decidable *iff there is a computable relation $f \subseteq \mathbb{N} \times \{0,1\}$ such that*

$$\textit{if } (x \in X \wedge y = 1) \textit{ then } \langle x, y \rangle \in f;$$

- semi-decidable *iff pre-semi-decidable and f is a function.*

Lemma 1. *If a set is decidable it is semi-decidable. If a set is pre-decidable it is pre-semi-decidable.*

Proof. Let $S \subseteq \mathbb{N}$ be a decidable set. Then S is pre-decidable and there is a computable function $f \colon \mathbb{N} \to \{0,1\}$ such that if $(x \in S \wedge y = 1) \vee (x \notin S \wedge y = 0)$ then $\langle x, y \rangle \in f$. Since any logic we consider will agree that $(A \vee B) \to C$ implies $(A \to C) \wedge (B \to C)$, it follows, by $\wedge$-elimination, that if $(x \in S \wedge y = 1)$ then $\langle x, y \rangle \in f$, so S is pre-semi-decidable, and since f is a function, S is semi-decidable. The same argument shows that if S is pre-decidable then it is pre-semi-decidable, with the only difference that this time f is a computable relation only. $\square$

Theorem 2 (*Weak Kleene's theorem*). *A set (or relation) S is pre-decidable iff both it and its complement $S^{\complement}$ are semi-decidable. But any inconsistent sets that are semi-decidable and have semi-decidable complements, are not decidable.*

Proof. If a set is decidable, then both it and its complement $S^{\complement}$ are semi-decidable, by the previous Lemma. Suppose both S and $S^{\complement}$ are semi-decidable. Then there are computable relations f_S and $f_{S^{\complement}}$. Putting these together $f_S \cup f_{S^{\complement}} = f$ shows that S is pre-decidable. Now suppose S is decidable—that is, suppose f is a function.

$$f(n) = \begin{cases} 1 & \text{if } n \in S \\ 0 & \text{if } n \notin S \end{cases}$$

If there is an inconsistent non-empty subset of $\mathbb{N}$, i.e., some S such that some $n \in S$ and $n \notin S$, then it would be that $f(n) = 0$ and $f(n) = 1$, which is impossible. So for any such S, f is not a function. Inconsistent computable sets are pre-decidable, but not decidable. $\square$

This suggests that the function problem can be *useful* in reductios. What we mean by this is that in some reductios ending in $0 = f(x) = 1$, instead of concluding that some substantial assumption was wrong (like the decidability of some set or the existence of some Turing machine), all we may have determined is that f is not a function. Perhaps the assumptions made are still compatible with f being a computable relation instead. Functionality is now one of the moving parts, a variable parameter, in the overall picture of computability. This does not tell us what a computable relation *is*, if not a relation with a recursive characteristic function. But it suggests that recasting familiar definitions to rethink results is a worthwhile strategy trying to answer our GUIDING QUESTIONS. This is a start.

7 Conclusion: Which Logic(s) for Paraconsistent Computable Functions?

Starting from **R** as a mid-point for a spectrum of logics for arithmetic, we see a choice—turning to the stronger logics like **RM3** and maximally paraconsistent systems, or turning to logics significantly weaker, like **DL** and other depth-relevant systems. How do these choices fare in light of our GUIDING QUESTIONS? We summarize our findings:

	$\mathbf{R}^\sharp$	$\mathbf{R}^{\sharp\sharp}$	$\mathbf{Q_R}$	$\mathbf{RM3}^i$	$\mathbf{Q_{PAC}}$	**DKA**	$\mathbf{MC}^\sharp$	$\mathbf{DL2Q}^{t,fc}$
Non-trivial	✓	✓	✓	✓	✓	✓	✓	?
Negation consistent	✓	✓	✓	×	?	?	✓	?
Negation complete	×	×	×	✓	?	?	×	?
Decidable	×	×	×	✓	?	?	×	?
Contractive	✓	✓	✓	✓	✓	×	×	×

The question marks indicate that there is a good deal still to uncover. As elsewhere in the relevant programme, there is no clear 'winner'.

Here are some concluding observations.

- If one works with the stronger, or even maximal, systems, there are many benefits—the most apparent being staying close to classical arithmetic and computability theory. The class of computable functions is unchanged from the classical case. This benefit, though, is also the most significant cost, in that the resulting computability theory will not significantly advance on previous attempts. While in some cases decidability is regained (*Entscheidung!*), proving completeness via 'internal' proofs of soundness are as out of reach as they have been since the mid-twentieth century. *Ignoramus et ignorabimus.*

- If one works with weaker systems, there are many costs—from the nigh-intractability of the function problem, to more mundane difficulties in deriving even the simplest of results (our guiding Q5). These costs, though, are only costs if one is holding out for a non-classical approach that somehow is not different from the classical approach. A complete, decidable system that represents its own proof relation and expresses its own soundness and consistency is going to be unlike classical arithmetic, which does none of these things.

The choice, as we noted from the start, may turn on the answer to the question: *paraconsistent* or *inconsistent?*

An inconsistent approach anticipates inconsistent subsets of $\mathbb{N}$. If these exist, either because of naïve set comprehension, or some inconsistency from Gödel's theorem (cf. [56]), or something else, they go hand in hand with weaker logics. On this approach, $\circ$ and equivalent operators should give us profound pause. They cannot reckon with Gödel's or Turing's limits any better than classical logic.

A simply paraconsistent approach not concerned with the ghosts of departed inconsistent sets can work much more flexibly with logics like **R** and its extensions. Strong logics close to classical—some might say *extremely* close to classical—are easy to work with and think about. They support familiar (but not exactly the same) arithmetic, as well as models of inconsistent theories with nice properties.[14]

Insofar as Löb's theorem is an incompleteness, an inability to express something manifestly true, then stronger logics are terminally incomplete. They are easy to work with and have some good-making features, but ultimately do not—will not—answer Hilbert's call. On that count, if the aim is to find a genuine solution to the function problem and with it a class of *properly* non-classical computable operations that cross the Church–Turing barrier, something more radical, only yet half-imagined, is still needed.

Appendix 1: Axiomatizations for R and Other Systems

RQ is given by the following axiom schemata and rules.

Axioms:

1. $A \to A$

2. $(A \wedge B) \to A$

3. $(A \wedge B) \to B$

---

[14] And they can describe, and can be represented by, hardware you can literally build this afternoon if you wanted to [15].

4. $((A \to B) \land (A \to C)) \to (A \to (B \land C))$

5. $A \to (A \lor B)$

6. $B \to (A \lor B)$

7. $((A \to C) \land (B \to C)) \to ((A \lor B) \to C)$

8. $(A \land (B \lor C)) \to ((A \land B) \lor (A \land C))$

9. $\neg\neg A \to A$

10. $(A \to \neg B) \to (B \to \neg A)$

11. $(A \to B) \to ((B \to C) \to (A \to C))$

12. $A \to ((A \to B) \to B)$

13. $(A \to (A \to B)) \to (A \to B)$

14. $\forall x\, A \to A[y/x]$ (where y is free for x in A)

15. $\forall x\, (A \to B) \to (A \to \forall x\, B)$ (where x is not free in A)

16. $\forall x\, (A \lor B) \to (A \lor \forall x\, B)$ (where x is not free in A)

Rules:

17. $\vdash A, \vdash A \to B \;\therefore\; \vdash B$

18. $\vdash A, \vdash B \;\therefore\; \vdash A \land B$

19. $\vdash A \;\therefore\; \vdash \forall x\, A$

Other logics discussed in this paper may be obtained from **RQ** as follows:

    RMQ $=$ **RQ** $+\, A \to (A \to A)$.
    RM3Q $=$ **RMQ** $+\; A \lor (A \to B)$.
    DKQ $=$ **RQ** $-\,11. -\,12. -\,13. +\, A \lor \neg A \,+\, ((A \to B) \land (B \to C)) \to (A \to C) \,+$
        $\vdash A \to B, \vdash C \to D \;\therefore\; \vdash (B \to C) \to (A \to D)$.
    MCQ $=$ **DKQ** $-\,8. -\, A \lor \neg A -\,16. +$ the following metarules:

- If $\vdash A, \vdash B \;\therefore\; \vdash C$ then $\vdash D \lor A, \vdash D \lor B \;\therefore\; \vdash D \lor C$.

- If $\vdash A, \vdash B[y/x] \;\therefore\; \vdash C[y/x]$ then $\vdash A, \vdash \exists x\, B \;\therefore\; \vdash \exists x\, C$ where y does not occur in A, B or C, and where 19. does not generalize on any free variable in the premises A and $B[y/x]$ in the derivation $\vdash A, \vdash B[y/x] \;\therefore\; \vdash C[y/x]$.

    MCQ$^-$ $=$ **MCQ** $-\,15. +\, A \to \forall x\, A$ (where x does not occur in A).
    DLQ $=$ **DKQ** $-\, A \lor \neg A \,+\, (A \to \neg A) \to \neg A$.
    DLQ^t $=$ **DLQ** $+\; t + \vdash A \;\therefore\; \vdash t \to A$.

$\mathbf{DL2Q}^{t,fc} = \mathbf{DLQ}^t + \exists F^n \, \forall y_1 \ldots \forall y_n \, (A \leftrightarrow A(F^n/B(y_1, \ldots, y_n)))$ where F does not occur free in B and B is free for F in A. Regarding 14.–16. and 19., quantification can be either first- or second-order.

See the appendix in [62] for an axiomatization of **PAC** (dubbed **A3** therein).

Appendix 2: Some Features of RM

An **RM**-frame $\langle W, w_0, \leqslant \rangle$ has an ordering that is reflexive, transitive, and connected. That is, **RM** has a linearly ordered infinite characteristic matrix, making it particularly easy to think about. Following Dunn, **RM**-models are obtained by adding valuations $v \colon \mathtt{prop} \times W \longrightarrow \{\{1\}, \{0\}, \{1,0\}\}$ that obey heredity (if $w_x \leqslant w_y$ then $v_{w_x}(p) \subseteq v_{w_y}(p)$) and the semantic clauses for $\neg$ and $\wedge$,

$$
\begin{aligned}
1 \in v_w(\neg A) \quad &\text{iff} \quad 0 \in v_w(A) \\
0 \in v_w(\neg A) \quad &\text{iff} \quad 1 \in v_w(A) \\
1 \in v_w(A \wedge B) \quad &\text{iff} \quad 1 \in v_w(A) \text{ and } 1 \in v_w(B) \\
0 \in v_w(A \wedge B) \quad &\text{iff} \quad 0 \in v_w(A) \text{ or } 0 \in v_w(B)
\end{aligned}
$$

The conditional has truth condition

$$
1 \in v_{w_x}(A \to B) \quad \text{iff} \quad \forall y \geqslant x((1 \in v_{w_y}(A) \text{ only if } 1 \in v_{w_y}(B)) \\
\wedge \, (0 \in v_{w_y}(B) \text{ only if } 0 \in v_{w_y}(A))
$$

This is nice, but unlike Routley–Meyer's $Rxyz$, it is hard for $A \to A$ to fail "honestly" [20, p. 172]. The falsity condition is disjunctive:

$$
0 \in v_{w_x}(A \to B) \quad \text{iff} \quad 1 \notin v_{w_x}(A \to B) \vee (1 \in v_{w_x}(A) \wedge 0 \in v_x(B))
$$

Dunn calls the first disjunct the "escape clause".[15]

Dunn [20] finds a countable family of weakenings of **RM**. Let $A \circ B = \neg(A \to \neg B)$. Then the **RM** axiom is $(A \circ A) \to A$.[16]
This in hand, we can recursively define

$$
A^1 = A \qquad A^{n+1} = A^n \circ A
$$

---

[15] See [8, p. 29] and [10, p. 126] for this and other very useful background. If we wrote the falsity condition as a conditional, it says:

$$
0 \in v(A \to B) \; \leftrightarrow \; (1 \in v(A \to B) \to (1 \in v(A) \wedge 0 \in v(B)))
$$

which we think is funny, and can be permuted/contraposed to say all sorts of things.

[16] By the following reasoning: $\neg(A \to \neg A) \to A$ implies $\neg A \to (A \to \neg A)$ by contraposition, which implies $A \to (\neg A \to \neg A)$ by permutation, which implies $A \to (A \to A)$ by contraposition and transitivity.

Then the **RM**s are successive weakenings of $A^{n+1} \to A^n$ or idempotence $(a \circ a)^{n+1} = (a \circ a)^n$ in the algebra.

Proof of Linearity from Mingle (reconstructed from [2, §29.5], "Why we don't like mingle"):

1. $\neg A \to (\neg A \to \neg A)$ (mingle)

2. $\neg A \to (A \to A)$ (1, contrapose)

3. $A \to (\neg A \to A)$ (2, permute)

4. $((A \to A) \wedge (B \to B)) \to (\neg((A \to A) \wedge (B \to B)) \to ((A \to A) \wedge (B \to B)))$ (instance of 3)

5. $(\neg(A \to A) \vee \neg(B \to B)) \to ((A \to A) \wedge (B \to B))$ (from 4, fiddling)

6. $\neg(A \to A) \to (B \to B)$ (5, axioms 5 and 3, transitivity)

7. $B \to (\neg(A \to A) \to B)$ (6, permute)

8. $B \to (A \circ \neg A \to B)$ (7, definition)

9. $B \to (\neg A \circ A \to B)$ (8, commute)

10. $B \to (\neg A \to (A \to B))$ (9, export)

11. $B \to (\neg(A \to B) \to A)$ (10, contrapose)

12. $\neg(A \to B) \to (B \to A)$ (11, permute)

13. $(A \to B) \vee (B \to A)$ (12, implication to disjunction)

Now, the above derivation does use steps that are disputable. If you are used to working in weaker systems, it feels very dirty! As Dunn says,

> One not used to relevance logic might be surprised at the author's delicious sense of forbidden pleasure derived from the use of the material conditional here and throughout the metalanguage [19, footnote 4].

So too with such casual permutation, among other moves, above.[17]

References

[1] Juan Agudelo and Walter Carnielli. Paraconsistent machines and their relation to quantum computing. *Journal of Logic and Computation*, 20(2):573–595, 2010.

---

[17]If instead of adding the Mingle axiom to **R** we add it to the weaker logic **B**, where many of these steps are not available, the situation may be significantly different; thanks to Andrew Tedder for pointing this out. And thanks to the editor for pointing out that since ticket entailment **T** (see [9]) does not have full permutation either, there is interest in adding Mingle there, too; cf. [34].

[2] Alan Ross Anderson and Nuel D. Belnap. *Entailment: The Logic of Relevance and Necessity*, volume 1. Princeton University Press, Princeton, 1975.

[3] Ofer Arieli, Arnon Avron, and Anna Zamansky. Ideal paraconsistent logics. *Studia Logica*, 99:31–60, 2011.

[4] F. G. Asenjo. Toward an antinomic mathematics. In Priest et al. [51], pages 394–414.

[5] Arnon Avron. On an implication connective of RM. *Notre Dame Journal of Formal Logic*, 27(2):201–209, 1986.

[6] Arnon Avron. Natural 3-valued logics—characterization and proof theory. *Journal of Symbolic Logic*, 56(1):276–294, 1991.

[7] Nuel D. Belnap. A useful four-valued logic. In J. M. Dunn and G. Epstein, editors, *Modern Uses of Multiple-Valued Logics*, pages 8–37. Reidel, Dordrecht, 1977.

[8] Katalin Bimbó. On the origins of gaggle theory. In I. Sedlár, editor, *Logica Yearbook 2021*, pages 19–36. College Publications, London, 2022.

[9] Katalin Bimbó and J. Michael Dunn. On the decidability of implicational ticket entailment. *Journal of Symbolic Logic*, 78(1):214–236, 2013.

[10] Katalin Bimbó and J. Michael Dunn. Entailment, mingle and binary accessibility. In Y. Weiss and R. Birman, editors, *Saul Kripke on Modal Logic*, volume 30 of *Outstanding Contributions to Logic*, pages 121–150. Springer, 2024.

[11] Ross Brady. Starting the dismantling of classical mathematics. *Australasian Journal of Logic*, 15(2):280–300, 2018.

[12] Ross T. Brady. The consistency of arithmetic, based on a logic of meaning containment. *Logique et Analyse*, 55(219):353–383, 2012.

[13] Ross T. Brady. Arithmetic formulated in a logic of meaning containment. *Australasian Journal of Logic*, 18(5):447–472, 2021.

[14] Ross T. Brady and Andrea Meinander. Distribution in the logic of meaning containment and in quantum mechanics. In *Paraconsistency: Logic and Applications*, pages 223–255. Springer, 2012.

[15] Fernando Cano-Jorge. Gates and circuits via Dunn semantics. Presented at the Seventh World Congress of Paraconsistency 2024, under consideration, 202+.

[16] Walter Carnielli, João Marcos, and Sandra De Amo. Formal inconsistency and evolutionary databases. *Logic and Logical Philosophy*, 8:115–152, 2000.

[17] Seungrak Choi. Is there an inconsistent primitive recursive relation? *Synthese*, 200(5):418, 2022.

[18] Ana Cruz, Alexandre Madeira, and Luís Soares Barbosa. A logic for para-consistent transition systems. *Electronic Proceedings in Theoretical Computer Science*, pages 270–284, 2022.

[19] J. Michael Dunn. A Kripke-style semantics for R-Mingle using a binary accessibility relation. *Studia Logica*, 35(2):163–172, 1976.

[20] J. Michael Dunn. R-Mingle and beneath. Extensions of the Routley–Meyer semantics for R. *Notre Dame Journal of Formal Logic*, 20(2):369–376, 1979.

[21] J. Michael Dunn. A theorem in 3-valued model theory with connections to number theory, type theory, and relevant logic. *Studia Logica*, 38(2):149–169, 1979.

[22] J. Michael Dunn. Relevant Robinson's arithmetic. *Studia Logica*, 38:407–418, 1980.

[23] J. Michael Dunn. R-Mingle is nice, and so is Arnon Avron. In O. Arieli and A. Zamansky, editors, *Arnon Avron on Semantics and Proof Theory of Non-classical Logics*, volume 21 of *Outstanding Contributions to Logic*, pages 141–165. Springer Nature, Switzerland, 2021.

[24] Herbert Enderton. *Computability Theory: An Introduction to Recursion Theory*. Academic Press, Elsevier, 2011.

[25] Luis Estrada-González and Manuel Tapia-Navarro. "A smack of irrelevance" in inconsistent mathematics? *Australasian Journal of Logic*, 18(5):503–523, 2021.

[26] Harvey Friedman and Robert K. Meyer. Whither relevant arithmetic? *Journal of Symbolic Logic*, 57:824–31, 1992.

[27] Patrick Girard and Koji Tanaka. Paraconsistent dynamics. *Synthese*, 193(1): 1–14, 2016.

[28] David Hilbert. Mathematical problems. *Bulletin of the American Mathematical Society*, 8:437–479, 1902. Translated from the 1900 address by Mary Winston Newton.

[29] Shay Logan. Nondeterministic and nonconcurrent computational semantics for BB+ and related logics. *Journal of Logic and Computation*, pages 1–20, 2023.

[30] Shay Logan and Graham Leach-Krouse. On not saying what we shouldn't have to say. *Australasian Journal of Logic*, 18(5):524–568, 2021.

[31] Shay A. Logan and Francesca Boccuni. Frege meets Belnap: Basic law V in a relevant logic. In I. Sedlár, S. Standefer, and A. Tedder, editors, *New Directions in Relevant Logic*, pages 401–423. Springer Nature, Switzerland, 2025.

[32] Francisco N. Martínez-Aviña. Changing the logic without changing the subject: The case of computability. *Journal of Logic and Computation*, 2024.

[33] Toby Meadows and Zach Weber. Computation in non-classical foundations? *Philosophers' Imprint*, 16:1–17, 2016.

[34] José Méndez, Gemma Robles, and Francisco Salto. Ticket entailment plus the mingle axiom has the variable-sharing property. *Logic Journal of the IGPL*, 20 (1):355–364, 2012.

[35] Robert K. Meyer. Kurt Gödel and the consistency of $R^{\#\#}$. In P. Hájek, editor, *Gödel '96: Logical Foundations of Mathematics, Computer Science and Physics—Kurt Gödel's Legacy, Brno, Czech Republic, August 1996, Proceedings*, pages 247–256. Springer-Verlag, Berlin, 1996.

[36] Robert K. Meyer. $\supset$E is admissible in "true" relevant arithmetic. *Journal of Philosophical Logic*, 27:327–351, 1998.

[37] Robert K. Meyer. Arithmetic formulated relevantly. *Australasian Journal of Logic*, 18(5):154–288, 2021. Typescript circa 1976.

[38] Robert K. Meyer. The consistency of arithmetic. *Australasian Journal of Logic*, 18(5):289–379, 2021. Typescript circa 1976.

[39] Robert K. Meyer and J. Michael Dunn. E, R and γ. *Journal of Symbolic Logic*, 34:460–474, 1969.

[40] Robert K Meyer and Chris Mortensen. Inconsistent models for relevant arithmetics. *The Journal of Symbolic Logic*, 49(3):917–929, 1984.

[41] Chris Mortensen. *Inconsistent Mathematics*. Kluwer Academic Publishers, Dordrecht; New York, 1995.

[42] Chris Mortensen. Implication principles in Routley arithmetic. In Z. Weber, editor, *Ultralogic as Universal? The Sylvan Jungle-Volume 4*, pages 185–194. Springer, 2019.

[43] Tore Fjetland Øgaard. Skolem functions in non-classical logics. *Australasian Journal of Logic*, 14(1):181–225, 2017.

[44] Hitoshi Omori and Zach Weber. Just true? On the metatheory for paraconsistent truth. *Logique et Analyse*, 248:415–433, 2019.

[45] Hitoshi Omori, Graham Priest, and Zach Weber. The (in)consistency of consistency. In H. Antunes, A. Freire, and A. Rodrigues, editors, *Walter Carnielli on Reasoning, Paraconsistency, and Probability*. Springer, 202x. to appear.

[46] Jeff B. Paris and Alla Sirokofskich. On LP-models of arithmetic. *Journal of Symbolic Logic*, 73:212–226, 2008.

[47] Graham Priest. The logic of paradox. *Journal of Philosophical Logic*, 8:219–241, 1979.

[48] Graham Priest. Inconsistent models of arithmetic, II: The general case. *Journal of Symbolic Logic*, 65:1519–29, 2000.

[49] Graham Priest. *In Contradiction: A Study of the Transconsistent*. Oxford University Press, Oxford, 2nd edition, 2006.

[50] Graham Priest and Richard Routley. *On Paraconsistency*. Research School of Social Sciences, Australian National University, 1983. Reprinted as introductory chapters in [51].

[51] Graham Priest, Richard Routley, and Jean Norman, editors. *Paraconsistent Logic: Essays on the Inconsistent*. Philosophia Verlag, Munich, 1989.

[52] Michael O. Rabin and Dana Scott. Finite automata and their decision problems. *IBM Journal of Research and Development*, 3(2):114–125, 1959.

[53] Gemma Robles. The quasi-relevant 3-valued logic RM3 and some of its sublogics lacking the variable sharing property. *Reports on Mathematical Logic*, 51:105–131, 2016.

[54] Richard Routley. *Ultralogic as Universal?* Sylvan Jungle volume 4, edited by Z. Weber, Synthese Library, Springer, 2019. (First appeared in two parts in the *Relevance Logic Newsletter* 2(1):51–90, January 1977 and 2(2):138–175, May

1977; reprinted as appendix to *Exploring Meinong's Jungle and Beyond* 1980, pp. 892–962).

[55] Richard Routley and Robert K. Meyer. Dialectical logic, classical logic, and the consistency of the world. *Studies in Soviet Thought*, pages 1–25, 1976.

[56] Stewart Shapiro. Incompleteness and inconsistency. *Mind*, 111:817–832, 2002.

[57] Michael Sipser. *Introduction to the Theory of Computation*. Cengage Learning, 3rd edition, 2013.

[58] Peter Smith. *An Introduction to Gödel's Theorems*. Cambridge University Press, 2007.

[59] Shawn Standefer. Variable sharing as relevance. In Igor Sedlár, Shawn Standefer, and Andrew Tedder, editors, *New Directions in Relevant Logic*, pages 97–117. Springer Nature, Switzerland, 2025.

[60] Richard Sylvan and Jack Copeland. Computability is logic relative. In D. Hyde and G. Priest, editors, *Sociative Logics and their Applications*, pages 189–199. Ashgate, 2000.

[61] Andrew Tedder. Axioms for finite collapse models of arithmetic. *Review of Symbolic Logic*, 3:529–539, 2015.

[62] Andrew Tedder. On consistency and decidability in some paraconsistent arithmetics. *Australasian Journal of Logic*, 18(5):473–502, 2021.

[63] Andrew Tedder. A note on R-Mingle and the danger of safety. *Australasian Journal of Logic*, 19(1), 2022.

[64] Kazushige Terui. Light affine set theory: A naive set theory of polynomial time. *Studia Logica*, 77:9–40, 2004.

[65] Alasdair Urquhart. The undecidability of entailment and relevant implication. *Journal of Symbolic Logic*, 49:1059–1073, 1984.

Received 30 December 2024